Symmetry in Mechanical Engineering: Properties and Applications

Symmetry in Mechanical Engineering: Properties and Applications

Guest Editors

Yunqing Tang
Kun Xu
Bing Yang

MDPI

Basel • Beijing • Wuhan • Barcelona • Belgrade • Novi Sad • Cluj • Manchester

Guest Editors

Yunqing Tang
School of Mechanical Engineering
Shandong University
Jinan
China

Kun Xu
College of Electrical Engineering
Henan University of Technology
Zhengzhou
China

Bing Yang
School of Mechanical Engineering
Shandong University of Technology
Zibo
China

Editorial Office
MDPI AG
Grosspeteranlage 5
4052 Basel, Switzerland

This is a reprint of the Special Issue, published open access by the journal *Symmetry* (ISSN 2073-8994), freely accessible at: https://www.mdpi.com/journal/symmetry/special_issues/CF9LDE6OBT.

For citation purposes, cite each article independently as indicated on the article page online and as indicated below:

Lastname, A.A.; Lastname, B.B. Article Title. *Journal Name* **Year**, *Volume Number*, Page Range.

ISBN 978-3-7258-7108-7 (Hbk)
ISBN 978-3-7258-7109-4 (PDF)
https://doi.org/10.3390/books978-3-7258-7109-4

Contents

About the Editors . . . . **vii**

Preface . . . . **ix**

Keyang Chen and Yunqing Tang
Research Progress on the Design of Surface Texture in Tribological Applications: A Mini-Review
Reprinted from: *Symmetry* **2024**, *16*, 1523, https://doi.org/10.3390/sym16111523 . . . . **1**

Joo-Mi Bahk, Sun-Hak Kim and Jong-Yun Yoon
System Modeling and Simulation for Investigating Dynamic Characteristics of Geared Symmetric System Based on Linear Analysis
Reprinted from: *Symmetry* **2023**, *15*, 1904, https://doi.org/10.3390/sym15101904 . . . . **22**

Fayong Wu, Jie Hong, Xueqi Chen and Yanhong Ma
Analysis of High-Speed Rotor Vibration Failure Due to Sudden Angular Deformation of Bolt Joints
Reprinted from: *Symmetry* **2023**, *15*, 1937, https://doi.org/10.3390/sym15101937 . . . . **56**

Fayong Wu, Jie Hong and Xueqi Chen
Distributed Rotational Inertia Load Excitation Model and Its Impact on High-Speed Jointed Rotor Dynamic Response
Reprinted from: *Symmetry* **2023**, *15*, 2009, https://doi.org/10.3390/sym15112009 . . . . **78**

Przemysław Snopiński
Effects of KoBo-Processing and Subsequent Annealing Treatment on Grain Boundary Network and Texture Development in Laser Powder Bed Fusion (LPBF) AlSi10Mg Alloy
Reprinted from: *Symmetry* **2024**, *16*, 122, https://doi.org/10.3390/sym16010122 . . . . **102**

Lovro Liverić, Wojciech Sitek, Przemysław Snopiński, Wojciech Maziarz and Tamara Holjevac Grgurić
A Microstructural Study of Cu-10Al-7Ag Shape Memory Alloy in As-Cast and Quenched Conditions
Reprinted from: *Symmetry* **2024**, *16*, 545, https://doi.org/10.3390/sym16050545 . . . . **117**

Jordy Josue Martinez Cardona, Manuel Cardona, Jorge I. Canales-Verdial and Jose Luis Ordoñez-Avila
Direct and Inverse Kinematics of a 3RRR Symmetric Planar Robot: An Alternative of Active Joints
Reprinted from: *Symmetry* **2024**, *16*, 590, https://doi.org/10.3390/sym16050590 . . . . **134**

Kun Xu, Zhongyou Lu, Li Wang and Yunqing Tang
Enhancing the Proportion of Sub-5 μm Atomized Droplet Size in Medical Air-Compression Nebulizer
Reprinted from: *Symmetry* **2024**, *17*, 6, https://doi.org/10.3390/sym17010006 . . . . **153**

Sergey Lychev, Alexander Digilov and Nikolay Djuzhev
Galerkin-Type Solution of the Föppl–von Kármán Equations for Square Plates
Reprinted from: *Symmetry* **2024**, *17*, 32, https://doi.org/10.3390/sym17010032 . . . . **168**

Zhicheng Huang, Yingjie Chen, Xingguo Wang and Fulei Chu
Bending Analysis of Symmetrical Porous Functionally Graded Sandwich Panels
Reprinted from: *Symmetry* **2025**, *17*, 327, https://doi.org/10.3390/sym17030327 . . . . **202**

Zhicheng Huang, Biao Xiong, Xingguo Wang and Fulei Chu
Effect of Laying Angle on the Stress Distribution and Stiffness Degradation of Symmetrically Cracked Laminates
Reprinted from: *Symmetry* **2025**, *17*, 495, https://doi.org/10.3390/sym17040495 . . . . **221**

Liqun Wang, Yunqing Tang, Jianbin Yu, Wei Qin, Yangyang Zhang, Guoyan Wang, et al.
Bound Optimization by Quadratic Approximation for Heat-Dissipation-Oriented Design of an Air-Cooled Lithium Battery Energy Storage Cabinet
Reprinted from: *Symmetry* **2026**, *18*, 107, https://doi.org/10.3390/sym18010107 . . . . **236**

About the Editors

Yunqing Tang

Yunqing Tang is currently a professor at Shandong University. He received his bachelor and Ph.D. degrees, respectively, from the China University of Mining and Technology in 2012 and Jiangsu University in 2017. He then worked at the University of Alberta as a post-doctoral fellow and joined Shandong University in 2023. His main research areas are wear and friction, the design and machining of structural materials, the reliability of electronic packaging, and micro-/nanoscale heat transfer. He has published more than 100 journal papers in high-impact journals such as *Sci Adv, CRPS, Acta Mater, Wear,* and *Carbon*. He has 15 authorized Chines patents and 1 PCT patent (authorized in Australia and South Africa).

Kun Xu

Kun Xu is currently an associate professor at Henan University of Technology. He received his Ph.D. degree in Mechanical Engineering from Jiangsu University in 2017. His main research interests focus on mechatronics technology, flexible electronics, flexible sensors, and semiconductor device physics. He has published more than 50 journal papers in high-impact journals such as *npj Flexible Electronics, Science China Materials, IEEE Transactions on Instrumentation and Measurement,* and *Food Chemistry*. He also holds 12 authorized Chinese patents and 3 authorized PCT patents in the United States.

Bing Yang

Bing Yang is currently a teacher at Shandong University of Technology. He received his bachelor and Ph.D. degrees, respectively, from Jinan University in 2015 and Jiangsu University in 2021. He then worked in at University of Technology in 2021. His main research areas are micro-/nanoscale heat transfer, thermal management of power devices, and laser micro/nanomanufacturing. He has published more than 40 journal papers in high-impact journals such as *Materials Today Physics, ACS Applied Materials & Interfaces, Applied Thermal Engineering,* and *International Communications in Heat and Mass Transfer*. He one authorized patent in China.

Preface

This Reprint, titled "Symmetry and Asymmetry in Mechanical Engineering," assembles twelve peer-reviewed contributions that collectively illustrate how symmetry and its intentional modification function as powerful levers in modern mechanical engineering. Symmetry has long been a cornerstone of engineering design because it supports balanced load transfer, predictable deformation patterns, repeatable manufacturing, and stable dynamic behavior. Yet contemporary demands for higher efficiency, multi-function integration, and operation under increasingly extreme or constrained conditions are pushing designers beyond classical balanced layouts. In many advanced systems, carefully introduced asymmetry in geometry, material distribution, interfaces, or surface features is not a flaw to be avoided but a purposeful strategy to realize directional responses, localized performance enhancement, or improved controllability. The subject of this Reprint is therefore not symmetry alone but the design logic that connects symmetric principles with engineered departures from symmetry to achieve measurable performance gains.

The scope of this Reprint spans structural mechanics, dynamics, thermal- and fluid-related performance, mechanisms and robotics, advanced materials, and tribology. Several papers focus on how symmetry governs stress redistribution, stiffness evolution, and load carrying capacity in structural components, including cracked laminates and porous functionally graded sandwich panels. Fundamental modeling is also represented through plate analyses and solution strategies for nonlinear governing equations, demonstrating how symmetry assumptions and boundary conditions shape analytical tractability and physical interpretation. In parallel, this Reprint includes work on symmetric mechanism design and kinematic formulation, where geometric balance is central to motion accuracy, workspace characteristics, and actuation strategies in planar robotic architectures.

A second major theme is the role of symmetry in dynamics and vibration, particularly in high-speed rotating machinery and jointed assemblies. Symmetry is often associated with stable rotor response and reduced sensitivity to excitation, but real systems inevitably include imperfections, joint compliance, and time-varying disturbances that break ideal symmetry. The studies included herein analyze excitation modeling, vibration response, and failure mechanisms related to sudden angular deformation and bolted joint behavior, offering perspectives that connect modeling assumptions with practical reliability concerns. By addressing both idealized symmetric systems and realistic symmetry-breaking factors, the Reprint provides readers with a clearer pathway from theoretical concepts to diagnostic and design decisions for rotating equipment.

Thermal management and fluid-related performance form another important component of the collection. The Reprint features work oriented toward heat dissipation-driven optimization in air-cooled energy storage cabinets, highlighting how geometric arrangement, flow path design, and optimization methodology can be combined to improve thermal safety margins. It also includes research on atomized droplet size enhancement in medical nebulization, where symmetric or near symmetric flow structures, nozzle design, and operating conditions influence atomization efficiency and distribution characteristics. These topics underscore that symmetry considerations are not limited to load-bearing structures but extend to transport phenomena and system level performance targets.

The purpose of this Reprint is threefold. First, it offers a curated snapshot of how symmetry and asymmetry are currently being studied across diverse subfields of mechanical engineering, using a range of methods that includes analytical derivation, numerical simulation, optimization, and microstructural characterization. Second, it provides a shared vocabulary for designers and researchers who may approach symmetry from different angles, whether through structural topology, material architecture, kinematic constraints, or surface engineering. Third, it encourages a more nuanced design mindset; rather than treating symmetry as an unconditional ideal, it invites readers to evaluate when symmetry should be preserved to ensure robustness and manufacturability, and when controlled asymmetry can be introduced to enable directional functions, localized enhancement, or improved system integration.

The motivations behind assembling this Reprint arise from both enduring engineering fundamentals and emerging technological pressures. Symmetry remains essential for many applications because it reduces complexity and supports predictable performance. However, the rapid growth of high-power-density energy systems, high-speed rotating machinery, lightweight composite structures, and miniaturized devices is increasing the importance of targeted, multi-objective performance tuning.

Yunqing Tang, Kun Xu, and Bing Yang
Guest Editors

symmetry

MDPI

Review

Research Progress on the Design of Surface Texture in Tribological Applications: A Mini-Review

Keyang Chen [1,2,3,4] and Yunqing Tang [1,2,3,4,5,*]

[1] School of Mechanical Engineering, Shandong University, Jinan 250061, China; 202320658@mail.sdu.edu.cn
[2] State Key Laboratory of Advanced Equipment and Technology for Metal Forming, Shandong University, Jinan 250061, China
[3] Key Laboratory of High Efficiency and Clean Mechanical Manufacture of Ministry of Education, Jinan 250061, China
[4] Key National Demonstration Center for Experimental Mechanical Engineering Education, Jinan 250061, China
[5] Guangxi Key Laboratory of Manufacturing System & Advanced Manufacturing Technology, Guilin University of Electronic Technology, Guilin 541004, China
* Correspondence: tangyunqing@sdu.edu.cn

Abstract: Surface texturing technology, as an advanced method to improve surface tribological properties of friction pairs, has been widely used in many fields. In this work, the influence of surface texture parameters on tribological properties of friction pair surfaces are reviewed. For the currently most developed surface textures with symmetry and simple geometries and distributions, it is found that they could help reduce friction mainly by enhancing their dynamic pressure lubrication capability, storing abrasive debris and lubricants for dynamic lubrication or promoting the formation of friction films on surfaces of friction pairs. The dominant design parameters of surface textures influencing their tribological performance are found to be shape, geometry and density, while working condition, including contact mode and lubrication situation, also has a significant influence on the performance of surface textures with specific features. Asymmetric textures and multi-scale composite textures also show great tribological performance, while the coupling mechanism across different factors is still unclear, which makes it a challenge to maximize the advantage of asymmetric or multi-scale composite textures. The development of machine learning provides promising approaches for the multi-parameter optimization of surface textures, which is expected to promote and accelerate the design of advanced surface textures.

Keywords: surface texture; wear resistance; friction; symmetry and asymmetry

1. Introduction

Friction and wear are common phenomena in production and daily life, and scientists have been making efforts to minimize the cost induced by friction and wear for a long time [1–5]. In modern industry, around 60–80%of the occurrence failure of mechanical components are caused by wear, and the energy consumed by friction globally accounts for nearly 1/4 of the total energy used [6,7]. Apart from developing wear-resistant materials, lubricating media and surface treatment technology [8–10], controlling surface structure is also an effective way to control the friction and wear of materials. It has been observed that non-smooth surfaces do not always lead to more friction and wear [11], and this phenomenon has attracted great attention [12–17]. Experimental results show that surfaces with specific textures could result in less friction and wear compared to untextured surfaces [18–20]. Although the anti-friction mechanisms of surface textures are still lacking a unified theory, the effect of surface textures in reducing friction and wear has been widely accepted, and textures on the surface of friction pairs have been used more and more in tribological applications [20–24].

Surface textures are specific structures on the surface of friction pairs with a specific distribution arrangement [25–27]. Surface textures have been applied in many fields such as cutting tools and human joints for the purpose of friction reduction [28–30]. Surface textures exhibit varying anti-friction properties under different tribological conditions, so establishing a correlation between surface texture parameters and their performance across different working conditions is crucial for advancing the design of sophisticated surface textures. Generally speaking, factors such as shape, geometric shape, distribution and density are significant in determining the performance of surface textures, and most of the current research on surface texture design focuses on the optimization these features.

Here, we reviewed the research progress on the design of surface texture to look into the influence of surface texture design parameters on the tribological properties of surfaces. We discussed the effects of the design parameters of surface textures on the tribological properties of friction pair surfaces and illustrated friction reduction mechanisms for different types of surface textures under different working conditions. The advantages of machine learning in dealing with complex parameter optimization are also introduced, which provides a new and effective approach for the design of surface textures. This work is expected to provide references for a more effective optimization of design parameters in surface textures.

2. The Effect of Symmetric Texture Parameters on Tribological Properties

The geometric dimensions of surface texture have been studied from micro to macro levels to suit different working conditions [31–33]. According to the relative height between surface texture and friction pair surface, surface texture could be classified into two types: the convex texture (Figure 1a) and the pit texture (Figure 1b). Among the two types of textures, the pit texture attracts more attention and is currently more commonly used, which is mainly because of its better manufacturability. The pit-textured structures could be obtained by different mechanical, physical or chemical approaches, which are more mature than the additive manufacturing approach used to fabricate convex-textured structures. The pit texture has been proven to have good performance in anti-friction and anti-wear properties, load-bearing capacity and self-cleaning properties, and it could meet the requests of multiple applications with a relatively simple optimization of parameters [10,11]. Therefore, this work mainly focuses on the design and development of the pit texture.

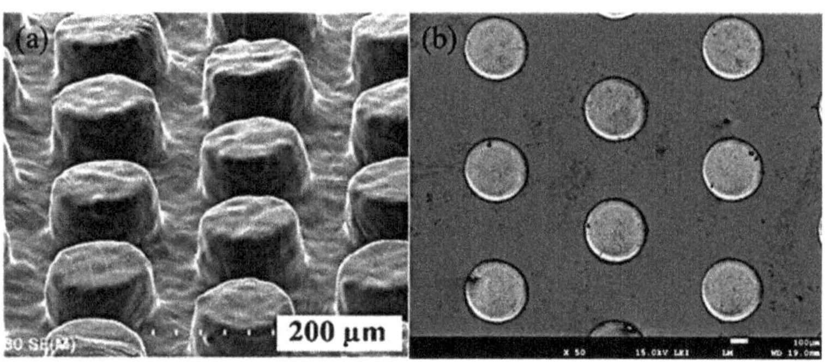

Figure 1. Texture Types (**a**) Convex texture (Reproduced with permission from [25], Copyright 2019 by the Elsevier). (**b**) Circular pit texture.

2.1. Surface Texture Contour Shape

Significant advances in computer graphics, coupled with a deeper understanding of the microstructure of biological body surfaces, have greatly facilitated the development of surface texture design techniques. These technological advances have not only simplified the design process but also improved the accuracy of the design results [34]. The excellent surface properties of biological body surfaces and the working mechanisms behind them have been an important source of inspiration for surface texture design. However, not all of these naturally generated surface structures can be directly applied to engineering practice. Therefore, performing graphical optimization becomes a crucial step in translating these bio-inspired structures into practical designs [35]. Zhang et al. [36] used a genetic algorithm to simulate the biological evolution process in order to investigate the effect of surface texture contour shape on tribological performance for shape contour optimization. After 28 iterations of the genetic algorithm, two types of textures, fish-shaped and bullet-shaped, were finally obtained. In order to verify the tribological properties of optimized texture shapes, three shapes of textures were prepared on GCr15 discs (Figure 2), in which all design parameters were kept the same apart from the shape. Unidirectional sliding experiments were carried out on the UMT-2 friction and wear tester; the average friction coefficient is shown in Figure 2d. It can be observed that the fish-shaped texture has the lowest friction coefficient. However, the longer boundary perimeter was found to result in more severe stress concentration, so the fish-shaped texture would fail earlier during relative friction. The experimental results show that the bullet-shaped textures have lower friction coefficients under all operating conditions compared to the circular textures, indicating that the bullet shape is more suitable for textures under the unidirectional sliding condition than the circular shape. Therefore, the optimal contour shape for all considerations was found to be the bullet shape.

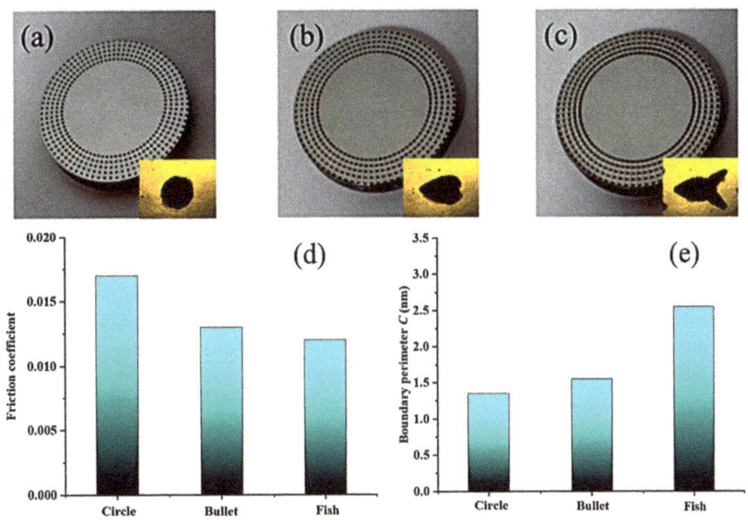

Figure 2. Photographs of disks with textures in shapes of circle (**a**), bullet (**b**) and fish (**c**); (**d**,**e**) are the friction coefficients and boundary perimeters of textures in different shapes. (Reproduced with permission from [36], Copyright 2019 by Elsevier).

Guo et al. [37] found that the texture shape of the material surface has a significant impact on the lubrication effect. They studied the effects of three types of convex textures on the surface of ultra-high-molecular-weight polyethylene materials under water lubrication. The three textures are cylindrical, cubic and rectangular. Due to the limitations of convex texture processing methods, the texture size is in the millimeter range. Friction and wear experiments were designed based on the service conditions of the materials. The average friction coefficient is shown in Figure 3. The coefficient of friction at a load of 0.3 MPa is obviously larger than the other cases, which is determined by the characteristics of the ultra-high-molecular-weight polyethylene materials. The textured surfaces have fewer scratches and material transfer compared to non-textured surfaces, and the cubic texture shows the best wear performance. Khan et al. [38] prepared three types of textures, namely straight grooves, curved grooves and circular pits, on the surface of tungsten carbide cutting tools using lasers. Compared to non-textured surfaces, the wear of these textured cutting tools is significantly reduced, which is attributed to the presence of textured structures effectively reducing the friction coefficient and controlling the generation of heat. It should be noted that although the wear resistance of the circular pit texture is the worst among the three textures, as is reflected in the service life of the tool, it still has a longer service life than tools without textured surfaces. This indicates that even textures with relatively low wear resistance can improve the durability of cutting tools to a certain extent, providing important references for tool design and material optimization.

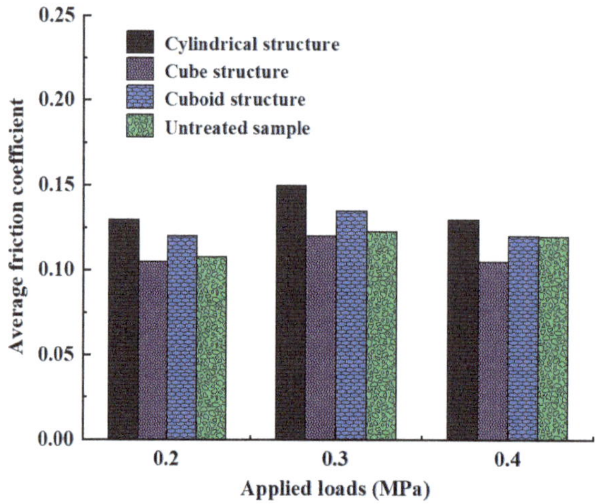

Figure 3. Average friction coefficients under different conditions.

The improvement of surface tribological properties by lubrication is significant, and different shapes of textures exhibit different friction reduction mechanisms under lubricated conditions. Morris et al. [39] combined numerical simulations and experimental tests to ascertain the positive effects of pressure perturbations induced by microfluidic mechanisms on the entrapment of lubricant in cavities of textures, and the improvement in micro-wedge flow dynamics was also estimated. Reciprocal friction tests on a V-shaped textured surface showed that the reciprocating sliding lubricant was entrained into the texture pits for storage, leading to varying degrees of effectiveness of the micro-wedge effect. The reduction in friction coefficient was most pronounced when the angle of the V-shaped texture was 80° and the depth was 3 μm. Higher texture density can store more lubricant, but the texture density should be designed carefully considering the air leakage problem when the piston ring is in contact with the cylinder liner. Wang et al. [40] ascribed the friction reduction mechanism of pit textures under oil lubrication to the variable cavitation effects induced by different shapes of pits. The friction coefficients of the three types of textures studied under lubrication conditions are shown in Figure 4. The square texture shows the lowest friction coefficient due to its larger compression area and superior dynamic pressure-bearing capacity. The dynamic lubrication effect provided by the texture has been confirmed in both simulation and experimental analysis and is currently one of the accepted texture friction reduction mechanisms.

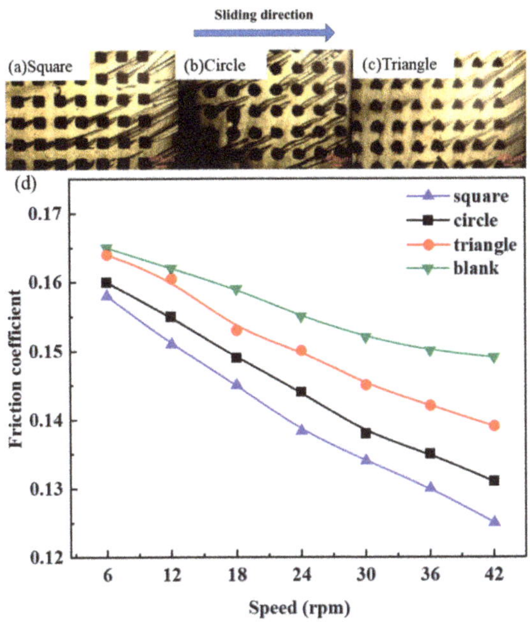

Figure 4. Photographs of surface with textures in shapes of square (**a**), circle (**b**) and triangle (**c**); (**d**) is the friction coefficient between the textured surface and the original surface. (Reproduced with permission from [40], Copyright 2020 by the Wiley).

In addition to the texture shape, there is a correlation between the dynamic pressure-bearing capacity and the depth and occupancy of pits. In the study by Nie et al. [41], a detailed simulation analysis was conducted on the relationship between surface texture parameters and bearing capacity; the results obtained are shown in Figure 5. H is the ratio of oil film thickness to texture depth, and W is the ratio of texture width to texture unit width. When $W < 0.5$, the load-bearing capacities of textured surfaces are lower than those of the non-textured surface, while when $W > 0.5$, the load-bearing capacities of textured surfaces is significantly higher than those of the non-textured surface. In the study by Qiu et al. [42], the geometric shapes of six commonly used textures were optimized, and it was found that contours with curves had a greater advantage in reducing the friction coefficient than straight edge contours, which was almost opposite to the results obtained by Wang et al. [40]. The main reason is that the lubrication condition of Qiu et al. [42] was gas lubrication and that only simulation analysis was performed, while the experimental results of Wang et al. [40] were obtained in oil lubrication. Babu et al. [43] combined simulation and experiment to study the frictional performance of several shape textures under boundary lubrication conditions. From the experimental results in Figure 6, it can be found that among different texture shapes, the elliptical texture shows the lowest friction coefficient at different slip velocities, whereas the circular texture shows the highest friction coefficient.

From the above discussion, one may see that the texture shape has a significant effect on the reduction of surface friction coefficient. Based on the analysis on friction coefficient and wear rate, the performance of different texture shapes under the same working conditions can be effectively compared, and the optimal texture design can be determined. It is worth noting that with changes in lubrication conditions, the same texture shape may exhibit vastly different performance. This is mainly due to differences in friction reduction mechanisms. Under dry friction conditions, surface texture achieves anti-friction ability mainly by reducing contact area and storing debris, while under lubricated conditions, the primary anti-friction mechanism becomes the enhancement of load-bearing capacity. Therefore, different lubrication states may lead to changes in texture properties. Specifically, a circular texture may not be comparable to other textures in terms of friction reduction effect, but its design and processing are relatively simple, and it can still provide a certain degree of friction reduction compared to surfaces without a texture [10]. This allows circular textures to still play a role in practical applications, especially in situations where high processing requirements are needed.

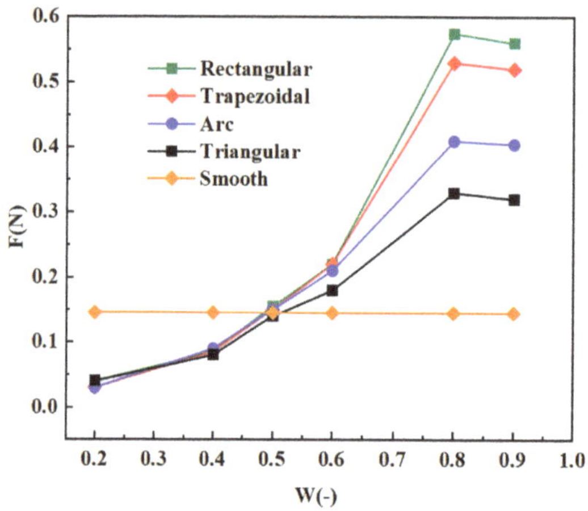

Figure 5. Oil film load capacity of texture units with different shapes (H = 1.0).

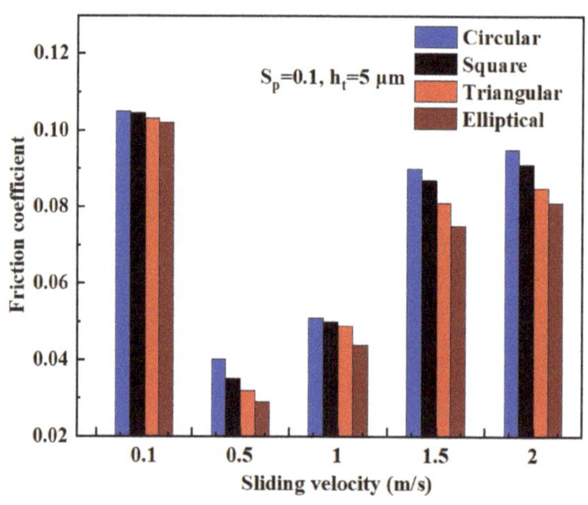

Figure 6. Friction coefficients of various types of textures under different sliding velocity.

2.2. Geometric Parameters

The optimization of parameters is a crucial step in the design of products. For surface textures with specific shapes, the active optimization of geometric parameters is essential to improving their tribological properties [11,14,25]. Huang et al. [44] were inspired by the surface microstructure of tree frogs and used the response method to optimize the geometric size of the texture. A biomimetic hexagonal texture design was carried out on the surface of AISI 4140 steel. The average friction coefficient was reduced by 20% after the optimization; this is ascribed to the uniform lubricating film generated on the optimized textured surface, which can help decrease the direct contact between friction pairs.

Zhang et al. [45] experimentally verified that the friction coefficient of groove-textured surfaces is lower than that of non-textured surfaces. In the state of oil lubrication, the wear resistance of textured surfaces with different parameters also shows differences. The effect of groove width and depth on wear rate is shown in Figure 7. The results show that the wear rate of textured surfaces with larger depth and width values is lower. The SEM characterization indicates that the observed phenomenon is attributed to the presence of abrasives in the operational environment. The grooves with increased width and depth can capture more abrasives, which reduces residual abrasives on the surface and turns the wear type from abrasive wear to mild abrasive wear and scratches. Patel et al. [46] also found that deeper and wider textures could significantly decrease the friction coefficient.

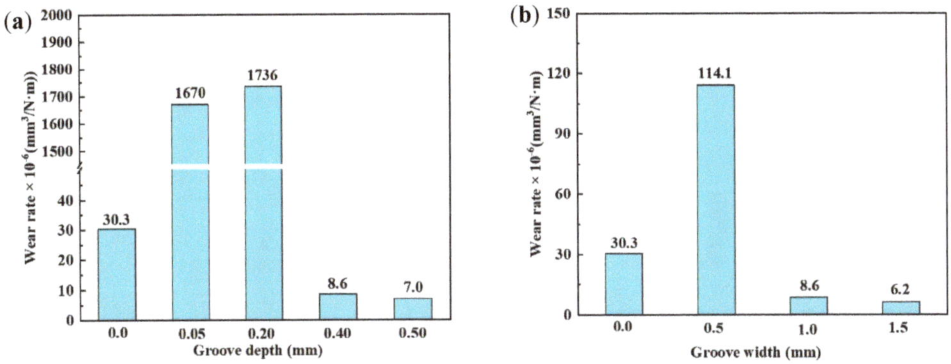

Figure 7. Effect of texture parameters: (**a**) depth, (**b**) width.

Xu et al. [47] studied the effects of full oil lubrication and oil-deficient lubrication on the formation of lubricating oil film and the wear performance of elliptical textures. The presence of a friction film is key to reducing surface friction wear. Under the same lubrication conditions, textured surfaces have thicker lubrication films. XPS analysis shows that carbon, oxygen and sulfur are the primary elements composing the friction film. With sufficient lubrication, the contents of carbon and oxygen in the film are more than with starved lubrication. If the ratio of the long axis to the short axis of an ellipse is too high, it will increase the contact stress and damage the friction film, forming wear debris. M. Varman et al. [48] studied the effect of different diameters and depths of textures under lubrication conditions on the tribological properties of DLC (diamond-like carbon) coatings. Figure 8 shows the influence of texture diameter and depth on the wear volume. It is found that the pit texture with a depth to diameter ratio of 0.06 has the lowest wear loss. Friction reduction on textured surfaces is not always better than on untextured surfaces, especially if the texture is not well-designed. The depth and diameter of the texture largely determine the lubricant storage capacity. Well-designed texture pits can effectively store lubricant and thus reduce friction. However, if the pit is too deep or too large, it may cause the oil film to rupture and increase friction. However, shallow or small pits may not be able to efficiently store lubricants, leading to an impact on the lubrication effect.

The design process of surface texturing is complex and involves numerous influencing factors, making it extremely challenging to reveal the relationship between texture parameters and surface functionality solely through simple statistical analysis. An adequate understanding and effective control of related influencing factors are the key to achieving optimal textured surfaces. During the design stage, it is essential to take different factors into consideration, including properties of materials and morphological features of textures such as shape, size and distribution density. Zeng et al. [49] proposed a method in the article that utilizes 3D surface texture parameters and statistical functions to establish a connection between surface geometric features and surface functionality, which is more accurate and effective than 2D parameters. The research results also reveal that establishing the relationship between the geometric shape of surface texture and its function cannot rely solely on a single or a few parameters. Given the numerous parameters involved and the complexity of experimental conditions, relying solely on human labor for statistics and analysis is impractical. In the context of the big data era, with the increasing improvement of computer technology and artificial intelligence algorithms, the use of machine learning algorithms to predict the relationship between texture parameters and surface properties has shown promising prospects. Machine learning can effectively establish potential connections in complex data, build multidimensional models and enable researchers to optimize texture design with higher accuracy and efficiency. This method has been applied in material design and surface texture design [50–52]. Li et al. [53] proposed a method based on machine learning for designing texture parameters, which trains and evaluates a large number of collected experimental parameters to reverse predict the relationship between texture parameters and performance functions. They found that the coefficient of friction is mainly affected by the texture geometry and the predictions from machine learning algorithms have a reliability of more than 90%, which effectively accelerates the design and optimization of surface texture parameters.

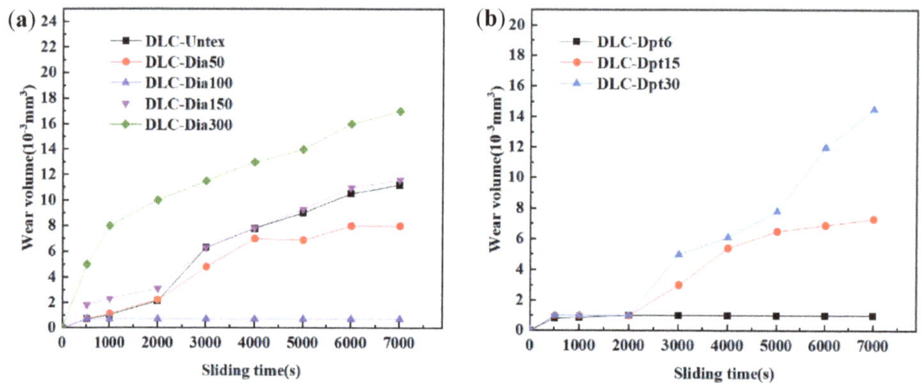

Figure 8. Wear volume (**a**) the effect diameter on wear (**b**) the effect of depth on wear.

2.3. Texture Density

The surface texture could lower the friction coefficient by reducing the contact area between friction pairs. Both convex and pit textures can effectively reduce the contact area and thus lower the friction coefficient. Therefore, a reasonable design of the proportion of the textured area on the whole surface is essential to achieve the desired tribological performance. Due to the generally small size of surface texture, higher texture density can increase processing difficulty and have adverse effects on surface quality. Therefore, it is necessary to optimize the texture density for different application surfaces.

Li et al. [20] prepared a V-shaped texture on the surface of hard alloys to achieve the purpose of reducing friction. The experimental environment was lubricated with lubricating oil, and the friction direction was consistent with the V-shaped convergence direction. Figure 9 shows the friction coefficients for different texture densities. The results shows that the texture density of 9.5% has the lowest and most stable friction coefficient under different loads. If the density is too high, it will actually increase the friction coefficient, while if the density is too low, it will not provide sufficient bearing capacity. Additionally, for lubrication conditions, Zhang [45] et al. concluded in their study that a larger texture density significantly enhances the wear resistance of the surface, and this conclusion is based on the fact that a larger texture density can capture the abrasive material present between the surfaces to a greater extent and minimize the three-body abrasive wear. Nui et al. [54] found in their study that increasing the density and diameter of pits appropriately is beneficial for reducing friction and wear when the texture depth is small. With the same number of textured units, increasing the diameter of the pits results in an increase in the density of the texture, so the increase in both density and diameter can enhance the ability of textures to capture abrasive particles, thereby reducing the wear rate.

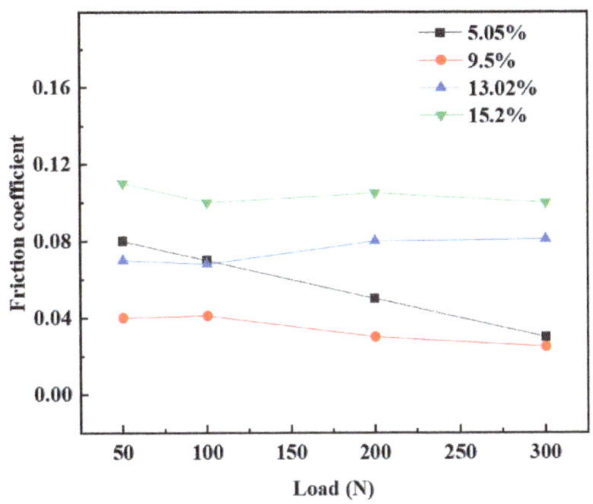

Figure 9. Friction coefficient of 9.5% texture density samples with changes in test force.

Yuan et al. [55] prepared pit textures on the surfaces of four metal materials and used variance analysis to comprehensively analyze and determine the optimal texture density for each material. However, the optimal texture density for different materials is different. Therefore, optimizing texture density cannot completely rely on literature references or simulations. Instead, for each material, texture parameters must be optimized through a systematic design of experiments and statistical analyses to ensure optimal tribological properties. The surface texture not only stores abrasive debris to reduce three-body abrasive wear but also stores some tiny solid lubricants, which diffuse out from the inside of the texture pits to lubricate the friction partner when it moves relative to each other. The surface texture not only captures debris and prevents abrasive wear but also serves as a reservoir for small solid lubricants. These lubricants gradually diffuse from the texture to the surface, providing lubrication during the relative motion of the friction pair. The study by Zhang et al. [56] confirms this phenomenon, but it requires the surface texture density to be within a reasonable range. From Figure 10, it can be concluded that the wear rate is lowest when the surface texture density is 36% and increases when it is greater than or less than 36%. The highest wear rate occurs on the non-textured surface. Electron probe microanalysis and EDS analysis results indicate that the presence of surface texture results in the diffusion of solid lubricants to the worn surface, thereby reducing direct contact between friction pairs. When the texture density is 36%, the solid lubricant has the best synergistic effect with the surface texture and exhibits the best tribological performance.

The lubricant plays a crucial role in reducing friction and wear between friction pairs. In addition to conventional lubricants and greases, water-based lubricants have been found to be of practical value in specific application scenarios. In particular, there are industrial processes where the operating environment of the friction pair is inherently wet, such as in underwater operations, hydraulic machinery and marine engineering. Zou et al. [57] confirmed under water lubrication conditions that there is a positive correlation between texture-bearing capacity and wear resistance, where texture-bearing capacity is also related to texture density. The best bearing capacity is achieved at a texture density of 2%, and the wear volume is also the smallest, as shown in Figure 11. Li et al. [58] also pointed out the considerable significance of texture density on anti-wear performance. Their results show that modifying the texture density can obviously enhance the tribological properties of the textured surface. The key factors contributing to improved anti-wear performance in friction pairs are the enhancement of dynamic pressure lubrication capabilities and the formation of friction film.

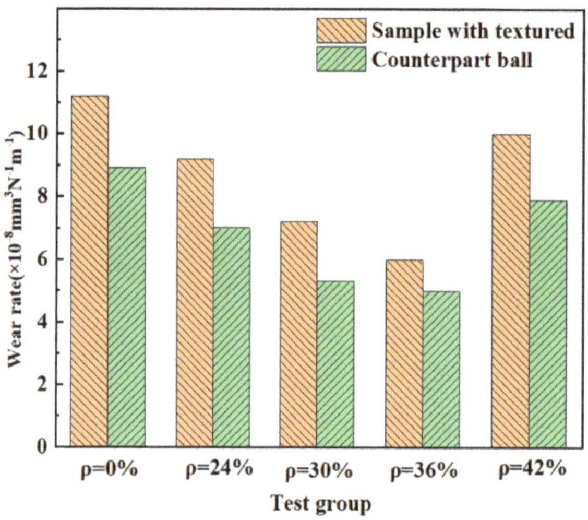

Figure 10. Average friction coefficients of texture surfaces with different textured densities (ρ = 0%, 24%, 30%, 36% and 42%).

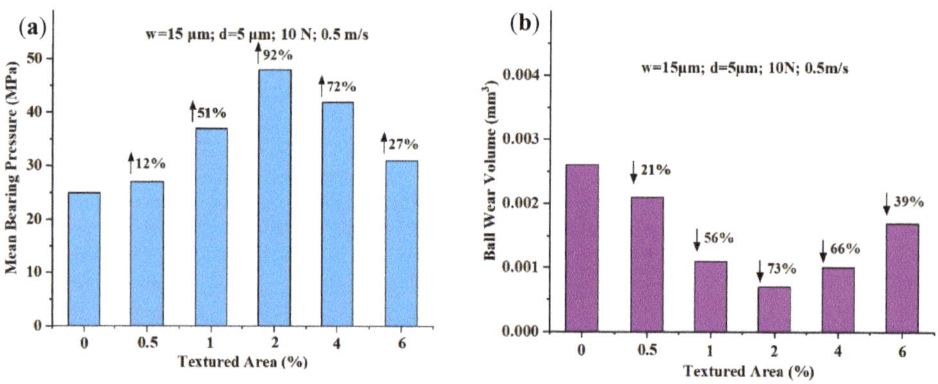

Figure 11. (**a**) mean scar contact pressure. (**b**) Si_3N_4 ball-wear volume as a function of the texture depths.

2.4. Other Factors

Apart from shape, dimensions and density, there are many other factors influencing the design of surface textures. For example, the distribution arrangement was observed to largely affect the performance of surface textures. However, there is a lack of systematic research on the arrangement of surface textures, and it is unclear how the texture arrangement influences wear behaviors. Shen et al. [59] studied the differences in the arrangement of V-shaped textures on the improvement of bearing capacity. Compared to V-shaped textures with the parallel arrangement shown in Figure 12a, the V-shaped arrangement shown in Figure 12b shows improved lubrication performance and load-bearing capacity, which is ascribed to the more effective convergence of the lubricating oil film with the V-shaped arrangement [11]. Li et al. [60] observed significant influences of texture shape and lubricant flow direction on the fluid dynamic pressure effects. The reverse-arranged gourd-shaped texture and the circular pit texture are found to have close friction coefficients, while the forward-arranged gourd-shaped texture has a significantly reduced friction coefficient. Therefore, the convergence of the texture in the direction of lubricant flow is considered to be an important factor affecting the hydrodynamic pressure effect of textures. In some studies [10,11,43,54], it is believed that the ratio of texture depth to diameter has an impact on the surface of friction pairs, and there is a certain correlation between the friction coefficient and the ratio of texture depth to diameter. The ratio of depth to diameter affects the secondary lubrication ability of lubricants [61,62].

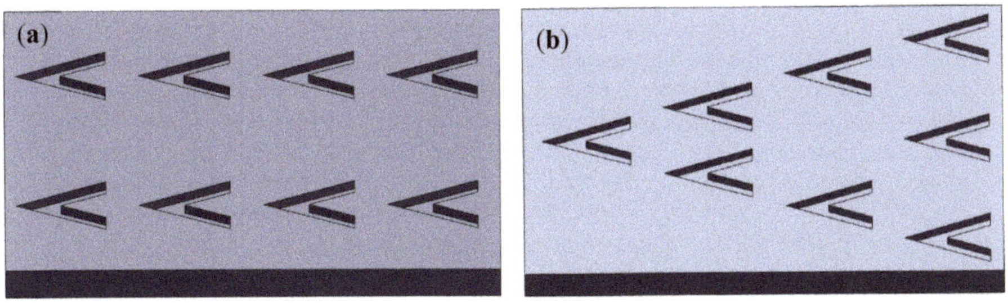

Figure 12. Texture arrangement (**a**) Parallel arrangement (**b**) V-shaped arrangement.

The wear products would affect the tribological interface and the subsequent frictional processes. In some cases, abrasive debris generated during tribological processes would result in three-body abrasive wear and exacerbate material loss. In contrast, some materials may form friction films during tribological processes, and in some cases the wear products could also act as a lubricant to reduce the coefficient of friction and reduce wear losses. Kong et al. [63] prepared hetero-structured NiCr-WS_2-Ti self-lubricating composites, and the sulfide products after the interfacial friction and oxidation reaction acted as a solid lubricant, which facilitated the formation of a protective layer on the sliding surfaces. At elevated temperatures, titanium would be firstly oxidized due to its high activity, which promotes the spontaneous formation of tiny textures on the surface of material, the textures can store the self-lubricating material debris, and the oxides could serve as a protective glaze layer on the contact surface. Srivyas et al. [64] actively designed the texture parameters and prepared the textured surface using lasers. The synergistic lubrication between the textured surface and lubricant was investigated, and the results showed that the coefficient of friction and wear of the textured surface were lower than those of the normal surface under all the design conditions, which was attributed to the synergistic action between the textured surface and the lubricant in the formation of a self-lubricating layer, which reduces the coefficient of friction and the wear loss.

When the friction pair surface has hydrophobicity, it can effectively reduce the liquid in contact with the surface, reduce the erosion of liquid molecules on the surface and reduce adhesive wear and corrosion wear [65,66]. Shum et al. [67] prepared textures with different geometric parameters on the surface using laser surface texturing technology, and they then deposited DLC coatings on the surface by magnetron sputtering to form a textured DLC surface. The surface contact angle of the textured DLC coating was observed to be greater than 130°. The wear experiment was conducted using a reciprocating ball plane wear tester in a simulated body fluid environment. When the texture density was 10%, the distance between adjacent texture units was most conducive to reducing wear, which was related to the surface hydrophobicity and anti-adhesion caused by the texture units. Wang et al. [68] found a correlation between the change in contact angle caused by surface texture and the friction coefficient. Hydrophobic textures were prepared on the surface of YG8 hard alloy cutting tools using a femtosecond laser. Contact angle measurements showed a 123% increase in specific surface contact angle compared to ordinary surfaces. Friction and wear experiments showed a 61.154% decrease in surface friction coefficient compared to ordinary surfaces. SEM and EDS results showed that the micro-textures can significantly reduce adhesive wear, promote the separation of abrasive particles and prevent the aggravation of wear.

2.5. Summary

Surface texture plays an important role in controlling friction and wear by reducing friction coefficients and wear loss during relative motion, which greatly benefits the surface life and reliability of components. The reduction of the friction coefficient improves the energy utilization efficiency of mechanical systems, reducing energy waste and costs, and the decrease in wear loss increases the surface life and reduces maintenance costs. Although it is still hard to give a quantitative techno-economic evaluation for the surface texture technology at the current stage, the development of manufacturing technology and continuously improving performance makes the application of texture technology promising in terms of both efficiency and economy. The main mechanisms contributing to the anti-friction properties of surface textures are found to be enhancing the dynamic pressure lubrication capability, storing abrasive debris and lubricants for dynamic lubrication and promoting the formation of friction films [45,48,54]. At present, the surface texture is more common in circular, oval, square and other symmetrical texture shapes, and there are also a variety of shape composite surface textures [36,43,47]. Compared with non-textured surfaces, traditional types of textured surfaces typically exhibit superior wear resistance [47]. However, in specific application scenarios, specially designed irregular textures can provide better wear resistance due to their unique geometric features (Table 1). Currently, compared to other preparation techniques, laser processing technology is the most commonly used method for preparing surface textures due to its flexibility and high accuracy. By adjusting the frequency, power and scanning strategy of the laser, the physical features of textures can be precisely controlled, and with the advancement of technology, the cost of laser-textured surfaces is expected to be lowered. However, the geometric dimensions of surface texture are still limited to the micrometer level. The surface occupancy rate of the texture in current research typically ranges from 2% to 20%, and in some special cases the occupancy rate may come up to 30–40% [47]. At present, there is a lot of research on textured surfaces of hard alloy materials, titanium alloys, stainless steel and other materials. This is because these materials are widely used in wear resistance and corrosion resistance, and textured surfaces further improve the frictional properties of wear-resistant material surfaces [20,38,44,55]. The formation of surface lubricating film is related to the surface material and lubricant. Textured surfaces and lubricated surfaces are more likely to form stable lubricating films in the later stage of friction compared to non-textured surfaces. Furthermore, a reasonable texture unit design can dynamically release lubricants, store abrasive particles and ensure the continuous existence of the lubricating film. The main factors influencing lubricating films include the type of lubrication, the self-lubricating properties of the surface material and the dynamic pressure lubrication capability of the texture unit [39,44,47,63,64]. The anti-friction effect of surface texture involves numerous complex influencing factors (Table 1), which require a comprehensive consideration of the interaction of these factors in the design stage. In order to achieve optimal tribological performance, a systematic approach must be adopted to comprehensively analyze and evaluate the effects of various parameters. Machine learning, as an advanced data analysis tool, provides an effective means to handle and optimize multi-parameter problems in surface texture design. By training algorithms to identify complex patterns and trends, machine learning can predict the contributions of different texture parameters to tribological properties and guide rational design optimizations. It is worth noting that when designing the texture shape, the actual working conditions in the application should be taken into account. Radial bearings usually use surface textures with circular pits. However, due to the problem of combustion pressure leakage, circular textures are not suitable for piston rings.

3. The Effect of Asymmetric Texture and Multi-Scale Composite Texture on Tribological Performance

3.1. Asymmetric Texture

Asymmetric design plays an important role in nature, science and engineering. Asymmetry is an important concept in the field of topology in mathematics. In material design, asymmetry can be used to enhance the mechanical properties of materials, such as improving material strength and toughness through asymmetric microstructures [69,70]. Asymmetric design also exists in the design of surface textures to achieve the improvement of surface properties. Shen et al. [71] proposed a numerical method for optimizing surface texture shape based on the SQP algorithm, which eliminates the limitation on the occupancy rate of texture area in the optimization algorithm and can maximize the finding of the optimal texture shape. After optimization, it is found that the optimal texture geometry is similar to the V-shape texture but with a flat front end and an asymmetric structure, as shown in Figure 13. Such an asymmetric texture has much better load-bearing capacity compared to the texture with a symmetric V-shape, which is consistent with the results from the experimental and theoretical work by Li et al. [72]. A similar concept is proposed by Han et al. [73], in which ultimate bearing capacity is used as an effective basis to evaluate the lubrication characteristics of textured surfaces. The research results indicate that asymmetric micro pits can achieve greater load-bearing capacity than symmetric rectangular micro pit textures, which will help improve surface lubrication performance. Asymmetric micro pit structures may become a highly promising candidate shape for future texture design.

Figure 13. Optimization results of surface texture shape based on SQP algorithm (Reproduced with permission from [71], Copyright 2015 by the Elsevier).

The effect of asymmetric texture on surface friction under dry friction conditions was studied by Tewelde et al. [74], in which the groove has an asymmetric cross section as shown in Figure 14. The results show that the sliding direction has a significant influence on the performance of asymmetric textures. The friction coefficient under clockwise rotation is 28.1% higher compared to that under counterclockwise rotation. This is related to the variation in the sharpness of the asymmetric texture edges under different relative motions. Jiang et al. [75] proposed a numerical method based on a genetic algorithm to explore the optimal type of texture under reciprocating sliding conditions. The results showed that the asymmetric texture optimized by numerical methods has the ability to improve overall performance and achieve directional friction control. It also indicates that the design of texture types should be based on actual working conditions, as the optimal texture is not necessarily symmetrical or asymmetric. Schuh et al. [76] successfully achieved the preparation of asymmetric texture on the friction pair surface through end milling and experimentally verified the effect of asymmetric texture surface on reducing the friction coefficient. It shows that there exists an optimal value of β for reducing the friction coefficient of the asymmetric texture, the optimal value of β depends on the height of the operating gap and the optimal value in the study was obtained from the analysis of the friction coefficient, as shown in Figure 15.

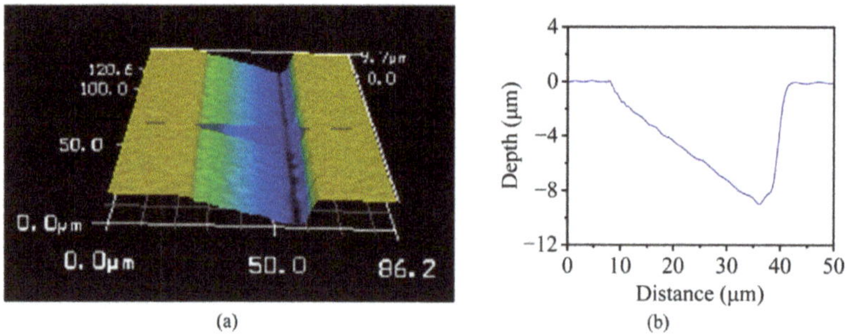

Figure 14. Morphology of a single micro-groove, (**a**) stereoscopic profile image (**b**) sectional profile view (Reproduced with permission from [74], Copyright 2023 by Elsevier).

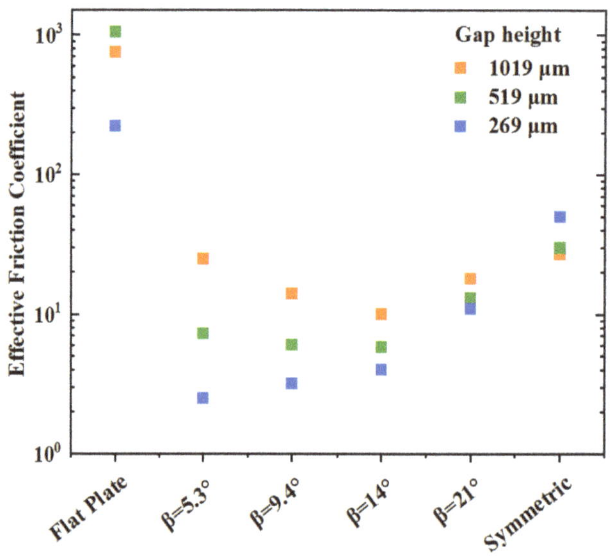

Figure 15. Effective friction coefficient for the S600 oil evaluated at Ω = 100 rad/s for the flat plate and five textures.

The asymmetric textures have shown unique advantages and great potential in directional sliding and enhanced dynamic lubrication compared to symmetric textures in specific cases [71,74]. However, the research on asymmetric surface textures is relatively limited. Through a statistical analysis of the relevant experimental literature, as shown in Table 1, it can be found that there are relatively few studies on asymmetric textures, which may be related to the limitations of their preparation methods. The preparation methods of asymmetric textures in Table 1 mainly rely on mechanical processing, and this method has a limited selection of design parameters. Moreover, the values of surface texture design parameters are at the mesoscale, which increases the difficulty of preparing asymmetric textured surfaces. In contrast, symmetric textures are more easily obtained owing to relatively mature preparation processes and simple geometries. Therefore, in order to promote the development of asymmetric surface texturing technology, future research focusing on innovative preparation methods are desired, which is expected to overcome the challenge in fabrication and minimize the potential benefits of asymmetric textures in tribological applications.

Table 1. Brief summary of experimental data from some references (d (depth), dia (diameter), h (heigh), w (width), CA (Crosshatch angle), β (β is the angle that the material was tilted to make the texture), s (spacing), r (ratio of major axis to minor axis of ellipse)).

	Profile Shape	Geometrical Parameter	Texture Density% or Spacing	Sliding Type	Lubrication Type	Tribological Performance	Friction Reduction Mechanism	Ref.
Laser	Chevron	d = 32 μm	5.05, 9.5, 13.02, 15.2	Unidirectional sliding	Liquid lubricant	The surface with 9.5% texture density shows the lowest friction coefficient.	Increase fluid dynamic pressure	[20]
Laser	Linear groove	d = 20–100 μm w = 20–50 μm	32	Reciprocating sliding	Dry sliding	Compared to non-textured surfaces, the friction coefficient is reduced by 32%, the wear rate is reduced by over 84%.	Form friction films	[35]
Laser	Circular, bullet, fish	d = 15 μm	15.2	Unidirectional sliding	Oil	Compared to circular texture, the friction coefficients of fish shaped and bullet shaped materials are reduced by 21.1% and 30.1%, respectively.	Increase fluid dynamic pressure	[36]
Turning	Cylindrical, cube, cuboid	d = 1.0 mm	/	Unidirectional sliding	Water	Compared to non-textured surfaces, the decrease in friction coefficient is not significant, and the wear loss is reduced by 55%, 63% and 58% for cylindrical, cube and cuboid textures, respectively.	Store debris	[37]
Laser	Chevron	d = 3 μm	9.1 11.5	Reciprocating sliding	Oil	The surface with texture density of 11.5% has the lowest friction coefficient.	Increase the thickness of the lubricating film	[39]
Laser	Square, Circular, triangle	d = 90 μm	8	Unidirectional sliding	Oil	Compared to non-textured surfaces, the friction coefficients are reduced by 17%, 13% and 6% for square, circular and triangle textures, respectively.	Increase fluid dynamic pressure	[40]
Chemical Etching	Square	h = 10, 40, 70, 100 μm	10–50	Unidirectional sliding	Oil	Compared to non-textured surfaces, the lower the height and texture density, the more significant the decrease in friction coefficient: The maximum friction reduction rate of a surface with a height of 10 μm and a texture density of 10% is 80.6%.	Increase fluid dynamic pressure	[43]
Laser	Hexagonal (bionic)	d = 490 μm	s = 360 μm	Reciprocating sliding	Solid lubricant	Compared to non-textured surfaces, the friction coefficient is decreased by 20.82%, and the wear loss is decreased by 65.65%.	Form friction films	[44]
Milling	Linear groove	d = 400 μm w = 1.5 mm	34	Unidirectional sliding	Lubricating oil contains sands	The wear resistance is 4.9 times that of non-textured surfaces.	Store silica sand	[45]
photoetching	Ellipse	r = 1.5 (ST), 2 (MT), 3 (LT)	s = 2 mm	Reciprocating sliding	Starved oil lubrication	Compared to the non-textured surface, the wear loss is decreased by 10.1% and 23.4% for ST and MT surfaces and increased by 35.5 for LT surface.	Form friction films	[47]

Table 1. *Cont.*

	Profile Shape	Geometrical Parameter	Texture Density% or Spacing	Sliding Type	Lubrication Type	Tribological Performance	Friction Reduction Mechanism	Ref.
Laser	Dimple	d = 6–30 μm dia = 50–300 μm	20	Reciprocating sliding	Oil	The samples with texture diameters of 100 μm and depths of 6 μm show enhanced tribological properties.	Store lubricants and remove wear particles	[48]
Laser	Dimple	d = 10 μm	5, 10, 15	Unidirectional sliding	Starved oil lubrication	Compared to the non-textured surface, the friction coefficient is decreased by 51%, 55% and 48% for texture densities of 5%, 10% and 15%, respectively.	Store debris and form lubricating films	[54]
EDM wire cutting	Rectangular Micro-grooves	d = 500 μm w = 300 μm	24, 30, 36, 42	Reciprocating sliding	Grease	Compared to the non-textured surface, the wear rate is the most reduced by 38.6% with the texture density of 36%.	Store lubricating grease	[56]
Laser	Dimple	dia = 15 μm	0.5, 1, 2, 4, 6	Unidirectional sliding	Deionized water	Compared to the non-textured surface, the wear loss is decreased by 21%, 56%, 73%, 66% and 39%, for texture densities of 0.5%, 1%, 2%, 4% and 6%, respectively.	Increase fluid dynamic pressure	[57]
Laser	Dimple	dia = 90 μm d = 10 μm	2, 4, 6, 8, 10	Unidirectional sliding	Graphene/5CB suspension lubrication	The surface with texture density of 8% shows the lowest friction coefficient.	Store lubricating oil, reduce contact area, secondary lubrication	[58]
photoetching	Micro-grooved crosshatch	CA = 20°–60° W = 40 μm	20	Unidirectional sliding	Paraffin oil	The friction coefficient first increases and then decreases, with the highest value at 30°, which is about 6 times higher than that at 60°.	Promote fluid flow	[62]
Laser	Square	d = 10 μm w = 50 μm	s = 100 μm	Reciprocating sliding	Dry sliding, lubrication (virgin PAO-4, SAE20W50)	Compared to the non-textured surface, the friction coefficient of the textured surface is decreased by 63.88%, 70.53% and 44.99%, and the wear loss is decreased by 32.13%, 60.30% and 86.03% under conditions of dry sliding, virgin PAO-4 lubrication and SAE20W50 lubrication, respectively.	Form transfer layers	[64]
Turning	Asymmetric depth of groove	d = 4 μm	20, 30, 40, 70	Opposite bidirectional rotational sliding	Dry sliding	The sharpness of the groove outlet edge increases when rotated clockwise, resulting in a higher friction coefficient. The increase in friction coefficient is 61.3% and 28.1% for texture densities of 20% and 70%, respectively.	Reduce contact area	[74]
Milling	Asymmetric-depth-profile textures	β = 5.3°, 9.4°, 14°, 21.7°	s = 8.953 mm	Bidirectional relative motion	Newtonian fluids	The asymmetric textures result in lower friction coefficient than symmetric textures. The asymmetric texture with a base angle β = 5.3 ° shows the lowest effective friction coefficient.	Reduce shear stress and generate normal load	[76]

3.2. Multi-Scale Composite Texture

Considering the fact that a single texture has shown the capacity to improve surface tribological properties, it is attractive to explore the design and application of multiscale composite textures. There are generally two types of multiscale composite textures. One is to combine multiple textures of different shapes on the same surface, and the other one is to combine textures of the same shape but different sizes on the same surface [77–79]. Segu et al. [80] demonstrated in their study that a combination of circular and elliptical surface textures was used to prepare multi-scale textures with specific formula arrays, as shown in Figure 16. Friction and wear experiments were conducted on surfaces with multi-scale composite textures and non-textured surfaces on a unidirectional friction and wear tester. It was found that the surface with multi-scale texture has a lower and more stable friction coefficient than the non-textured surfaces, which is attributed to the enhanced hydrodynamic lubrication. Meng et al. [81] studied the effect of multi-scale composite pit texture on the frictional performance of sliding bearings, and the results showed that compared to a single texture, a multi-scale composite texture performed excellently in improving load-bearing capacity and reducing the friction coefficient. However, it is not necessarily the case that multi-scale composite textures are superior to single textures or smooth surfaces in improving surface tribological performance. The research of Zou et al. [57] shows that multi-scale composite textures do not significantly improve tribological performance under certain conditions. Therefore, whether it is a single texture or a multi-scale composite texture, it is necessary to fully consider the parameters and experimental conditions of the texture itself when using it.

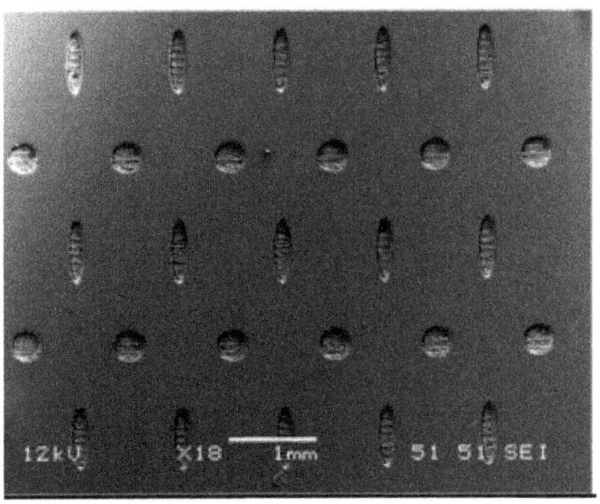

Figure 16. The SEM micrographs of multi-scale textured surface. (Reproduced with permission from [80], Copyright 2013 by the Elsevier).

3.3. Summary

Asymmetric textures and multi-scale composite textures have shown unique advantages in improving tribological properties compared to single-texture and non-textured surfaces. Due to the fact that there is no consensus on the mechanism of the tribological performance of single and symmetric textures, how multi-scale composite or asymmetric textures contribute to the reduction of friction and wear is not clear, and related research deep into the mechanism is limited at the current stage. More influencing factors such as the complex combination of shape, geometry and distribution of textures and the complexity of geometric features of asymmetric textures makes it hard to establish a systemic approach for the design of asymmetric or composite textures. However, the development of processing techniques makes it easier to access asymmetric and composite textures with complex parameters, which promotes the design, discovery and understanding of complex textures in tribological applications.

4. Summary and Prospect

In this work, we review the research progress on the design of surface textures in tribological applications. The influence of surface texture parameters on the tribological properties of friction pairs is discussed from the perspective of texture parameters. It can be concluded that the shape, geometry and density of textures have a significant impact on tribological performance while the distribution of textures has a smaller impact on tribological performance. The main anti-friction mechanism of surface textures includes enhancing the dynamic pressure lubrication capability, storing abrasive debris and lubricants for dynamic lubrication or promoting the formation of friction films on surfaces of friction pairs. A texture with a unidirectional convergence shape, such as a V-shape, in unidirectional lubrication conditions has good tribological properties. The formation of friction film is related to the surface material and lubrication state. Reasonable surface texture design can promote the formation of friction film and reduce the friction coefficient. Under dry friction conditions, a reasonable design of texture depth can store more debris and reduce abrasive wear. Asymmetric textures and multi-scale composite textures also show great anti-friction performance and have potential in specific applications, but they still lack systematic research. The main factor hindering their development is the limited manufacturing approach. Surface texturing technology has been proven to effectively reduce the friction coefficient and energy consumption through the optimization of texture parameters, but considering the significant influence of working conditions on the performance of surface textures, the current design of surface textures still largely relies on case studies. Therefore, some future research directions are proposed based on the review:

1. The development of a general mathematical method for designing optimal texture parameters based on the anti-friction mechanism and experimental design parameters;
2. The development of standardized testing methods and computational models to better understand and predict the performance of surface textures;
3. The development of preparation methods for asymmetric textures;
4. The development of a multi-parameter optimization method based on machine learning, which could be used to analyze and predict the tribological properties of textured surfaces, evaluate the weight of different parameters and optimize the design of surface textures.

Author Contributions: K.C.: writing—original draft preparation; Y.T.: review and editing, supervision, funding acquisition. All authors have read and agreed to the published version of the manuscript.

Funding: This work was supported by National Natural Science Foundation of China (No. 52305203), "Qilu Young Scholar" program of Shandong University (No. 31360082363167), Shandong Excellent Young Scientists Fund Program (Overseas) (No. 2024HWYQ-003), Young Taishan Scholars Program of Shandong Province (No. tsqn202306087), Guangxi Key Laboratory of Manufacturing System & Advanced Manufacturing Technology (No. 24354S012), and Central Guidance on Local Science and Technology Development Fund of Shandong Province (YDZX2024084), during the course of this work.

Data Availability Statement: No new data were created or analyzed in this study. Data sharing is not applicable to this article.

Conflicts of Interest: The authors declare no conflict of interest.

References

1. Zhao, J.; Sheng, W.; Li, Z.M.Q.; Zhang, H.; Zhu, R.P. Effect of lubricant selection on the wear characteristics of spur gear under oil-air mixed lubrication. *Tribol. Int.* **2022**, *167*, 107382. [CrossRef]
2. Somers, A.E.; Howlett, P.C.; MacFarlane, D.R.; Forsyth, M. A Review of Ionic Liquid Lubricants. *Lubricants* **2013**, *1*, 3–21. [CrossRef]
3. Meng, Y.G.; Xu, J.; Jin, Z.M.; Prakash, B.; Hu, Y.Z. A review of recent advances in tribology. *Friction* **2021**, *8*, 221–300. [CrossRef]
4. Liao, C.; Erbaugh, J.T.; Kelly, A.C.; Agrawal, A. Clean energy transitions and human well-being outcomes in lower and middle income countries: A systematic review. *Renew. Sustain. Energy Rev.* **2021**, *145*, 111036. [CrossRef]

5. Poudyal, R.; Loskot, P.; Nepal, R.; Parajuli, R.; Khadka, S.K. Mitigating the current energy crisis in Nepal with renewable energy sources. *Renew. Sustain. Energy Rev.* **2019**, *116*, 109338. [CrossRef]
6. Yu, R.F.; Chen, W. Research progress and prospect of surface texturing in industrial tribology. *J. Mech. Eng.* **2017**, *53*, 100–110. [CrossRef]
7. Holmberg, K.; Erdemir, A. Influence of tribology on global energy consumption, costs and emissions. *Friction* **2017**, *5*, 263–284. [CrossRef]
8. Zhao, X.D.; Zhang, Y.M.; Gao, S.Z. Evaluation and analysis of abrasive wear resistance of hybrid roller bearings under lubricant contamination. *Wear* **2024**, *558–559*, 205570. [CrossRef]
9. Tang, Y.Q.; Pan, H.B.; Li, D.Y. Contribution of cold-work to the wear resistance of materials and its limitation—A study combining molecular dynamics modeling and experimental investigation. *Wear* **2021**, *476*, 203642. [CrossRef]
10. Roy, S.; Speed, L., Jr.; Viola, M.; Luo, H.M.; Leonard, D.; Qu, J. Oil miscible phosphonium-phosphate ionic liquid as novel antiwear and antipitting additive for low-viscosity rear axle lubricants. *Wear* **2021**, *466–467*, 203588. [CrossRef]
11. Mohit, V.; Paras, K.; Qasim, M. Surface texturing techniques to enhance tribological performance: A review. *Surf. Interfaces* **2021**, *27*, 101463.
12. Andriy, K.; Oyelayo, A.; Ali, E.; George, F.; Izhak, E. The effect of laser surface texturing on transitions in lubrication regimes during unidirectional sliding contact. *Tribol. Int.* **2005**, *38*, 219–225.
13. Wang, X.L.; Koshi, A.; Katsunori, O.; Koji, K. Optimization of the surface texture for silicon carbide sliding in water. *Appl. Surf. Sci.* **2006**, *253*, 1282–1286. [CrossRef]
14. Chen, K.Y.; Yang, X.F.; Lv, G.J.; Zhang, Y.F.; Yang, H.; Gao, Y.L. Research progress of improving surface friction properties by surface texture technology. *Int. J. Adv. Manufcturing Technol.* **2021**, *116*, 2797–2821. [CrossRef]
15. Chen, K.Y.; Yang, X.F.; Zhang, Y.F.; Lv, G.J.; Yang, H.; Gao, Y.L. Research progress of surface modification and solid-liquid lubrication synergistic friction reduction and wear resistance. *Int. J. Adv. Manufcturing Technol.* **2022**, *122*, 1115–1141. [CrossRef]
16. Dean, B.; Bhushan, B. Shark-skin surfaces for fluid-drag reduction in turbulent flow: A review. *Philos. Trans. A Math. Phys. Eng. Sci.* **2010**, *368*, 5737.
17. Wang, Z.Q.; Ye, R.H.; Xiang, J.B. The performance of textured surface in friction reducing: A review. *Tribol. Int.* **2023**, *177*, 108010. [CrossRef]
18. Hamilton, D.B.; Walowit, J.A.; Allen, C.M.A. Theory of lubrication by micro asperities. *J. Basic Eng. ASME* **1966**, *88*, 177–185. [CrossRef]
19. Xiang, P.; Cao, J.; Long, Q.H.; Wang, S.M.; Wu, J.X.; Abbas, Z. Tribological properties of PTFE/PI-PAI composite coatings filled by GO/ ZrO2 on the surfaces of different micro-textures. *Colloids Surf. A Physicochem. Eng. Asp.* **2024**, *686*, 133412. [CrossRef]
20. Li, D.; Yang, X.F.; Lu, C.Y.; Cheng, J.; Wang, S.R.; Wang, Y.J. Tribological characteristics of a cemented carbide friction surface with chevron pattern micro-texture based on different texture density. *Tribol. Int.* **2020**, *142*, 106016.
21. Kuriachen, B. Influence of surface texture of electric discharge machined Ti6Al4V on the surface wettability. *Precis. Eng.* **2024**, *86*, 317–330. [CrossRef]
22. Magnus, C.; Gulenc, I.; Rainforth, W.M. Ambient dry sliding friction and wear behaviour of laser surface textured (LST) Ti_3SiC_2 MAX phase composite against hardened steel and alumina. *Wear* **2022**, *490–491*, 204184. [CrossRef]
23. Meng, Y.; Deng, J.X.; Zhang, Y.; Wang, S.J.; Li, X.M.; Yue, H.Z.; Ge, D.L. Tribological properties of textured surfaces fabricated on AISI 1045 steels by ultrasonic surface rolling under dry reciprocating sliding. *Wear* **2020**, *460–461*, 203488. [CrossRef]
24. Guo, Q.G.; Zheng, L.; Zhong, Y.H.; Wang, S.K.; Ren, L.Q. Numerical simulation of hydrodynamic lubrication performance for continuous groove-textured surface. *Tribol. Int.* **2022**, *167*, 107411. [CrossRef]
25. Lena, A.; Kari, M.; Suvanto, M. Precise fabrication of microtextured stainless steel surfaces using metal injection moulding. *Precis. Eng.* **2020**, *62*, 89–94.
26. Koshy, P.; Tovey, J. Performance of electrical discharge textured cutting tools. *CIRP Ann.—Manuf. Technol.* **2011**, *60*, 153–156. [CrossRef]
27. Wang, W.H.; Yuan, W.; Guo, Q.J.; Wang, N.N.; Chi, B.T.; Yu, J. Effect of picosecond laser surface texturing under Babbitt coating mask on friction and wear properties of GCr15 bearing steel surface. *Eng. Fail. Anal.* **2024**, *157*, 107878. [CrossRef]
28. Wang, B.; Liu, Z.Q.; Cai, Y.K.; Luo, X.C.; Ma, H.F.; Song, Q.H.; Xiong, Z.H. Advancements in material removal mechanism and surface integrity of high speed metal cutting: A review. *Int. J. Mach. Tools Manuf.* **2021**, *166*, 103744. [CrossRef]
29. Hu, Y.H.; Chen, X.N.; Chen, J.; Zhang, C.Q.; Fu, W.H. The influence of crescent texture parameters on the axial force when drilling bone. *Med. Eng. Phys.* **2021**, *87*, 87–94. [CrossRef]
30. Borghi, A.; Gualttieri, E.; Marchetto, D.; Moretti, L.; Valeri, S. Tribological effects of surface texturing on nitriding steel for high-performance engine applications. *Wear* **2008**, *265*, 1046–1051. [CrossRef]
31. Lu, P.; Wood, R.J.K. Tribological performance of surface texturing in mechanical applications—A review. *Surf. Topogr. Metrol. Prop.* **2020**, *8*, 043001. [CrossRef]
32. Chen, L.X.; Liu, Z.Q.; Shen, Q. Enhancing tribological performance by anodizing micro-textured surfaces with nano-MoS_2 coatings prepared on aluminum-silicon alloys. *Tribol. Int.* **2018**, *122*, 84–95. [CrossRef]
33. Tan, N.; Li, Y.; Lou, L.Y.; Zhang, G.L.; Xing, Z.G.; Wang, H.D. Influence of micro-nano multiscale surface texture on wettability of Ni-based droplets at high temperature. *Surf. Coat. Technol.* **2021**, *418*, 127103. [CrossRef]

34. Zhang, L.W.; Liu, G.; Guo, Y.R.; Wang, Y.; Zhang, D.Y.; Chen, H.W. Bioinspired functional surfaces for medical devices. *Chin. J. Mech. Eng.* **2022**, *35*, 43. [CrossRef]
35. Li, X.; Deng, J.X.; Yue, H.; Ge, D.; Zou, X. Wear performance of electrohydrodynamically atomized WS_2 coatings deposited on biomimetic shark-skin textured surfaces. *Tribol. Int.* **2019**, *134*, 240–251. [CrossRef]
36. Zhang, H.; Hua, M.; Dong, G.Z.; Zhang, D.Y.; Chen, W.J.; Dong, G.N. Optimization of texture shape based on Genetic Algorithm under unidirectional sliding. *Tribol. Int.* **2017**, *115*, 222–232. [CrossRef]
37. Guo, Z.W.; Xie, X.; Yuan, C.Q.; Bai, X.Q. Study on influence of micro convex textures on tribological performances of UHMWPE material under the water-lubricated conditions. *Wear* **2019**, *426–427*, 1327–1335. [CrossRef]
38. Khan, M.A.; Gupta, K. A study on machinability of nickel based superalloy using micro-textured tungsten carbide cutting tools. *Mater. Res. Express* **2020**, *7*, 016537. [CrossRef]
39. Morris, N.; Leighton, M.; De la Cruz, M.; Rahmani, R.; Rahnejat, H.; Howell-Smith, S. Combined numerical and experimental investigation of the micro-hydrodynamics of chevron-based textured patterns influencing conjunctional friction of sliding contacts. *Proc. Inst. Mech. Eng. Part J J. Eng. Tribol.* **2015**, *229*, 316–335. [CrossRef]
40. Wang, J.; Yan, Z.; Fang, X.; Shen, Z.; Pan, X. Observation and experimental investigation on cavitation effect of friction pair surface texture. *Lubr. Sci.* **2020**, *32*, 404–414. [CrossRef]
41. Nie, T.; Yang, K.; Zhou, L.; Wu, X.; Wang, Y. CFD analysis of load capacity of journal bearing with surface texture. *Energy Rep.* **2022**, *8*, 327–334. [CrossRef]
42. Qiu, M.F.; Minson, B.R.; Bart, R. The effect of texture shape on the friction coefficient and stiffness of gas-lubricated parallel slider bearings. *Tribol. Int.* **2013**, *67*, 278–288. [CrossRef]
43. Babu, P.; Syed, I.; Ben, B. Experimental and numerical studies of positive texture effect on friction reduction of sliding contact under mixed lubrication. *Proc. Inst. Mech. Eng. Part J J. Eng. Tribol.* **2020**, *208–210*, 1994–1996. [CrossRef]
44. Huang, Q.P.; Shi, X.L.; Xue, Y.W.; Zhang, K.P.; Wu, C.H. Optimization of bionic textured parameter to improve the tribological performance of AISI 4140 self-lubricating composite through response surface methodology. *Tribol. Int.* **2021**, *161*, 107104. [CrossRef]
45. Zhang, D.Y.; Zhao, F.F.; Wei, X.; Gao, F.; Li, P.Y.; Dong, G.N. Effect of texture parameters on the tribological properties of spheroidal graphite cast iron groove-textured surface under sand-containing oil lubrication conditions. *Wear* **2019**, *428–429*, 470–480. [CrossRef]
46. Patel, D.S.; Singh, A.; Balani, K.; Ramkumar, J. Topographical effects of laser surface texturing on various time-dependent wetting regimes in Ti6Al4V. *Surf. Coat. Technol.* **2018**, *349*, 816–829. [CrossRef]
47. Xu, Y.F.; Zheng, Q.; Abuflaha, R.; Olson, D.; Furlong, O.; You, T.; Zhang, Q.Q.; Hu, X.G.; Tysoe, W.T. Influence of dimple shape on tribofilm formation and tribological properties of textured surfaces under full and starved lubrication. *Tribol. Int.* **2019**, *136*, 267–275. [CrossRef]
48. Arslan, A.; Masjuki, H.H.; Varman, M.; Kalam, M.A.; Quazi, M.M.; Mahmud, K.A.H.A.; Gulzar, M.; Habibullah, M. Effects of texture diameter and depth on the tribological performance of DLC coating under lubricated sliding condition. *Appl. Surf. Sci.* **2015**, *365*, 1135–1149. [CrossRef]
49. Zeng, Q.R.; Qin, Y.; Chang, W.L.; Luo, X.C. Correlating and evaluating the functionality-related properties with surface texture parameters and specific characteristics of machined components. *Int. J. Mech. Sci.* **2018**, *149*, 62–72. [CrossRef]
50. Butler, K.T.; Davies, D.W.; Cartwright, H.; Isayev, O.; Walsh, A. Machine learning for molecular and materials science. *Nature* **2018**, *559*, 547–555. [CrossRef]
51. Li, B.C.; Deng, B.L.; Shou, W.; Oh, T.H.; Hu, Y.M.; Luo, Y.Y.; Shi, L.; Matusik, W. Computational discovery of microstructured composites with optimal stiffness-toughness trade-offs. *Sci. Adv.* **2024**, *10*, eadk4284. [CrossRef] [PubMed]
52. Zhu, B.; Zhang, W.X.; Zhang, W.S.; Li, H.X. Generative design of texture for sliding surface based on machine learning. *Tribol. Int.* **2023**, *179*, 108139. [CrossRef]
53. Li, Z.S.; Li, J.Q.; An, B.; Li, R. The design method for surface texture of sliding friction pairs based on machine learning under mixed lubrication. *Tribol. Int.* **2024**, *194*, 109563. [CrossRef]
54. Niu, Y.X.; Pang, X.J.; Yue, S.W.; Shangguan, B.; Zhang, Y.Z. The friction and wear behavior of laser textured surfaces in non-conformal contact under starved lubrication. *Wear* **2021**, *476*, 203723. [CrossRef]
55. Yuan, S.; Lin, N.M.; Wang, W.H.; Zhang, H.X.; Liu, Z.Q.; Yu, Y.; Zeng, Q.F.; Wu, Y.C. Correlation between surface textural parameter and tribological behaviour of four metal materials with laser surface texturing (LST). *Appl. Surf. Sci.* **2022**, *583*, 152410. [CrossRef]
56. Zhang, K.P.; Shi, X.L.; Xue, Y.W.; Huang, Q.P.; Wu, C.H. Effects of composite textured surface on friction characteristics of 42CrMo steel under grease lubrication. *Wear* **2022**, *504–505*, 204419. [CrossRef]
57. Zou, H.B.; Yan, S.; Shen, T.; Wang, H.J.; Li, Y.; Chen, T.Y.; Meng, Y.Q.; Men, S.C.; Zhang, Z.J.; Sui, T.Y.; et al. Efficiency of surface texturing in the reducing of wear for tests starting with initial point contact. *Wear* **2021**, *482–483*, 203957. [CrossRef]
58. Li, S.S.; Chen, H.; Luo, T.; Xiao, G.C.; Yi, M.D.; Chen, Z.Q.; Zhang, J.J.; Xu, C.H. Tribological properties of laser surface texturing modified GCr15 steel under graphene/5CB lubrication. *J. Mater. Res. Technol.* **2022**, *18*, 3598–3611. [CrossRef]
59. Shen, Z.J.; Wang, F.C.; Chen, Z.G.; Ruan, X.P.; Zeng, H.H.; Wang, J.H.; An, Y.R.; Fan, X.L. Numerical simulation of lubrication performance on chevron textured surface under hydrodynamic lubrication. *Tribol. Int.* **2021**, *154*, 106704. [CrossRef]

60. Li, J.L.; Chen, P.; Shao, T.M. The Effect of gourd-shaped surface texture on tribological performance of stainless steel. *Tribology* **2016**, *36*, 207–214.
61. Shinkarenko, A.; Kligerman, Y.; Etsion, I. The effect of surface texturing in soft elastohydrodynamic lubrication. *Tribol. Int.* **2009**, *42*, 284–292. [CrossRef]
62. Suh, M.; Chae, Y.; Kim, S. Effect of geometrical parameters in micro-grooved crosshatch pattern under lubricated sliding friction. *Tribol. Int.* **2010**, *43*, 1508–1517. [CrossRef]
63. Kong, X.; Sun, W.Y.; Wang, Q.C.; Chen, M.H.; Zhang, T.; Wang, F.H. Improving high-temperature wear resistance of NiCr matrix self-lubricating composites by controlling oxidation and surface texturing. *J. Mater. Sci. Technol.* **2022**, *131*, 253–263. [CrossRef]
64. Srivyas, P.D.; Wani, M.F.; Sehgal, R.; Bisht CS, S.; Charoo, M.S.; Raina, A.; Haq MI, U. Synergetic effect of surface texturing and graphene nanoplatelets on the tribological properties of hybrid self-lubricating composite. *Tribol. Int.* **2022**, *168*, 107434. [CrossRef]
65. Liu, J.Q.; Rong, H.; Yang, R.L.; Zhang, Y.B.; Hu, Y.; Dong, A.J.; Wang, X.F.; Wang, Y.; Chen, D.M.; Liu, K.Z.; et al. A novel antibacterial and wear-resistant strategy based on metal surface micro-texture nitriding coupled functional mesoporous silicon coating. *Surf. Coat. Technol.* **2024**, *492*, 131220. [CrossRef]
66. Lo, C.C.; Chen, L.J. Comparative wetting behavior of condensed and sessile drops on single micro-scale textured hydrophobic surfaces: Physical insights for condenser design. *Surf. Interfaces* **2024**, *48*, 10427. [CrossRef]
67. Shum, P.W.; Zhou, Z.F.; Li, K.Y. To increase the hydrophobicity, non-stickiness and wear resistance of DLC surface by surface texturing using a laser ablation process. *Tribol. Int* **2014**, *78*, 1–6. [CrossRef]
68. Wang, Y.; Peng, W.X.; Tong, H.; Sun, Y.B.; Liu, Z.J.; Wu, F.H. A biomimetic micro-texture based on shark surface for tool wear reduction and wettability change. *J. Manuf. Process.* **2024**, *129*, 202–214. [CrossRef]
69. Taali, S.; Moazzen, P.; Toroghinejad, M.R.; Chen, G.H.; Mola, J. Microstructure, texture, and mechanical properties of asymmetrically cold-rolled $Ni_{1.5}FeCrCu_{0.5}$ high-entropy alloy. *J. Mater. Res. Technol.* **2022**, *21*, 3489–3501. [CrossRef]
70. Xie, G.M.; Cui, H.B.; Luo, Z.A.; Misra, R.D.K.; Wang, G.D. Asymmetric distribution of microstructure and impact toughness in stir zone during friction stir processed a high strength pipeline steel. *Mater. Sci. Eng. A* **2017**, *704*, 401–411. [CrossRef]
71. Shen, C.; Khonsari, M.M. Numerical optimization of texture shape for parallel surfaces under unidirectional and bidirectional sliding. *Tribol. Int.* **2015**, *82*, 1–11. [CrossRef]
72. Li, Y.L.; Zhang, Z.H.; He, Y.Y.; Luo, J.B. Optimization of surface texture distribution on the thrust bearing using the mass-conserving cavitation boundary condition: Theory and experiments. *Tribol. Int.* **2024**, *194*, 109551. [CrossRef]
73. Han, J.; Fang, L.; Sun, J.P.; Wang, Y.Q.; Ge, S.R.; Zhu, H. Hydrodynamic Lubrication of Surfaces with Asymmetric Microdimple. *Tribol. Trans.* **2011**, *54*, 607–615. [CrossRef]
74. Tewelde, F.B.; Zhou, T.F.; Zhou, J.; Guo, W.J.; Zhao, B.; Ge, X.Y.; Wang, W.Z.; Wang, X.L.; Wang, X.B. Asymmetric surface texturing for directional friction control under dry sliding condition. *Tribol. Int.* **2023**, *181*, 108321. [CrossRef]
75. Jiang, S.; Liu, Q.; Wang, J.W.; Ji, H.; Dong, G.N. Numerical optimization of asymmetric surface texturing under reciprocating sliding conditions. *Tribol. Int.* **2023**, *180*, 108310. [CrossRef]
76. Schuh, J.K.; Ewoldt, R.H. Asymmetric surface textures decrease friction with Newtonian fluids in full film lubricated sliding contact. *Tribol. Int.* **2016**, *97*, 490–498. [CrossRef]
77. Annadi, R.R.; Syed, I. Impact of multi-scaled surface textures on tribological performance of parallel sliding contact under lubricated condition. *Tribol. Int.* **2023**, *183*, 108415. [CrossRef]
78. Marian, M.; Zambrano, D.F.; Rothammer, B.; Waltenberger, V.; Boidi, G.; Krapf, A.; Merle, B.; Stampfl, J.; Rosenkranz, A.; Gachot, C.; et al. Combining multi-scale surface texturing and DLC coatings for improved tribological performance of 3D printed polymers. *Surf. Coat. Technol.* **2023**, *466*, 129682. [CrossRef]
79. Tang, X.; Tian, Y.; Tian, X.W.; Li, W.; Han, X.; Kong, T.T.; Wang, L.Q. Design of multi-scale textured surfaces for unconventional liquid harnessing. *Mater. Today* **2021**, *43*, 62–83. [CrossRef]
80. Segu, D.Z.; Choi, S.G.; Choi, J.; Kim, S.S. The effect of multi-scale laser textured surface on lubrication regime. *Appl. Surf. Sci.* **2013**, *270*, 58–63. [CrossRef]
81. Meng, F.M.; Zhang, L.; Liu, Y.; Li, T.T. Effect of compound dimple on tribological performances of journal bearing. *Tribol. Int.* **2015**, *91*, 99–110. [CrossRef]

symmetry MDPI

Article

System Modeling and Simulation for Investigating Dynamic Characteristics of Geared Symmetric System Based on Linear Analysis

Joo-Mi Bahk, Sun-Hak Kim and Jong-Yun Yoon *

Department of Mechatronics Engineering, Incheon National University, (Songdo-dong) 119 Academy-ro, Yeonsu-gu, Incheon 22012, Republic of Korea; bahkjm@inu.ac.kr (J.-M.B.); sunhak@inu.ac.kr (S.-H.K.)
* Correspondence: yoon3932@inu.ac.kr; Tel.: +82-32-835-8682

Abstract: Complex vibrational phenomena, such as gear impacts and mesh stiffness excitations, often require a significant amount of effort to be revealed using nonlinear analytical methods. However, key parameters for addressing vibrational problems can often be identified through simplified approaches based on linear analysis models. In light of these considerations, this study aimed to propose linear analytical methods to investigate the influences of various key parameters within symmetric systems. To achieve the main goal of this study, system modeling and eigensolutions were first implemented, focusing on a specific manual transmission with a front-engine/front-wheel configuration. Second, analytical techniques to reduce the number of degrees of freedom from the original symmetric system were suggested, and the reduced model was validated. Third, the system responses in the time domain were examined, along with key system parameters, such as gear mesh stiffness and clutch dampers, using state–variable equations. As a result, the findings from the linear system model demonstrated the fundamental dynamic characteristics of the torsional system within specific frequency regimes relevant to noise and vibration problems. Furthermore, the reduced lumped linear model employing the state–variable formula established its reliability in determining key parameters for mitigating noise and vibration problems.

Keywords: geared system; manual transmission; driveline; linear analysis; eigensolutions; system reduction; symmetrically torsional system; state–variable equation

1. Introduction

In the field of system dynamics, which focuses on torsional vibrations occurring in a vehicle driveline with symmetric conditions, numerous studies have been conducted to improve noise, vibration, and harshness (NVH) qualities [1–22]. These studies have aimed to identify the main causes of NVH problems, such as gear rattle and whine noise, by comprehending the basic mechanisms of drivelines with gear pairs and their dynamic characteristics through both linear and nonlinear analyses [10–13]. To determine the key parameters related to the specific noise and vibration issues, linear analysis can be performed as a fundamental step to reveal the critical system parameters relevant to the NVH problems in question [1–7]. When it comes to examining system responses, the analytical approach is commonly more efficient for diagnosing and improving vehicle driveline problems under symmetric conditions than the experimental method, as experiments generally require more significant costs and time than simulations [16–19].

In a review of prior studies, Yoon and Singh investigated the dynamic properties of gear rattle under wide-open throttle (WOT) and coast conditions. They proposed a simplified nonlinear model employing drag torque estimation [1]. Guo et al. suggested a prediction model for gear rattle incorporating various key parameters, such as backlash, time-varying mesh stiffness, nonlinear oil film force between teeth, and drag torque. In this study, the developed model was compared with experimental methods, revealing the influence of each parameter on gear rattle [2]. Zhou et al. conducted experimental studies on the rattle behavior of a complete dual-clutch transmission (DCT) system, both on a rattle test bench and in a vehicle [3]. Through this study, the authors investigated the gear rattle sensitivity and the dynamic evolution of the geartrain in a DCT. Diez-Ibarbia et al. proposed examining the lubricant effect between teeth in contact to understand its role in non-stationary phenomena. In this context, the gear transmission model was used to calculate the dynamic transmission error (DTE), with a specific focus on the influence of fluid viscosity on the system's dynamic behavior [4]. Bozca proposed the method to reduce the gear-rattle noise of the five-speed manual gearbox with the optimization of gearbox geometric design parameters [5]. Rigaud and Perret-Liaudet investigated the nonlinear dynamic behavior of a rattling spur gear pair induced by vibro-impacts between gear teeth. They suggested their own experimental set-up to visualize the contact zone and identify the successive impacts [6]. Donmez and Kahraman proposed a gear rattle prediction model by incorporating manufacturing errors such as eccentricities and tooth indexing errors [7]. Singh et al. proposed design guidelines for automotive manual transmissions and discussed gear rattle criteria based on clutch, flywheel, and drag torque [8]. Pizzolante et al. proposed an analytical procedure for the generalization of the rattle index in any type of ordinary transmission layout. To determine the rattle index, they conducted numerical studies to examine the effectiveness and reliability of the introduced index [9]. Guo et al. suggested a comprehensive nonlinear model to investigate the effects of influencing parameters such as drag torque, nonlinear oil film force, lubricant viscosity, time-varying meshing stiffness, friction force, and gear backlash. This study demonstrated a nonlinear dynamic model of gear rattling that considered the combined effects of the aforementioned parameters and their severity [10]. Shangguan et al. presented a nonlinear torsional model for a driveline system with four degrees of freedom to study gear rattle under idling conditions [11]. In this study, they developed experimental and calculation methods to reduce gear rattle in a generic transmission at idle. Trochon suggested a lumped parameter model of an automotive manual transmission corresponding to a nonlinear system and compared neutral rattle phenomena under driving and coast conditions [12]. Idehara et al. conducted and verified a system model of a powertrain with gear pairs to reduce gear impacts by changing clutch disc parameters [13]. Beinstingel et al. investigated parametrically excited vibrations and parametric instabilities arising from time-varying gear mesh stiffness, resulting in gear whine [14]. In this study, the authors analyzed the nonlinear model using the Newmark integration scheme and the harmonic balance approach to identify parametric instabilities. Pizzolante et al. proposed an analytical formulation capable of forecasting the main overall direction and magnitude of bearing reaction forces on an idler gear under quasi-static conditions [15]. In this study, they aimed to examine the geartrain formed by three gears to observe the modification of the ellipse shape concerning the excitation frequency and different modal shapes. Palermo et al. introduced a transmission error (TE) measurement with respect to the traditional direct method to demonstrate that low-cost digital encoders can be successfully used together with the elapsed time method [16]. To verify this measurement, the authors exploited it to compare the theoretical and experimental methods. Mughal et al. introduced a gear tribodynamics model using the potential energy method to estimate time variable meshing stiffness (TVMS) [17]. In this research, the authors suggested a fully analytical and time-efficient model for lubricated contact stiffness based on transitions in the regimes of lubrication. Yoo et al. proposed a hybrid metal composite gear to reduce gear whine noise and evaluated dynamic transmission error (DTE) [18]. In this study, they investigated the effect of composite material in reducing DTE in a hybrid metal composite gear in which the tooth and hub were steel-based and the blank was composed of a composite material. Barthod et al. suggested an experimental gearbox model for characterizing the rattle threshold and noise caused by engine torque fluctuations, investigating the rattle threshold, influence of excitation, and geometrical gearbox parameters [19]. Cui et al. proposed an analytical algorithm for meshing stiffness relative to the gear peeling position based on the energy method, comparing the effects on peeling length, depth, and width [20]. Chen et al. proposed matching models between the dual-mass flywheel (DMF) and power transmission system based on sensitivity and the natural frequency range, adjusting structural and resultant parameters to understand the dynamic characteristics using the torsional stiffness of the DMF [21]. Ren et al. suggested a high-speed vehicle model to obtain the gear meshing forces and investigated their effect on the gear transmission under high-speed conditions [22].

According to the summaries of prior research, there appears to be a lack of fundamental approaches for conducting linear analysis in symmetric driveline systems, even though Singh et al. suggested a linearized design guideline to address gear impact problems [8]. Furthermore, the reviewed articles conducted their work using nonlinear analytical methods and their own experimental setups to resolve gear impact vibration issues. However, these methods require a significant amount of effort and cost. Thus, to overcome the complexity associated with nonlinear analysis, this study proposes the use of linear analytical methods to investigate and determine gear impact problems and their relevant key parameters. Therefore, the specific objectives of this study were as follows: (1) to propose mathematical techniques for reducing the number of degrees of freedom (DOFs) by comparing eigensolutions and frequency response functions (FRFs) derived from the original system with the results obtained from the reduced model; (2) to investigate linear system responses in the time domain using state variable equations, which were compared with results obtained from key factors, such as magnitude and phase, in the FRFs. This approach aimed to present efficient methods for determining key parameters to reduce the specific noise and vibration problems with less effort compared with nonlinear analysis. To achieve these specific objectives, this study focused on a specific manual transmission with a front-engine/front-wheel driveline assembly. In addition, Figure 1 illustrates the analytical approach using linear analysis. For example, the system modeling process was conducted in the first step, followed by the development of the original and reduced system models. The verification of the reduced model, involving a comparison of dynamic characteristics, such as eigensolutions and FRFs, led to the state variable equations to reveal the system responses and the effectiveness of key parameters in the time domain. In the case study, the primary key parameters, such as K_c, I_{o5}, and K_g, were the focus, and their designations are announced later.

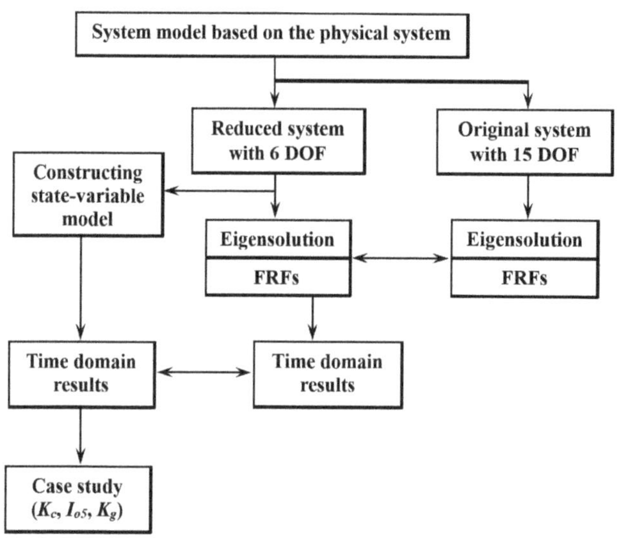

Figure 1. Flowchart of analytical approach for this study.

2. Problem Formulation

In this study, a specific manual transmission was analyzed under 3rd gear engaged and 5th gear unloaded condition within a front-engine/front-wheel drive configuration. Two models with 15 and 6 degrees of freedom (DOFs) were used for analysis [1]. Figure 2a depicts the physical driveline system with 15 DOFs. The power generated by the engine is transmitted from the flywheel and clutch to the tires through the transmission, as shown in Figure 2.

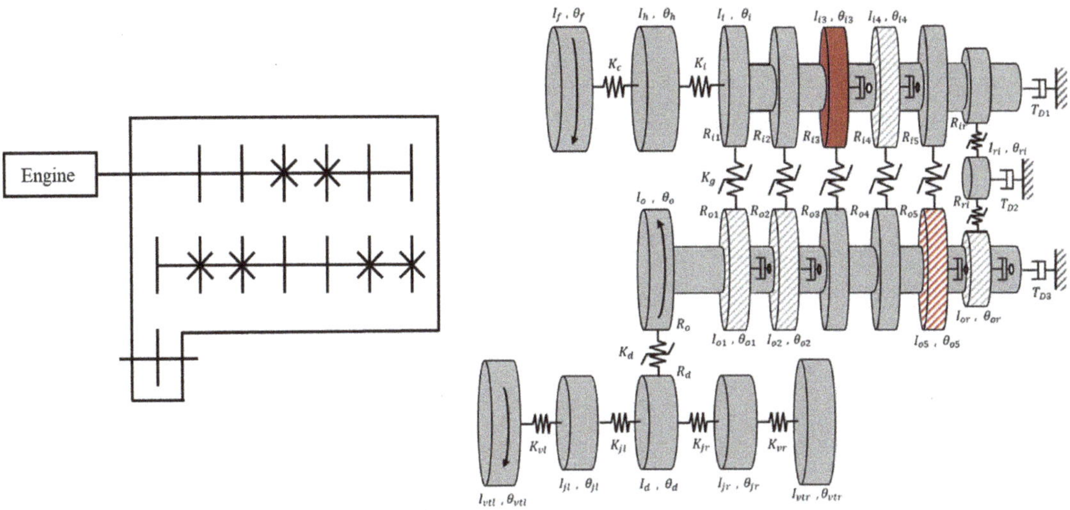

Figure 2. Schematic of driveline system with 15 DOFs: (**a**) physical driveline system [1]; (**b**) schematic diagram.

The manual transmission plays a crucial role in adapting the transferred torque and speed to suit various driving conditions. The transmission consists of three major components: input and output shafts, speed gears, and synchronizers. In Figure 2a, the symbols marked with a cross represent the speed gears, which remain unloaded and freely rotate unless the synchronizer is engaged. The straight lines in Figure 2a represent the fixed gears welded to the input (or output) shaft. Figure 2b provides a schematic diagram based on the physical system shown in Figure 2a. Each element in the schematic diagram is represented by lumped inertia or stiffness. The synchronizers and fixed gears are assumed to be lumped into the input and output shafts. The symbols and their descriptions for the lumped inertia and stiffnesses are summarized in Tables 1 and 2. To fulfill the objectives of this study, several assumptions were made: first, nonlinearities such as clutch hysteresis and gear backlashes were disregarded in the linear analysis. This omission removed the energy loss effect caused by dry friction and practical gear impact phenomena, respectively; second, gear pairs were assumed to be continuously in contact on the driving side, which always ensures a gear rattle-free condition; third, only one clutch stiffness value was considered, even though the clutch dampers exhibited multi-staged properties. This limitation prevented the exploration of gear impact conditions caused by changes in stiffness.

Table 1. Inertia employed in simulations.

Inertia (Description)	Values (kg·m^2)	Inertia	Values (kg·m^2)
I_f (Flywheel)	1.38 × 10−1	I_{o5} (5th gear on output shaft)	5.23×10^{-4}
I_h (Clutch hub)	5.76×10^{-3}	I_{ri} (Idler)	4.35×10^{-4}
I_i (Input shaft)	3.10×10^{-3}	I_{or} (Reverse gear on output shaft)	1.33×10^{-3}
I_o (Output shaft)	25.33×10^{-3}	I_d (Differential)	2.15×10^{-2}
I_{i3} (3rd gear on input shaft)	5.80×10^{-4}	I_{jl} (Left cv joint)	3.91×10^{-3}
I_{i4} (4th gear on input shaft)	8.73×10^{-4}	I_{jr} (Right cv joint)	4.35×10^{-3}
I_{o1} (1st gear on output shaft)	2.60×10^{-3}	I_{vtl} (Left tire and vehicle)	23.9591
I_{o2} (2nd gear on output shaft)	1.39×10^{-3}	I_{vtr} (Right tire and vehicle)	23.9591

Table 2. Stiffness employed in simulations.

Stiffness (Description)	Values (N·m·rad^{-1})	Stiffness	Values (N·m^{-1})
K_i (Input shaft stiffness)	10,000	K_g (Gear mesh stiffness)	2.7×10^8
K_{jl}, K_{jr} (Stiffness between left or right drive shaft and differential)	10,000	K_d (Gear mesh stiffness between output shaft and differential)	2.7×10^8
K_{vl}, K_{vr} (Stiffness between left or right cv joint and vehicle)	10,000		
K_c (Clutch stiffness)	1838		

3. Modeling and Analysis of the Symmetric Driveline for Linear Analysis

3.1. Modeling of the Torsional System

The 15-DOF driveline system for this study consisted of several components, including the flywheel, clutch, driveshaft, joint, and vehicle, as shown in Figure 2b. Each of the elements depicted in Figure 2b represents lumped inertias and stiffnesses. The lumped mass, denoted I_f, is connected to the clutch hub, I_h, by the clutch stiffness, K_c. Using Newton's second law, Equation (1) can be derived as follows.

$$\sum_{i=1}^{n} T_i = -K_c\left(\theta_f - \theta_h\right) = I_f\ddot{\theta}_f \quad (1)$$

The torsional displacements of sub-systems connected with gear pairs are influenced by each gear ratio. The effect of these gear ratios should be incorporated into the mathematical formulation by considering the relative displacements in the gear pairs. The schematic for the clutch hub, input shaft, and speed gears is depicted in Figure 3a, where the inertia values of the 3rd gear and its relevant synchronizer are combined into the input shaft inertia, considering their engaged status. Figure 3b illustrates the basic mechanism with a free body diagram (FBD) that operates between gear pairs, such as the input shaft and speed gears on the output shaft, along with the torque flow. The torque transferred at the engaged gear pairs can be expressed as $R_{in}K_g\delta_n$. Here, $\delta_n = R_{in}\theta_{in} + R_{on}\theta_{on}$ represents the relative displacement of the speed gear pairs. The radius of each gear is summarized in Table 3.

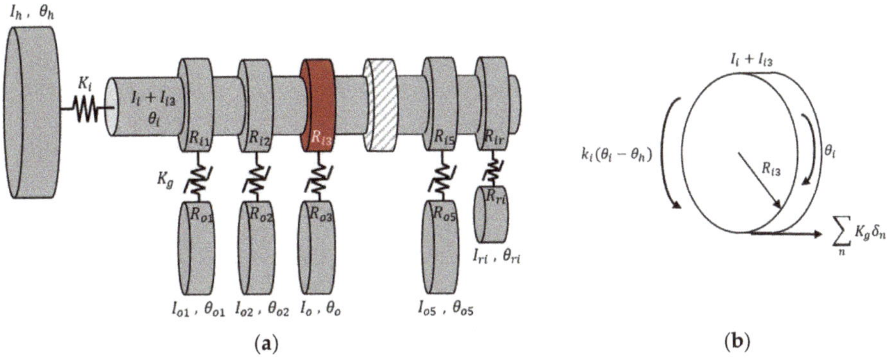

Figure 3. Schematic of the clutch hub, input shaft, and speed gears: (**a**) lumped masses with clutch hub, input shaft, and speed gears; (**b**) free body diagram of the input shaft.

Table 3. Gear radii employed in simulations.

Radius (Description)	Values (mm)	Radius	Values (mm)
R_{i1} (1st gear on input shaft)	18.30	R_{o2} (2nd gear on output shaft)	51.99
R_{i2} (2nd gear on input shaft)	29.51	R_{o3} (3rd gear on output shaft)	46.01
R_{i3} (3rd gear on input shaft)	35.50	R_{o4} (4th gear on output shaft)	39.55
R_{i4} (4th gear on input shaft)	41.95	R_{o5} (5th gear on output shaft)	35.58
R_{i5} (5th gear on input shaft)	45.92	R_{or} (Reverse gear on output shaft)	54.95
R_{ir} (Reverse gear on input shaft)	16.36	R_o (Final gear on output shaft)	26.63
R_{ri} (Idler)	40.14	R_d (Final gear on differential)	103.37
R_{o1} (1st gear on output shaft)	63.20		

As illustrated in Figure 3, the basic derivation of dynamic equations can be described as follows.

$$\begin{aligned}\sum_{i=1}^{n} T_i = -K_i(\theta_i - \theta_h) - K_g(R_{i1}\theta_{i1} + R_{o1}\theta_{o1})R_{i1} - K_g(R_{i2}\theta_{i2} + R_{o2}\theta_{o2})R_{i2} \\ -K_g(R_{i5}\theta_{i5} + R_{o5}\theta_{o5})R_{i5} - K_g(R_{ir}\theta_i + R_{ri}\theta_{ri})R_{ir} - K_g(R_{i3}\theta_i + R_{o3}\theta_o)R_{i3} \\ = (I_i + I_{i3})\ddot{\theta}_i\end{aligned} \quad (2)$$

Thus, the complete equation of motion with 15 DOFs can be established as follows. Here, the torque excitation and damping effects are neglected for the implementation of the linear real eigensolutions, and $\underline{\theta}(t)$, $\underline{\underline{M}}$, and $\underline{\underline{K}}$ represent the absolute torsional displacement, inertia, and stiffness matrices, respectively.

$$\underline{\underline{M}}\,\ddot{\underline{\theta}}(t) + \underline{\underline{K}}\,\underline{\theta}(t) = \underline{0}. \quad (3)$$

$$\underline{\theta} = \left[\theta_f\ \theta_h\ \theta_i\ \theta_o\ \theta_{01}\ \theta_{02}\ \theta_{05}\ \theta_{ri}\ \theta_{or}\ \theta_{i4}\ \theta_d\ \theta_{jl}\ \theta_{jr}\ \theta_{vlt}\ \theta_{vtr}\right]^T. \quad (4)$$

$$\underline{\underline{M}} = diag\left[I_f, I_h, (I_i + I_{i3}), I_o, I_{o1}, I_{o2}, I_{o5}, I_{ri}, I_{or}, I_{i4}, I_d, I_{jl}, I_{jr}, I_{vtl}, I_{vtr}\right]. \quad (5)$$

$$\underline{\underline{K}} = \begin{bmatrix} & k(i,j) & \end{bmatrix}. \tag{6}$$

$$k_{1,1} = k_c, \tag{7a}$$

$$k_{2,2} = k_c + k_i. \tag{7b}$$

$$k_{3,3} = k_i + k_g \left({R_{i1}}^2 + {R_{i2}}^2 + {R_{i3}}^2 + {R_{i5}}^2 + {R_{ir}}^2 \right). \tag{8}$$

$$k_{4,4} = k_g \left({R_{o3}}^2 + {R_{o4}}^2 \right) + k_d {R_o}^2. \tag{9}$$

$$k_{5,5} = k_g {R_{o1}}^2, \tag{10a}$$

$$k_{6,6} = k_g {R_{o2}}^2. \tag{10b}$$

$$k_{7,7} = k_g {R_{o5}}^2, \tag{11a}$$

$$k_{8,8} = 2k_g {R_{ri}}^2. \tag{11b}$$

$$k_{9,9} = k_g {R_{or}}^2, \tag{12a}$$

$$k_{10,10} = k_g {R_{i4}}^2. \tag{12b}$$

$$k_{11,11} = k_d {R_d}^2 + k_{jl} + k_{jr}. \tag{13}$$

$$k_{12,12} = k_{jl} + k_{vl}, \tag{14a}$$

$$k_{13,13} = k_{jr} + k_{vr}. \tag{14b}$$

$$k_{14,14} = k_{vl}, \tag{15a}$$

$$k_{15,15} = k_{vr}. \tag{15b}$$

$$k_{1,2} = k_{2,1} = -k_c, \tag{16a}$$

$$k_{2,3} = k_{3,2} = -k_i. \tag{16b}$$

$$k_{3,4} = k_{4,3} = k_g R_{i3} R_{o3}, \tag{17a}$$

$$k_{3,5} = k_{5,3} = k_g R_{i1} R_{o1}. \tag{17b}$$

$$k_{3,6} = k_{6,3} = k_g R_{i2} R_{o2}, \tag{18a}$$

$$k_{3,7} = k_{7,3} = k_g R_{i5} R_{o5}. \tag{18b}$$

$$k_{3,8} = k_{8,3} = k_g R_{ir} R_{ri}, \tag{19a}$$

$$k_{4,10} = k_{10,4} = k_g R_{i4} R_{o4}. \tag{19b}$$

$$k_{4,11} = k_{11,4} = k_g R_o R_d, \tag{20a}$$

$$k_{8,9} = k_{9,8} = k_g R_{ri} R_{or}. \tag{20b}$$

$$k_{11,12} = k_{12,11} = -k_{jl}, \tag{21a}$$

$$k_{11,13} = k_{13,11} = -k_{jr}. \tag{21b}$$

$$k_{12,14} = k_{14,12} = -k_{vl}, \tag{22a}$$

$$k_{13,15} = k_{15,13} = -k_{vr}. \tag{22b}$$

Here, the units K_g and K_d are in N·m^{-1}, and the torsional stiffness has the unit of N·m·rad^{-1}.

3.2. Modal Analysis of the Torsional System

In general, the dynamic characteristics of a system were examined on the basis of the modal analysis, in which the natural frequencies and mode shapes reveal the fundamental dynamic behaviors of the system when it is excited. Consequently, the results from modal analysis provide the practical information necessary for improving vibratory conditions resulting from various excitation forces or torques. The characteristic equation of this system can be obtained by multiplying Equation (3) by the inverse inertia matrix $\underline{\underline{M}}^{-1}$ as follows [23–29].

$$\underline{\underline{M}}^{-1}\underline{\underline{M}}\,\ddot{\underline{\theta}}(t) + \underline{\underline{M}}^{-1}\underline{\underline{K}}\,\underline{\theta}(t) = \underline{0}. \tag{23}$$

$$\ddot{\underline{\theta}}(t) + \underline{\underline{M}}^{-1}\underline{\underline{K}}\,\underline{\theta}(t) = \underline{0}. \tag{24}$$

$$\underline{\theta}(t) = \underline{U}e^{i(wt+\phi)}. \tag{25}$$

$$\ddot{\underline{\theta}}(t) = -w^2\underline{U}e^{i(wt+\phi)} = -\lambda\underline{U}e^{i(wt+\phi)}. \tag{26}$$

$$-\lambda\underline{U}e^{i(wt+\phi)} + \underline{\underline{M}}^{-1}\underline{\underline{K}}\,\underline{U}e^{i(wt+\phi)} = 0. \tag{27}$$

$$\left[-\lambda\underline{\underline{I}} + \underline{\underline{M}}^{-1}\underline{\underline{K}}\right]\underline{U}e^{i(wt+\phi)} = \underline{0}. \tag{28}$$

$$\underline{U} \neq 0,\ det\left|\underline{\underline{M}}^{-1}\underline{\underline{K}} - \lambda\underline{\underline{I}}\right| = 0. \tag{29}$$

$$(\lambda - \lambda_1)(\lambda - \lambda_2)\cdots(\lambda - \lambda_r)\cdots(\lambda - \lambda_{15}) = 0,\ \lambda_r = w_r{}^2,\ r = 1, 2, \cdots, r, \cdots 15. \tag{30}$$

$$\left[\underline{\underline{M}}^{-1}\underline{\underline{K}} - \lambda_r \underline{\underline{I}}\right]\underline{U}_r = \underline{0}. \tag{31}$$

$$\underline{\underline{K}}\,\underline{U}_r = \lambda_r \underline{\underline{M}}\,\underline{U}_r. \tag{32}$$

$$\left[\underline{\underline{U}}\right] = \left[\underline{U}_1\ \underline{U}_2 \cdots \underline{U}_r \cdots \underline{U}_{15}\right]. \tag{33}$$

In this context, the general solution of Equation (24) can be expressed as Equation (25), where $\underline{U}$ is an arbitrary constant vector with a size of 15 by 1, and ϕ represents an arbitrary phase. Here, $\lambda_r = w_r{}^2$ is an eigenvalue, and w_r is the corresponding natural frequency of the system. Eigenvectors corresponding to each eigenvalue are obtained from Equation (32), where $\underline{U}_r$ represents the rth eigenvector. The modal matrix $\underline{\underline{U}}$ consists of each eigenvector augmented, as described in Equation (33). Therefore, each column vector projects its relevant mode shape for each concerned natural frequency. Mode shapes are vectors that represent the dynamic motions for each natural frequency, and the systems positioned after the output shaft, including the idler, are influenced by the gear ratio. For instance, concerning the output shaft, the torsional displacement is affected by the radii of R_{i3} and R_{o3}, which are related to the engagement of the 3rd gear. Consequently, the modal vector component $U_{(4,i)}$, which is associated with the output shaft, should be normalized considering the gear ratio, resulting in a modified modal vector $U_{(4,i)} \times \frac{R_{o3}}{R_{i3}}$. Considering the influence of the gear ratio for the rest of the sub-systems and normalizing the eigenvectors with respect to their maximum vector components yields the normalized modal vectors. The first four natural frequencies and mode shapes are presented in Table 4 and Figure 4. This study utilized MATLAB to obtain the results. Here, the y- and x-axes of Figure 3 are the normalized values of the modal vector and their relevant subsystem, as described in Equations (4) and (5). Considering the typical range of vehicle speeds, they do not exceed 8000 RPM, and at this speed, the firing frequency is below 300 Hz. Therefore, natural frequencies higher than 300 Hz were ignored. Additionally, since this torsional system is semi-positive definite, a rigid body mode appears, which must be disregarded due to numerical instabilities in the simulations [12,23–29].

Table 4. First four natural frequencies of the 15-DOF system.

Mode	Description	Natural Frequency (Hz)
1	Hopping mode (f_1)	2.3
2	Driveline surging mode (f_2)	7.6
3	Clutch spring mode (f_3)	60.8
4	Clutch + input shaft mode (f_4)	272.5

3.3. *System Responses in the Frequency Domain*

To obtain the frequency response functions (FRFs) based on the results calculated through modal analysis, damping effects must be included, as the real eigensolution is conducted without considering damping effects. Since it can be challenging to directly measure physical damping under dynamic conditions, the damping matrix can be reasonably approximated from the decoupled modal domain [12,23–29]. The mathematical formulation $\underline{P}(t)$ as principal coordinates can be calculated by transferring the basic coordinates $\underline{\theta}(t)$, as follows.

$$\underline{\underline{M}}\,\ddot{\underline{\theta}}(t) + \underline{\underline{K}}\,\underline{\theta}(t) = \underline{0}, \tag{34a}$$

$$\underline{\theta}(t) = \underline{\underline{U}}\,\underline{P}(t). \tag{34b}$$

$$\underline{\underline{M}}\,\underline{\underline{U}}\,\ddot{\underline{P}}(t) + \underline{\underline{K}}\,\underline{\underline{U}}\,\underline{P}(t) = \underline{0}. \tag{35}$$

$$\underline{\underline{U}}^T\underline{\underline{M}}\,\underline{\underline{U}}\,\ddot{\underline{P}}(t) + \underline{\underline{U}}^T\underline{\underline{K}}\,\underline{\underline{U}}\,\underline{P}(t) = \underline{0}. \tag{36}$$

$$\underline{\underline{M_r}}\ddot{\underline{P}}(t) + \underline{\underline{K_r}}\,\underline{P}(t) = \underline{0}. \tag{37}$$

$$\begin{bmatrix} \ddots & & \\ & M_{ri} & \\ & & \ddots \end{bmatrix}\ddot{\underline{P}} + \begin{bmatrix} \ddots & & \\ & \lambda_i M_{ri} & \\ & & \ddots \end{bmatrix}\underline{P} = \underline{0},\ i = 1, 2, \cdots 15. \tag{38}$$

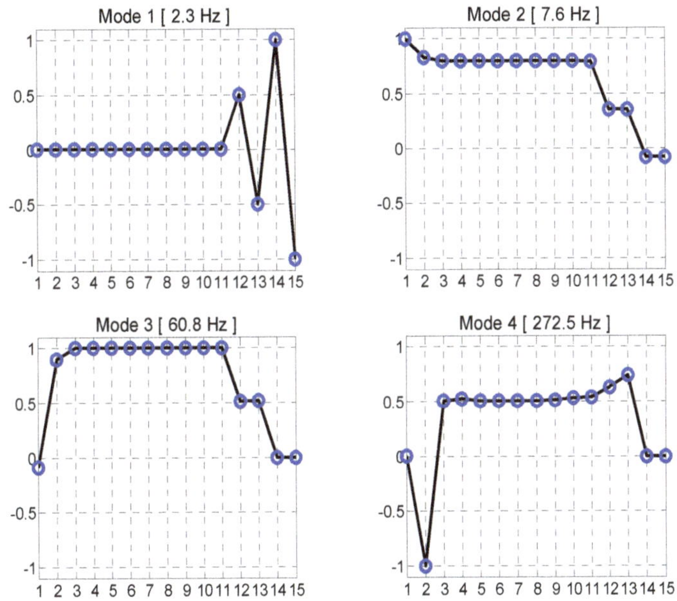

Figure 4. First four modes of the 15-DOF system.

Here, $\underline{\underline{M_r}}$ and $\underline{\underline{K_r}}$ represent the modal mass and stiffness matrices. Equation (38) describes the decoupled system, which is transformed into the modal coordinates $\underline{\eta}(t)$ from the physical coordinate system. From Equation (38), the normalized constant matrix $\underline{\underline{\beta_r}}$ is obtained as follows.

$$\underline{\underline{M_r}} = \begin{bmatrix} \ddots & & \\ & M_{ri} & \\ & & \ddots \end{bmatrix} = \begin{bmatrix} \ddots & & \\ & \sqrt{M_{ri}} & \\ & & \ddots \end{bmatrix}\begin{bmatrix} \ddots & & \\ & \sqrt{M_{ri}} & \\ & & \ddots \end{bmatrix}. \tag{39}$$

$$\underline{\underline{\beta_r}} = \underline{\underline{M_r}}^{\frac{1}{2}} = \begin{bmatrix} \ddots & & \\ & \sqrt{M_{ri}} & \\ & & \ddots \end{bmatrix}. \tag{40}$$

$$\underline{\underline{\beta_r}}^2\ddot{\underline{P}}(t) + \underline{\underline{K_r}}\,\underline{P}(t) = \underline{0},\ \underline{P}(t) = \underline{\underline{\beta_r}}^{-1}\underline{\eta}(t). \tag{41}$$

$$\underline{\underline{\beta_r}}^{-1}\underline{\underline{\beta_r}}\,\underline{\underline{\beta_r}}\,\underline{\underline{\beta_r}}^{-1}\ddot{\underline{\eta}}(t)+\underline{\underline{\beta_r}}^{-1}\underline{\underline{K_r}}\,\underline{\underline{\beta_r}}^{-1}\underline{\eta}(t)=\underline{0}. \tag{42}$$

$$\ddot{\underline{\eta}}(t)+\begin{bmatrix}\ddots & & \\ & {w_{ri}}^2 & \\ & & \ddots\end{bmatrix}\underline{\eta}(t)=\underline{0}. \tag{43}$$

Here, Equation (43) clearly shows that the system is decoupled by reflecting its natural frequencies of the 15-DOF system. To transform the physical coordinate $\underline{\theta}(t)$ into the modal coordinate $\underline{\eta}(t)$, two stages of calculations are performed, as described in Equations (34)–(43), using the principal coordinate $\underline{P}(t)$. However, this transformation can be simplified into a stage, as follows.

$$\frac{1}{\sqrt{M_{ri}}}\underline{U_i}^T\underline{\underline{M}}\,\underline{U_i}\frac{1}{\sqrt{M_{ri}}}=1,\ \mathrm{i}=1,2,3,\cdots,15. \tag{44}$$

$$\underline{\underline{\phi}}=\begin{bmatrix}\underline{U}_1 & \underline{U}_2 & \cdots & \underline{U}_{15}\end{bmatrix}\begin{bmatrix}\frac{1}{\sqrt{M_{r1}}} & & & \\ & \frac{1}{\sqrt{M_{r2}}} & & \\ & & \ddots & \\ & & & \frac{1}{\sqrt{M_{r15}}}\end{bmatrix}=\underline{\underline{U}}\,\underline{\underline{\beta}}. \tag{45}$$

$$\underline{\underline{M}}\,\ddot{\underline{\theta}}(t)+\underline{\underline{K}}\,\underline{\theta}(t)=\underline{0},\ \underline{\theta}(t)=\underline{\underline{\phi}}\,\underline{\eta}(t). \tag{46}$$

$$\underline{\underline{M}}\,\underline{\underline{\phi}}\,\ddot{\underline{\eta}}(t)+\underline{\underline{K}}\,\underline{\underline{\phi}}\,\underline{\eta}(t)=\underline{0}. \tag{47}$$

$$\underline{\underline{\phi}}^T\underline{\underline{M}}\,\underline{\underline{\phi}}\,\ddot{\underline{\eta}}(t)+\underline{\underline{\phi}}^T\underline{\underline{K}}\,\underline{\underline{\phi}}\,\underline{\eta}(t)=\underline{0}. \tag{48}$$

$$\ddot{\eta}_r+{w_r}^2\eta_r(t)=0, r=1,2,3,\cdots,15. \tag{49}$$

Here, $\underline{\underline{\phi}}$ is the normal modal matrix, and Equations (44)–(49) represent the series of techniques for transforming the physical coordinates into the modal domain. Using the normal modal matrix $\underline{\underline{\phi}}$, the viscous damping matrix can be approximated as follows.

$$\underline{\underline{\phi}}^T\cdot\underline{\underline{C}}\cdot\underline{\underline{\phi}}=\begin{bmatrix}\ddots & & \\ & C_r & \\ & & \ddots\end{bmatrix}=\begin{bmatrix}\ddots & & \\ & 2\zeta w_r & \\ & & \ddots\end{bmatrix}. \tag{50}$$

$$[\mathrm{C}]=\left[\phi^T\right]^{-1}\begin{bmatrix}\ddots & & \\ & 2\zeta_r w_r & \\ & & \ddots\end{bmatrix}[\phi]^{-1}. \tag{51}$$

$$\underline{\underline{M}}\,\ddot{\underline{\theta}}(t)+\underline{\underline{C}}\,\dot{\underline{\theta}}(t)+\underline{\underline{K}}\,\underline{\theta}(t)=\underline{T}(t). \tag{52}$$

$$\underline{T}(\mathrm{t})=\begin{bmatrix}T & 0 & 0 & 0 & 0 & 0 & 0 & 0 & 0 & 0 & 0 & 0 & 0 & 0 & 0\end{bmatrix}^T. \tag{53}$$

$$\underline{\theta}(t)=\tilde{\underline{\theta}}e^{i\omega t}, \tag{54a}$$

$$\dot{\underline{\theta}}(t)=i\cdot w\cdot\tilde{\underline{\theta}}e^{i\omega t}, \tag{54b}$$

$$\ddot{\underline{\theta}}(t) = -w^2 \cdot \underline{\tilde{\theta}} e^{i\omega t}. \tag{54c}$$

$$\left[\underline{\underline{K}} - \underline{\underline{M}} \cdot w^2 + i\underline{\underline{C}}w\right] \underline{\tilde{\theta}} e^{i\omega t} = \underline{\tilde{T}} e^{i\omega t}. \tag{55}$$

$$\underline{\tilde{\theta}}(w) = \left[\underline{\underline{K}} - \underline{\underline{M}}w^2 + i\underline{\underline{C}}w\right]^{-1} \underline{\tilde{T}}. \tag{56}$$

In this context, the damping matrix $\underline{\underline{C}}$ can be inversely derived reversely from the modal damping matrix C_r. In this study, a damping ratio ζ of 5% was employed, and Equation (52) represents the fundamental equation, incorporating the calculated physical damping matrix and its input torque vector. The excitation value for the input torque $\underline{T}$ is set to unity. From Equation (56), the FRFs of the 15-DOF system can be calculated. The FRFs in the 15-DOF system described in Figure 5 are defined as follows: $\left|Y_{ff}\right|$, FRFs of the flywheel; $\left|Y_{hf}\right|$, FRFs of the clutch hub; $\left|Y_{if}\right|$, FRFs of the input shaft. The resonance regions in the FRFs depicted in Figure 5 correlate well with the natural frequencies obtained from the modal analysis.

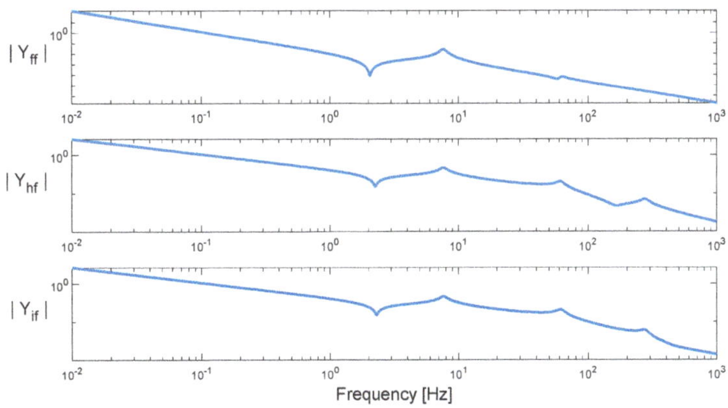

Figure 5. First three FRFs of 15-DOF model.

4. System Reduction and Validation with the Original System

The previously developed 15-DOF system model provides all dynamic characteristics, such as natural frequencies and resonance areas. However, the size of matrices and vectors in its mathematical formulations can lead to various issues in numerical simulations, including inefficient calculation times and convergence problems in numerical iterations, especially concerning sensitive nonlinearities [30]. To increase the reliability and efficiency of simulations, particularly when dealing with highly nonlinear elements such as gear impacts and parameter excitations in gear pairs, it is necessary to reduce the number of DOFs. To achieve the primary goal of this study, we focused on reducing the DOFs from 15 to 6 using a two-step process. First, the speed gears, which were not currently relevant for the examination, were lumped together with each driving input or output shaft. For instance, the 4th speed gear (or 1st, 2nd, and reverse gears) could be combined with the output shaft (or input shaft), as shown in Figures 2, 6 and 7 by considering their gear ratios, as illustrated in Figure 6 [31]. Second, the rotational system from the final gear to the wheel was simplified by distributing the inertia and stiffness values on the driveshaft equally, as illustrated in Figure 8 [32].

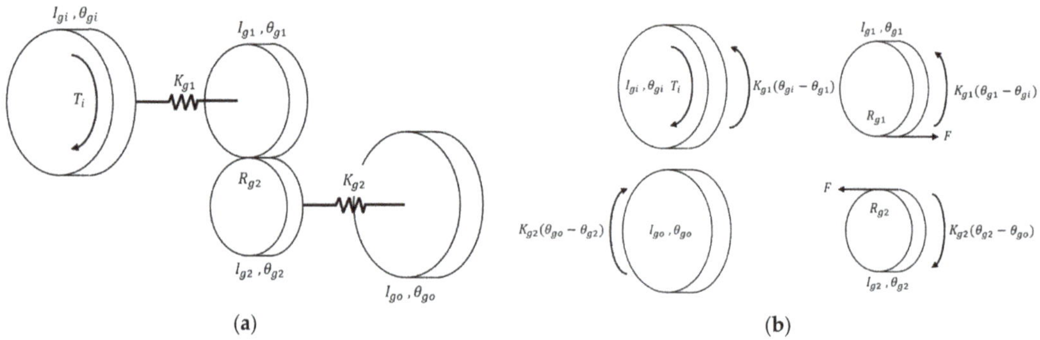

Figure 6. Geared system with 4 DOFs: (**a**) description of the torsional system with lumped inertias and springs; (**b**) free body diagram for the equation of motions.

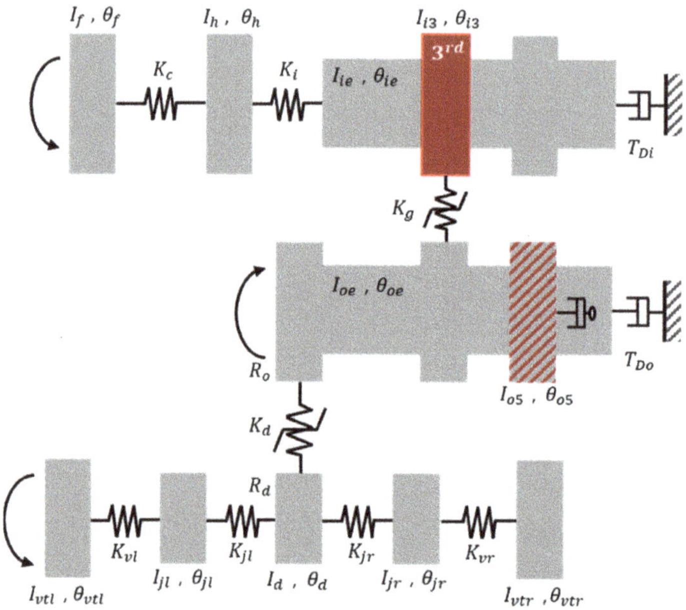

Figure 7. Lumped parameter system model for a 10−DOF torsional system.

To reduce the lumped model for the gear pairs, we can consider a simple geared system, as illustrated in Figure 6. Figure 6 shows a simple geared system with 4 DOFs and its FBD. In this diagram, I_{gi} and I_{go} represent the inertias of input and output gears, respectively, while I_{g1} and I_{g2} show the inertias of speed gear pairs. K_{g1} and K_{g2} present the assumed constant values of the gear mesh stiffness. θ_{gi}, θ_{go}, θ_{g1}, and θ_{g2} denote the angular displacements of each lumped mass, as presented in Figure 6. T_i represents the external torque operating on the input gear, and F is the internal force at the contact point between the gear pair. Basic dynamic motions can be derived using the illustrated FBD and Newton's second law, as follows.

$$I_{gi}\ddot{\theta}_{gi} + K_{g1}\theta_{gi} - K_{g1}\theta_{g1} = T_i. \quad (57)$$

$$I_{g1}\ddot{\theta}_{g1} + K_{g1}\theta_{g1} - K_{g1}\theta_{gi} = -(F \times R_{g1}). \quad (58)$$

$$I_{g2}\ddot{\theta}_{g2} + K_{g2}\theta_{g2} - K_{g2}\theta_{go} = (F \times R_{g2}). \tag{59}$$

$$I_{go}\ddot{\theta}_{go} + K_{g2}\theta_{go} - K_{g2}\theta_{g2} = 0. \tag{60}$$

$$\frac{I_{g1}\ddot{\theta}_{g1} + K_{g1}\theta_{g1} - K_{g1}\theta_{gi}}{I_{g2}\ddot{\theta}_{g2} + K_{g2}\theta_{g2} - K_{g2}\theta_{go}} = \frac{-(F \times R_{g1})}{(F \times R_{g2})} = -n. \tag{61}$$

$$\left(I_{g1} + n^2 I_{g2}\right)\ddot{\theta}_{g1} - K_{g1}\theta_{gi} + \left(K_{g1} + n^2 K_{g2}\right)\theta_{g1} - nK_{g2}\theta_{go} = 0. \tag{62}$$

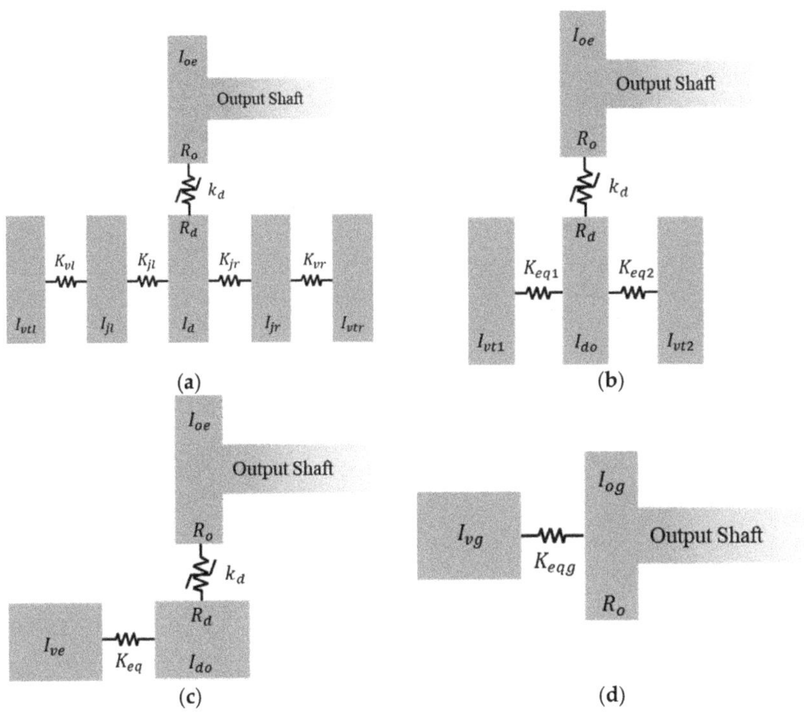

Figure 8. Schematics of the differential gear set: (**a**) schematic of the final gear drive with 10 DOFs; (**b**) schematic of the final gear drive with 8 DOFs; (**c**) schematic of the final gear drive with 7 DOFs; (**d**) schematic of the final gear drive with 6 DOFs.

Here, Equation (62) is derived from Equations (58) and (59) by utilizing the gear ratio $\frac{R_{g1}}{R_{g2}} = \frac{\theta_{g2}}{\theta_{g1}} = n$, as described in Equation (61). Consequently, the 4-DOF geared system depicted in Figure 6a is ultimately reduced to a 3-DOF system, in conjunction with Equations (58)–(62). Therefore, using the techniques suggested above, speed gears, such as 1st, 2nd, 4th, and reverse gears, as illustrated in Figure 2, can be combined with the input and output shafts. In summary, the original system with 15 DOFs, as shown in Figure 2, can be reduced to a 10-DOF geared system, as depicted in Figure 7. Thus, the reduced model for the practical system illustrated in Figure 7 is formulated with the reduced DOF using the following basic equations.

$$\underline{\underline{M}}\,\ddot{\underline{\theta}}(t) + \underline{\underline{K}}\,\underline{\theta}(t) = \underline{0}. \tag{63}$$

$$\underline{\theta} = \left[\theta_f \; \theta_h \; \theta_{ie} \; \theta_{05} \; \theta_{og} \; \theta_{vg}\right]^T. \tag{64}$$

$$\underline{\underline{M}} = diag\left[I_f, I_h, I_{ie}, I_{05}, I_{og}, I_{vg}\right]. \tag{65}$$

$$\underline{\underline{K}} = \left[\quad k(i,j) \quad\right]. \tag{66}$$

$$I_{ie} = I_{ie} + I_{i3} + \left(\frac{R_{i1}}{R_{o1}}\right)^2 I_{o1} + \left(\frac{R_{i2}}{R_{o2}}\right)^2 I_{o2} + \left(\frac{R_{ir}}{R_{ri}}\right)^2 \left\{I_{ri} + \left(\frac{R_{ri}}{R_{or}}\right)^2 I_{or}\right\}. \tag{67}$$

$$I_{oe} = I_o + \left(\frac{R_{o4}}{R_{i4}}\right)^2 I_{i4}. \tag{68}$$

The inertias of the reduced system were designated as follows: I_{ie}, input shaft lumped with 1st, 2nd, and reverse gears; I_{oe}, output shaft lumped with the 4th gear. In addition, the inertia and stiffness values on the driveshaft could be reduced by assuming equivalent stiffness values for the parallel and series connections of the spring. As illustrated in Figure 8, their equivalent properties were derived as follows.

$$K_{eqg} = (R_o/R_d)^2 K_{eq}, \tag{69a}$$

$$K_{eq} = K_{eq1} + K_{eq2}. \tag{69b}$$

$$K_{eq1} = K_{eq2} = \frac{K_{vr}K_{jr}}{K_{vr} + K_{jr}}. \tag{70}$$

$$I_{vg} = (R_o/R_d)^2 I_{ve}, \tag{71a}$$

$$I_{og} = I_{oe} + (R_o/R_d)^2 I_{do}. \tag{71b}$$

$$I_{ve} = I_{vt1} + I_{vt2}. \tag{72}$$

$$I_{vt1} = I_{vtl} + \frac{I_{jl}}{2}, \tag{73a}$$

$$I_{vt2} = I_{vtr} + \frac{I_{jr}}{2}. \tag{73b}$$

$$I_{do} = I_d + \frac{I_{jl} + I_{jr}}{2}. \tag{74}$$

In Figure 8c, K_{eq} is lumped into K_{eqg} in Figure 8d, as described in Equation (69a). The inertias and stiffnesses employed in the system reduction are designated as follows: I_{do}, differential gear lumped with the left and right cv joints; I_{vt1}, left tire and vehicle lumped with the left cv joint; I_{vt2}, right tire and vehicle lumped with the right cv joint; I_{ve}, whole inertia lumped with I_{vt1} and I_{vt2}; I_{og}, output shaft lumped with I_{do}; I_{vg}, inertia of the tire and vehicle affected by the gear ratio $\left(\frac{R_o}{R_d}\right)$; K_{eq1}, equivalent stiffness of the series connection of K_{vl} and K_{jl}; K_{eq2}, equivalent stiffness of the series connection of K_{vr} and K_{jr}; K_{eq}, equivalent stiffness of the parallel connection of K_{eq1} and K_{eq2}; K_{eqg}, differential stiffness lumped with K_{eq} by employing the gear ratio $\left(\frac{R_o}{R_d}\right)$. Finally, the number of DOFs for the system was reduced from 10 to 6, as illustrated in Figure 9, and all the components of the stiffness matrix $\underline{\underline{K}}$ from Equation (66) were obtained as follows.

$$K_{1,1} = K_c, \tag{75a}$$

$$K_{2,2} = K_c + K_i. \tag{75b}$$

$$K_{3,3} = K_i + \left(R_{i5}{}^2 + R_{i3}{}^2\right)K_g. \tag{76}$$

$$K_{4,4} = K_g R_{o5}{}^2. \tag{77}$$

$$K_{5,5} = K_g R_{o3}{}^2 + K_{eqg}, \tag{78a}$$

$$K_{6,6} = K_{eqg}. \tag{78b}$$

$$k_{1,2} = k_{2,1} = -k_c, \tag{79a}$$

$$k_{2,3} = k_{3,2} = -k_i. \tag{79b}$$

$$k_{3,4} = k_{4,3} = k_g R_{i5} R_{o5}, \tag{80a}$$

$$k_{3,5} = k_{5,3} = k_g R_{i3} R_{o3}. \tag{80b}$$

$$k_{5,6} = k_{6,5} = -k_{eqg}. \tag{81}$$

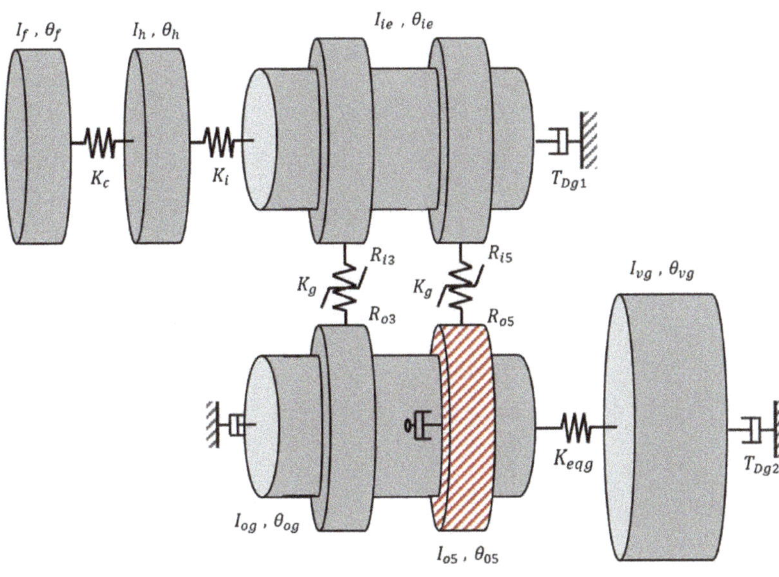

Figure 9. Reduced driveline system model with 6 DOFs based on the original torsional system.

The dynamic characteristics of the 6-DOF system can be obtained using modal analysis. Table 5 and Figure 10 show the first three natural frequencies and mode shapes for both models. Here, the y- and x-axes of Figure 10 are the normalized values of the modal vector and their relevant subsystem, as described in Equations (64) and (65). The differences between the natural frequencies of the 15-DOF and 6-DOF models are due to the reduction effect. The natural frequencies corresponding to the lumped inertia naturally disappeared as the relevant inertia was reduced. Therefore, the 1st natural frequency value of the 15-DOF model did not appear in the 6-DOF model. For example, the 1st mode shown in Figure 4 represents the hopping mode observed in the right and left wheels, which was removed when the differential gear, cv joint, vehicle, and tires were combined during the system reduction. Except for the hopping mode, which was not the main concern of this study, the first three mode shapes and natural frequencies presented in Figure 4, Figure 10 and Table 5 correlated well with each other.

Table 5. First four natural frequencies of the 15- and 6-DOF systems.

Natural Frequency (Hz)	6-DOF	15-DOF
f_1	7.6	2.3
f_2	60.6	7.6
f_3	272.8	60.8
f_4	1846.3	272.5

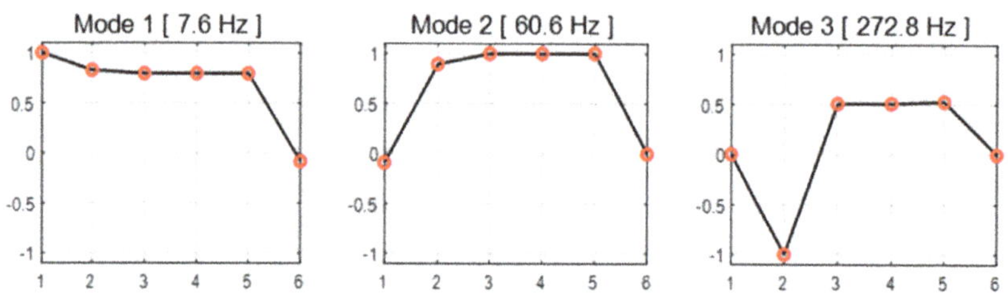

Figure 10. First three modes of the 6-DOF model.

To calculate the frequency response functions, the damping matrix was approximated, and unit torque excitation was considered, as described in Equations (50)–(56). The frequency response functions also correlated well between each model, as compared in Figure 11. The FRFs for both the 15- and 6-DOF system models described in Figure 11 are defined as follows: $|Y_1|$, FRFs of the flywheel; $|Y_2|$, FRFs of the clutch hub; $|Y_3|$, FRFs of the input shaft; $|Y_4|$, FRFs of the 5th gear on the output shaft; $|Y_5|$, FRFs of the output shaft; $|Y_6|$, FRFs of the tire and vehicle. The difference between $|Y_6|$, the FRFs of the vehicle in the 15- and 6-DOF system models, arose during the system reduction process, as the inertia and stiffness values for the vehicle components were modified with respect to the employed gear ratio.

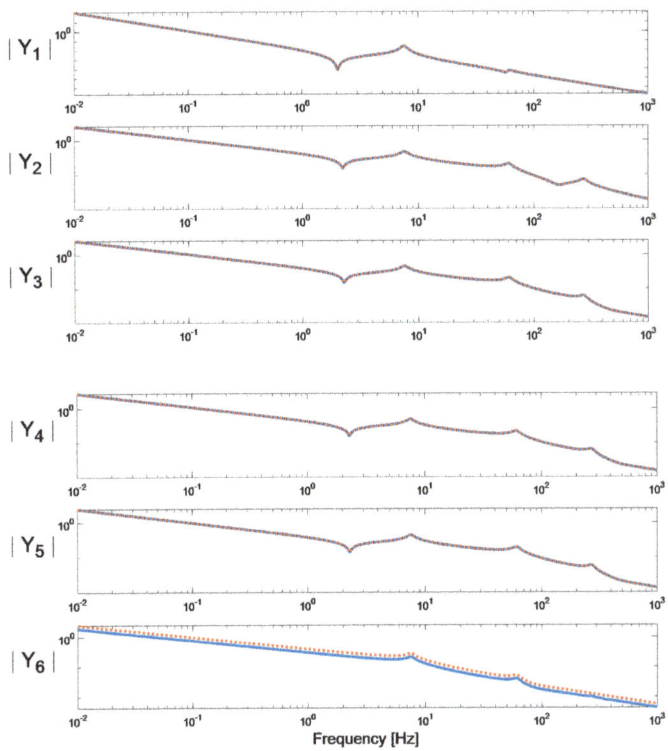

Figure 11. Comparisons of FRFs of 6-DOF model with 15-DOF model. Key: ———, 15-DOF; – – –, 6-DOF.

5. Linear Analysis in the Time Domain

From the results based on the modal analysis and FRFs, the linear system responses in the time domain can be investigated to find the main system parameters for improving the noise and vibration conditions. The time domain solutions are obtained with two different approaches. In general, the system responses consist of transient and steady-state terms. First, the basic equation model suggested previously can be used on the basis of the modal analysis and FRFs. For example, the modal domain and FRF results are simply used for investigating the transient and steady-state responses, respectively. Second, the state variable equations can be used as an alternative method which will lead to a simpler calculation procedure than the first approach, since both transient and steady-state responses are obtained simultaneously.

5.1. Time Responses Based on Modal Domain and FRFs

The transient responses can be directly calculated by transferring the modal domain coordinates $\underline{\eta}(t)$ to the physical domain for $\underline{\theta}(t)$, as described in Equation (46). Each coordinate can be estimated along with the initial conditions (ICs) as follows.

$$\underline{\underline{M}}\,\ddot{\underline{\theta}}(t) + \underline{\underline{C}}\,\dot{\underline{\theta}}(t) + \underline{\underline{K}}\,\underline{\theta}(t) = \underline{0}. \tag{82}$$

$$\ddot{\underline{\eta}}(t) + \begin{bmatrix} \ddots & & \\ & 2\zeta w_r & \\ & & \ddots \end{bmatrix} \dot{\underline{\eta}}(t) + \begin{bmatrix} \ddots & & \\ & w_{ri} & \\ & & \ddots \end{bmatrix} \underline{\eta}(t) = \underline{0}. \tag{83}$$

$$\underline{\eta}(t) = e^{-iw_{ni}t}[\underline{A}\cos(w_{di}t) + \underline{B}\sin(w_{di}t)],\ i = 1, 2, \cdots 6. \tag{84}$$

$$\underline{\eta}(0) = \underline{\underline{\phi}}^{-1}\underline{\theta}(0), \tag{85a}$$

$$\dot{\underline{\eta}}(0) = \underline{\underline{\phi}}^{-1}\dot{\underline{\theta}}(0). \tag{85b}$$

The 6-DOF model is a coupled system, as shown in Equation (82). Using the normal modal matrix $\underline{\underline{\phi}}$ from Equation (45), the coupled system of Equation (82) is transformed into the decoupled system, as derived in Equation (83). Overall, the transient responses of the torsional system can be expressed by Equation (84), where $\underline{A}$ and $\underline{B}$ are arbitrary constant vectors. These unknown constants are determined by substituting the modal initial values $\underline{\eta}(0)$ and $\dot{\underline{\eta}}(0)$. In this study, 0.1 (rad) and 0 (rad/s) were used as the initial torsional displacement and velocity, respectively. Finally, the transient responses of the torsional system in the physical domain were calculated by multiplying $\underline{\underline{\phi}}$ by $\underline{\eta}(t)$, and the results are shown in Figure 12. Along with the damping effects, the system responses converged to zero well. Here, the transient responses in the physical domain described in Figure 12 are defined as follows: X_{tr1}, transient response for the flywheel; X_{tr2}, transient response for the clutch hub; X_{tr3}, transient response for the input shaft; X_{tr4}, transient response for the 5th gear on the output shaft; X_{tr5}, transient response for the output shaft; X_{tr6}, transient response for the tire and vehicle.

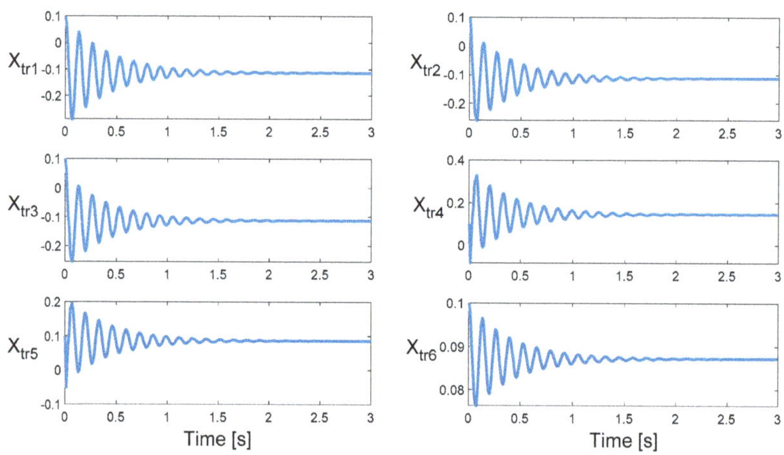

Figure 12. Transient response analysis based on the modal analysis.

From the whole system responses, combined with the transient and steady-state terms, only the steady-state responses are revealed after the transient responses converge to zero completely. To obtain the damped forced responses in a torsional system under sinusoidal excitation, the produced values such as the magnitude and phase from the FRFs, as seen in Figure 11, can be employed. The FRFs for the 6-DOF system can be derived as follows.

$$\underline{\underline{M}}\,\ddot{\underline{\theta}}(t) + \underline{\underline{C}}\,\dot{\underline{\theta}}(t) + \underline{\underline{K}}\,\underline{\theta}(t) = \underline{T}(t). \tag{86}$$

$$\underline{T}(t) = \begin{bmatrix} 1 & 0 & 0 & 0 & 0 & 0 \end{bmatrix}^T e^{i\omega t} = \tilde{\underline{T}} e^{i\omega t}. \tag{87}$$

$$\underline{\theta}(t) = \tilde{\underline{\theta}} e^{i\omega t}, \tag{88a}$$

$$\tilde{\theta}_n = \left|\tilde{\theta}_n\right| e^{i\varphi_n}, n = 1, 2, \cdots 6. \quad \text{(88b)}$$

$$\underline{\tilde{\theta}}(\omega) = \left[\underline{\underline{K}} - \underline{\underline{M}} \cdot \omega^2 + i \cdot \underline{\underline{C}} \cdot \omega\right]^{-1} \underline{\tilde{T}}. \quad \text{(89)}$$

$$\underline{\theta}(\mathrm{t}) = \left|\underline{\tilde{\theta}}(w_o)\right| e^{i\varphi(w_o)} e^{iw_o t} \text{ at } w = w_o. \quad \text{(90)}$$

The formula for calculating the FRFs in Equation (89) is derived by substituting Equations (88) and (87) into Equation (86). Since the results are generated as a function of frequency ω and are represented as complex values, their magnitudes and phases provide practical information regarding the amplitude and phase shift of the system responses. Thus, the dynamic behaviors in the time domain are obtained as described in Equation (90) for a specific excitation frequency value, ω_o. To focus on the linear analysis concerning specific noise and vibration problems, such as gear rattle, this study concentrated on the system responses at ω_o = 60 (Hz). Figure 13 displays the steady-state responses at ω_o = 60 (Hz) for the torsional system. In Figure 13, the responses from X_{ss1} to X_{ss6} represent, in order: steady-state responses based on the FRF components of the flywheel, clutch hub, input shaft, 5th speed gear on the output shaft, and vehicle.

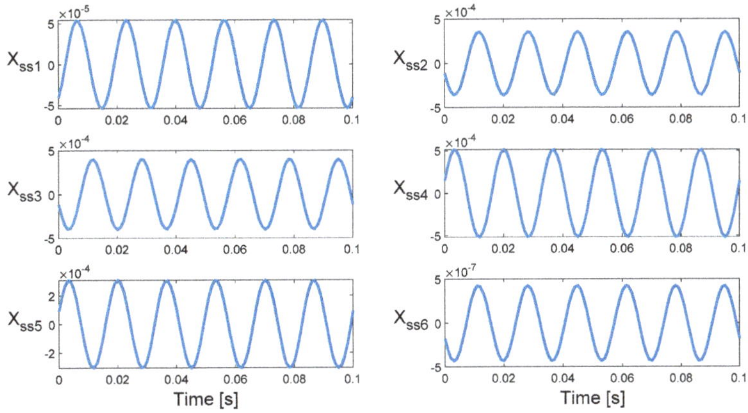

Figure 13. Steady-state responses based on the FRF components.

5.2. Time Responses Based on the State Variable Equation

The equations derived in Equations (82)–(86) are based on the 2nd-order ordinary differential equations (ODE). However, this method using the original forms described in Equations (82)–(86) becomes more complex as the order of the differential equation increases [33]. However, state variables are used to describe the mathematical state of a dynamic system by a set of 1st-order ODEs rather than by one or more nth-order ODEs. Thus, the simulations using the 1st-order ODEs can be more efficient in terms of modeling and analyzing the system responses, especially by including the MIMO (multi-input multi-output) system and time-varying and non-linear effects [25,34]. The state variable equation can be formulated by defining each state variable with the matrix and vector forms, as described in Equations (91) and (92).

$$\underline{\dot{x}}(t) = \underline{\underline{A}}\, \underline{x}(t) + \underline{\underline{B}}\, \underline{u}(t). \quad \text{(91)}$$

$$\underline{y}(t) = \underline{\underline{C}}\, \underline{x}(t) + \underline{\underline{D}}\, \underline{u}(t). \quad \text{(92)}$$

Equations (91) and (92) represent the state–space representation. In this representation, $\underline{x}(t)$ is the state vector containing the displacements and velocities, $\underline{\underline{A}}$ is the system matrix, $\underline{\underline{B}}$ is the input matrix, $\underline{\underline{C}}$ is the output matrix, $\underline{\underline{D}}$ is the constant matrix relevant to the input, $\underline{u}(t)$ is the input vector, and $\underline{y}(t)$ is the concerned output vector. The scope of this study was to investigate the system responses for the torsional dynamic behaviors of the physical system described earlier. Therefore, Equation (92) was not the main focus of this study; instead, Equation (91) was the primary focus for examining the system responses of the torsional system.

5.2.1. Transient Response

The system response in a zero-input case can be calculated using the state transition matrix $\underline{\underline{\Phi}}(t)$ and the initial values $\underline{x}(0)$. The general process for calculating the transient responses with the state transition matrix, based on the state variable equation, is described as follows [35].

$$\dot{\underline{x}}(t) = \underline{\underline{A}}\,\underline{x}(t). \tag{93}$$

$$\underline{x}(t) = \underline{b_0} + \underline{b_1}\mathrm{t} + \underline{b_2}t^2 + \underline{b_3}t^3 + \cdots + \underline{b_k}t^k \cdots. \tag{94}$$

$$\underline{b_1} + 2\underline{b_2}t + 3\underline{b_3}t^2 + \cdots + k\underline{b_k}t^{k-1} \cdots = \underline{\underline{A}}\left(\underline{b_0} + \underline{b_1}\mathrm{t} + \underline{b_2}t^2 + \underline{b_3}t^3 + \cdots + \underline{b_k}t^k \cdots\right). \tag{95}$$

$$\underline{b_k} = \frac{1}{k!}\underline{\underline{A}}^k\underline{b_0}\,, \tag{96a}$$

$$\underline{x}(0) = \underline{b_0}. \tag{96b}$$

$$\underline{x}(t) = \sum\nolimits_{k=0}^{\infty} \frac{\underline{\underline{A}}^k t^k}{k!} \tag{97a}$$

$$\underline{x}(0) = e^{\underline{\underline{A}}t}\underline{x}(0). \tag{97b}$$

$$\underline{x}(t) = \underline{\underline{\Phi}}(t)\underline{x}(0). \tag{98}$$

$$\underline{\underline{\Phi}}(t) = e^{\underline{\underline{A}}t} = \underline{\underline{u}}e^{\underline{\underline{\Lambda}}t}\underline{\underline{u}}^{-1}. \tag{99}$$

$$e^{\underline{\underline{\Lambda}}t} = \begin{bmatrix} e^{\lambda_1 t} & 0 & 0 & 0 \\ 0 & e^{\lambda_2 t} & 0 & 0 \\ 0 & 0 & \ddots & 0 \\ 0 & 0 & 0 & e^{\lambda_n t} \end{bmatrix}. \tag{100}$$

For this study, the initial values of 0.1 (rad) and 0 (rad/s) were used for the initial torsional displacement $\underline{x}(0)$ and velocity $\dot{\underline{x}}(0)$, respectively. The solution of the system can be assumed on the basis of the power series with arbitrary vectors $\underline{b_0}$ and $\underline{b_k}$ ($k = 1, 2, 3, \cdots$), as described in Equation (94). Thus, substituting Equation (94) into Equation (93) leads to a series expansion as stated in Equation (95). By equating the same t^k terms on both sides of Equation (95), the basic forms of constant vectors $\underline{b_k}$ ($k = 1, 2, 3, \cdots$) are expressed as Equation (96), where the initial torsional displacement is represented as $\underline{x}(0) = \underline{b_0}$. Thus, the basic solution of $\underline{x}(t)$ can be written as Equation (97) with the matrix exponential $e^{\underline{\underline{A}}t}$. Moreover, the matrix exponential $e^{\underline{\underline{A}}t}$, when the initial conditions are given at $t = 0$ in a linear system, is considered the state transition matrix $\underline{\underline{\Phi}}(t)$, as described in Equation (98).

In addition, $\underline{\underline{\Phi}}(t)$ can be calculated along with the eigenmatrix $\underline{u}$ and eigenvalue λ_n of the system matrix $\underline{\underline{A}}$. Figure 14 compares the results of transient responses based on the state variable equation with the ones obtained from the modal analysis, demonstrating a strong correlation between the two.

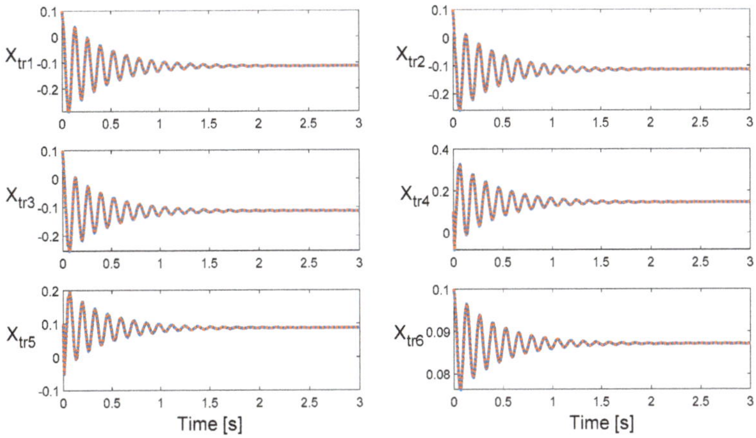

Figure 14. Comparisons of transient responses from two different analyses. Key: ▬▬▬, using the modal analysis; ▬ ▬ ▬, using the state variable equation.

5.2.2. Steady-State Responses for Semi-Definite System

The steady-state responses using the state variable equation under sinusoidal input can be calculated by employing the state-transition matrix and the convolution method [36,37]. The process of obtaining the steady-state responses using the state transition matrix $\underline{\underline{\Phi}}(t)$ and the initial condition $\underline{x}(0)$ is described as follows.

$$\dot{\underline{x}}(t) - \underline{\underline{A}}\,\underline{x}(t) = \underline{\underline{B}}\,\underline{u}(t). \tag{101}$$

$$e^{-\underline{\underline{A}}t}\dot{\underline{x}}(t) - e^{-\underline{\underline{A}}t}\underline{\underline{A}}\,\underline{x}(t) = e^{-\underline{\underline{A}}t}\underline{\underline{B}}\,\underline{u}(t). \tag{102}$$

$$\frac{d}{dt}\left\{e^{-\underline{\underline{A}}t}\underline{x}(t)\right\} = e^{-\underline{\underline{A}}t}\underline{\underline{B}}\,\underline{u}(t). \tag{103}$$

$$\int_0^t \frac{d}{d\tau}\left\{e^{-\underline{\underline{A}}\tau}\underline{x}(\tau)\right\}d\tau = \int_0^t e^{-\underline{\underline{A}}\tau}\underline{\underline{B}}\,\underline{u}(\tau)d\tau. \tag{104}$$

$$e^{-\underline{\underline{A}}t}\underline{x}(t) - \underline{x}(0) = \int_0^t e^{-\underline{\underline{A}}\tau}\underline{\underline{B}}\,\underline{u}(\tau)d\tau. \tag{105}$$

$$\underline{x}(t) = e^{\underline{\underline{A}}t}\underline{x}(0) + \int_0^t e^{\underline{\underline{A}}(t-\tau)}\underline{\underline{B}}\,\underline{u}(\tau)\,d\tau. \tag{106}$$

Here, the same initial condition, $\underline{x}(0)$, that was used for the transient responses was employed to calculate the steady-state responses. In addition, Equation (101) represents a revised formula achieved by transferring the term $\underline{\underline{A}}\,\underline{x}(t)$ to the left side of Equation (91), resulting in Equation (102) by multiplying $e^{-\underline{\underline{A}}t}$ on both sides of Equation (101). Furthermore, the left side of Equation (102) reflects the partial derivative of $e^{-\underline{\underline{A}}t}\underline{x}(t)$, which is summarized as Equation (103). Integrating both sides of Equation (103) over the range from 0 to t yields Equations (104)–(106). Using these final-form equations to calculate the steady-state responses, the dynamic behaviors of the system are shown in Figure 15. When comparing the results based on the FRF method and state variable equations, it is evident that they are well correlated for X_{ss2}, X_{ss3}, X_{ss4}, and X_{ss5}, as clearly seen in Figure 15. However, the simulated results for X_{ss1} and X_{ss6} exhibit significant discrepancies between the FRFs and state variable equation methods, primarily due to the rigid body mode observed at 0 Hz for the semi-definite system. This mode can lead to numerical issues that might mislead the integration routine by introducing an arbitrary constant, including infinity, into all solutions [12]. Thus, to thoroughly investigate the dynamic responses occurring between each coupled system, the absolute motions must be transformed into relative coordinates, thereby converting the semi-definite configuration into the definite system model [25].

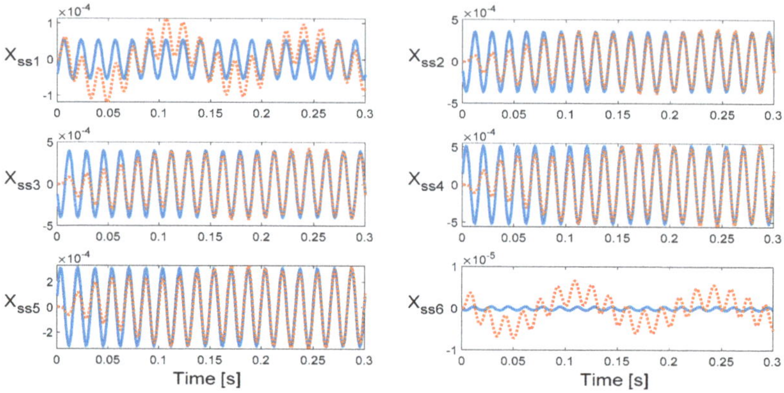

Figure 15. Examination of steady-state responses with absolution motions. Key: ▬▬▬, using the FRFs; ▬ ▬ ▬, using the state variable equation.

To transform the semi−definite system into a positive-definite system, new coordinates can be defined as $q_1 = \theta_f - \theta_h$, $q_2 = \theta_h - \theta_{ie}$, $q_3 = R_{i5}\theta_{ie} + R_{o5}\theta_{o5}$, $q_4 = R_{i3}\theta_{ie} + R_{o3}\theta_{og}$, and $q_5 = \theta_{og} - \theta_{vg}$. Thus, the relative angular displacement vector $\underline{q}$ and linear transformation matrix $\underline{\underline{P}}$ can be expressed as follows [12].

$$\underline{q} = \begin{bmatrix} q_1 \\ q_2 \\ q_3 \\ q_4 \\ q_5 \end{bmatrix} = \begin{bmatrix} 1 & -1 & 0 & 0 & 0 & 0 \\ 0 & 1 & -1 & 0 & 0 & 0 \\ 0 & 0 & R_{i5} & R_{o5} & 0 & 0 \\ 0 & 0 & R_{i3} & 0 & R_{o3} & 0 \\ 0 & 0 & 0 & 0 & 1 & -1 \end{bmatrix} \begin{bmatrix} \theta_f \\ \theta_h \\ \theta_{ie} \\ \theta_{o5} \\ \theta_{og} \\ \theta_{vg} \end{bmatrix} = \underline{\underline{P}}\,\underline{\theta}. \tag{107}$$

$$\underline{\underline{M}}^{*}\ddot{\underline{q}}(t) + \underline{\underline{C}}^{*}\dot{\underline{q}}(t) + \underline{\underline{K}}^{*}\underline{q}(t) = \underline{T}^{*}(t). \tag{108}$$

$$\underline{\underline{M}}^{*} = \underline{\underline{P}}\,\underline{\underline{M}}^{-1}\underline{\underline{M}}\,\underline{\underline{P}}^{*}, \tag{109a}$$

$$\underline{\underline{C}}^{*} = \underline{\underline{P}}\,\underline{\underline{M}}^{-1}\underline{\underline{C}}\,\underline{\underline{P}}^{*}. \tag{109b}$$

$$\underline{\underline{K}}^{*} = \underline{\underline{P}}\,\underline{\underline{M}}^{-1}\underline{\underline{K}}\,\underline{\underline{P}}^{*}, \tag{110a}$$

$$\underline{T}^{*}(t) = \underline{\underline{P}}\,\underline{\underline{M}}^{-1}\underline{T}(t). \tag{110b}$$

$$\underline{\underline{P}}^{*} = \underline{\underline{P}}^{t}\left(\underline{\underline{P}}\,\underline{\underline{P}}^{t}\right)^{-1}, \tag{111a}$$

$$\underline{\underline{P}}\,\underline{\underline{P}}^{t}\left(\underline{\underline{P}}\,\underline{\underline{P}}^{t}\right)^{-1} = \underline{\underline{I}}. \tag{111b}$$

$$\dot{\underline{x}}(t) = \underline{\underline{A}}^{*}\underline{x}(t) + \underline{\underline{B}}^{*}\underline{u}^{*}(t). \tag{112}$$

$$\underline{x}(t) = \left[q_1\ q_2\ q_3\ q_4\ q_5\ \dot{q}_1\ \dot{q}_2\ \dot{q}_3\ \dot{q}_4\ \dot{q}_5\right]^T, \tag{113a}$$

$$\underline{\underline{A}}^* = \begin{bmatrix} \underline{\underline{0}} & \underline{\underline{I}} \\ -\dfrac{\underline{\underline{K}}^*}{\underline{\underline{M}}^*} & -\dfrac{\underline{\underline{C}}^*}{\underline{\underline{M}}^*} \end{bmatrix}. \tag{113b}$$

$$\underline{\underline{B}}^* = \underline{\underline{0}},\ \underline{\underline{B}}^*(6,1) = 1, \tag{114a}$$

$$\underline{u}^*(t) = \underline{\underline{P}}\,\underline{\underline{M}}^{-1}\underline{u}(t). \tag{114b}$$

From Equations (107)–(114), $\underline{\underline{P}}$ is not generated as a square matrix. Therefore, to determine the inverse matrix with respect to $\underline{\underline{P}}$, the product of $\underline{\underline{P}}$ and $\underline{\underline{P}}^t$ can be introduced, as indicated in Equation (111b), where $\underline{\underline{P}}^* = \underline{\underline{P}}^t\left(\underline{\underline{P}}\,\underline{\underline{P}}^t\right)^{-1}$ of Equation (111a) is defined as a pseudo-inverse matrix. Overall, Equation (108) has new coordinates transferred from the absolute displacement $\underline{\theta}(t)$ to the relative displacement $\underline{q}(t)$, resulting in a positive-definite system. Using Equations (107)–(111), the state variable equation can be derived as expressed in Equations (112)–(114). In these equations, $\underline{x}(t)$, $\underline{\underline{A}}^*, \underline{\underline{B}}^*$, and $\underline{u}^*(t)$ represent state variables, the system matrix, input matrix, and input vector related to the system parameters based on $\underline{\underline{M}}^*$, $\underline{\underline{C}}^*$ and $\underline{\underline{K}}^*$. The steady-state responses with the new coordinate systems, based on Equations (107)–(114), are presented in Figure 16 and compared with the time responses based on FRFs. From the comparisons between two sets of results based on the state variable model and FRFs, it is evident that they are well correlated with each other. Furthermore, the rigid body modes within the nearly zero frequency ranges are completely eliminated.

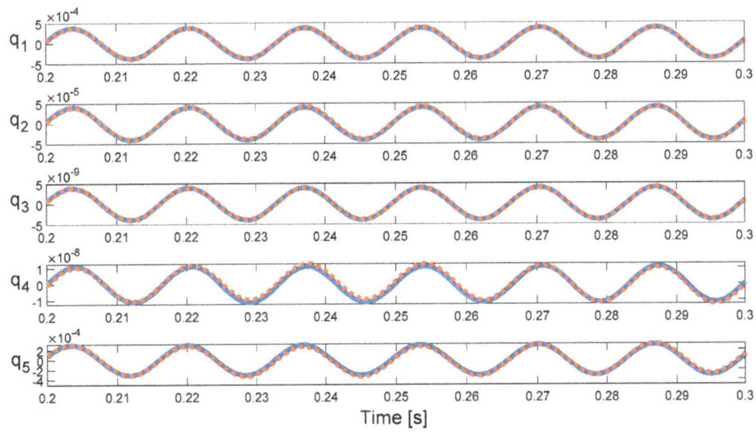

Figure 16. Steady-state responses based on the relative motions. Key: ▬▬▬, using the FRFs; ▬ ▬ ▬, using the state variable equation.

6. Case Study

Dynamic characteristics in a geared system can be investigated by focusing on several key parameters such as clutch stiffness K_c, unloaded gear inertia I_{o5}, and gear mesh stiffness K_g with respect to torsional vibration problems such as gear rattle and whine. To investigate the effects of these key parameters for resolving the aforementioned issues, the relevant parameters, such as K_c, I_{o5}, and K_g, were examined [1,5–8,14,20,22,38–40]. For example, to improve gear rattle conditions, prior research has suggested modifying clutch stiffness K_c and unloaded gear inertia I_{o5} [1,5–8,14]. On the other hand, gear whine is closely related to gear properties such as the gear involute and mesh stiffness. In general, K_c consists of multi-staged clutch dampers, as illustrated in Figure 17. Figure 17a illustrates the physical drawing of the entire clutch system with the clutch damper indicated as the clutch spring. Typically, multi-clutch spring components are located inside the clutch housing along with the clutch friction disk. Figure 17b presents an example of the clutch stiffness profile, featuring two staged clutch springs or clutch dampers [1]. In addition, K_g exhibits nonlinear dynamic characteristics along with gear rotational motions under the symmetric condition, as depicted in Figure 18. However, for the purposes of linear analysis concerning the effects of key parameters such as K_c, I_{05}, and K_g, this study employed various constant values to investigate their relationship with primary torsional vibratory problems, namely gear rattle and whine.

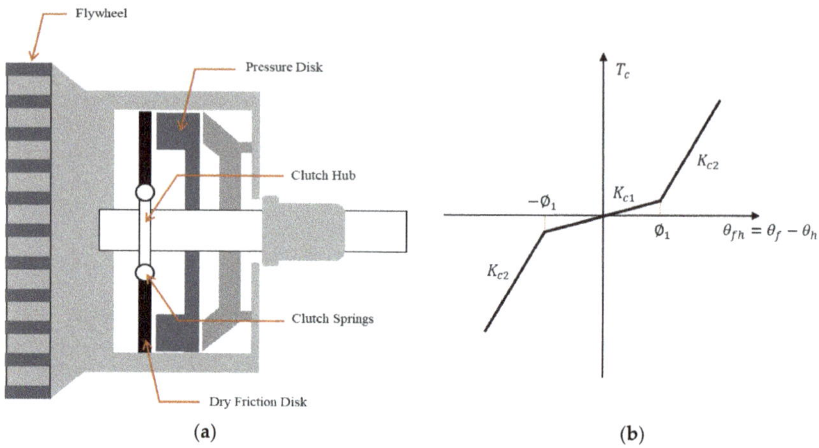

Figure 17. Clutch stiffness K_c: (**a**) schematic of clutch; (**b**) clutch torque induced by stiffness [1].

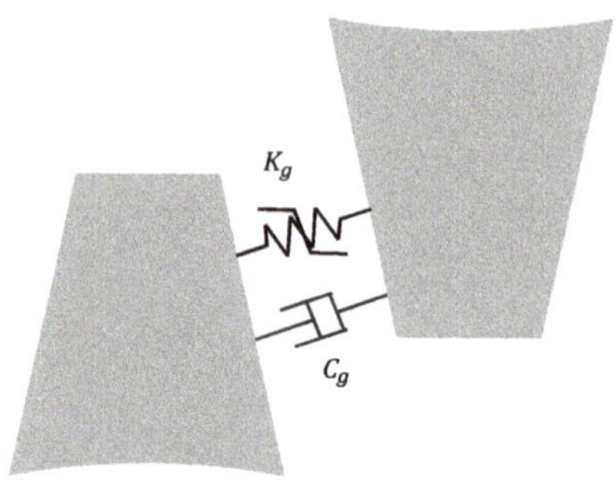

Figure 18. Gear mesh stiffness K_g.

6.1. Case Study I: Clutch Stiffness K_c

The natural frequencies of the system are compared for three different clutch stiffness values, as summarized in Table 6, where the values of K_{c1} and K_{c2} are given as 595.8 N·m·rad^{-1} and $(K_{c1} + K_c)/2$, respectively. Significant changes were observed in the second natural frequency because the dynamic behavior at the second mode was mostly affected by clutch stiffness. This effect was clearly observed in the mode shapes presented in Figures 3, 9 and 18a. When examining the second mode shapes (clutch spring modes) with various K_c values, as seen in Figure 19a, their basic mode shapes were nearly identical, except for the calculated natural frequencies, which are compared in Tables 6 and 7. In general, gear rattle was closely related to this clutch spring mode, which operates within the second resonance region [1,8,14]. Thus, based on the linearly simulated results presented in Figure 19a and Tables 6 and 7, clutch stiffness can be considered a key parameter for resolving gear impact problems.

Table 6. Comparison of natural frequencies along with clutch stiffnesses for the positive definite system.

Natural Frequency (Hz)	K_{c1} (595.8 N·m·rad^{-1})	K_{c2} (1216.9 N·m·rad^{-1})	Original K_c (1838 N·m·rad^{-1})
f_1	6.6	7.3	7.6
f_2	40.6	51.8	60.6
f_3	266.0	269.4	272.8
f_4	1845.7	1845.7	1846.3
f_5	4484.9	4484.9	4485.7

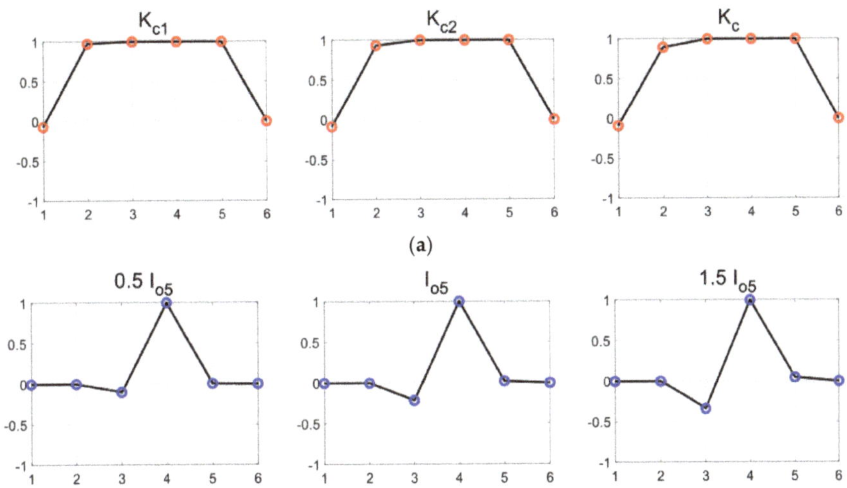

Figure 19. Mode shapes along with K_c and I_{o5} values: (**a**) second mode shapes with various K_c; (**b**) fifth mode shapes with various I_{o5}.

Table 7. Components of second modal vectors along with various K_c values.

2nd Mode Shape	K_{c1}	K_{c2}	K_c
1. flywheel	−0.0693	−0.0847	−0.0903
2. clutch hub	0.9742	0.9328	0.8930
3. input shaft	0.9999	0.9997	0.9992
4. unloaded gear	1	0.9999	0.9994
5. output shaft	0.9996	1	1
6. vehicle + wheel	−0.0032	−0.0020	−0.0014

The effect of K_c was also evident from the time domain analysis, as shown in Figure 20. To investigate the time domain responses, the relative motions in terms of $q_3(t)$ and $q_4(t)$ were the primary focus, as gear impacts and their associated torsional vibrations generally occur at the gear pairs. To simulate the system responses in the time domain, the engine torque T_e was employed as the excitation torque, as described below.

$$T_e = T_m + T_{pm}\cos(w_p t) \tag{115}$$

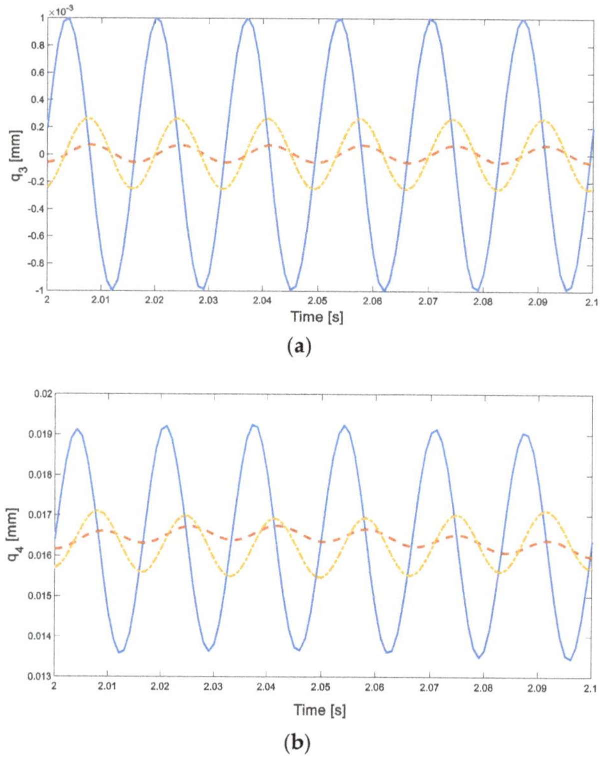

Figure 20. Comparisons of steady-state responses along with K_c values for the positive-definite system: (**a**) steady-state responses of q_3; (**b**) steady-state responses of q_4. Key: ■■■■, K_{c1}; ———, K_c; — · — · –, K_{c2}.

Here, T_m and T_{pm} represent the mean and alternating components of the engine torque, respectively, and w_p is the firing frequency, with the values used for T_m, T_{pm}, and w_p being 168.9 N·m, 251.5 N·m, and 60 Hz, respectively [1]. Figure 20 compares the time responses for q_3 and q_4, which represent the relative motions between unloaded and loaded gear pairs, respectively. Figure 20a illustrates the system responses with various K_c values. It is worth noting that the amplitude of q_3 decreased as the value of K_c was reduced. For instance, when smaller values such as K_{c1} (595.8 N·m·rad^{-1}) and K_{c2} (1216.9 N·m·rad^{-1}) were employed, the amplitude of q_3 decreased compared with the response for K_c (1838 N·m·rad^{-1}). This reduction in amplitude was further confirmed when comparing the peak-to-peak values, as summarized in Table 8. Furthermore, Figure 20b and Table 8 demonstrate that q_4 was also significantly influenced by the key parameter K_c. In this context, q_3 can be expected to reflect the gear rattle status, which occurs at the unloaded gear pair. In addition, q_4 can be considered the vibratory conditions transferred from the loaded gear pair to the rest parts of the driveline. Consequently, these results provide valuable insights into improving the vibration conditions associated with gear rattle, as compared in Figure 20 and Tables 6–8, even though they do not explicitly depict the gear contact motions. In summary, it can be concluded that clutch stiffness values have a significant influence on the dynamic behavior of the gears, and reducing clutch stiffness improves the vibrational problems caused by gear impacts.

Table 8. Comparison of peak-to-peak value along with various K_c values.

	Peak-to-Peak Value q_3 (mm)	Peak-to-Peak Value q_4 (mm)
K_{c1} (595.8 N·m·rad^{-1})	1.2723×10^{-4}	3.9210×10^{-4}
K_{c2} (1216.9 N·m·rad^{-1})	5.1967×10^{-4}	15×10^{-4}
Original K_c (1838 N·m·rad^{-1})	20×10^{-4}	56×10^{-4}

6.2. Case Study II: Unloaded Gear Inertia I_{o5}

Once again, the natural frequencies of the system, along with the unloaded gear inertia I_{o5}, are listed in Table 9. In this case, $0.5I_{o5}$ (1.30×10^{-4} kg·m^2) and $1.5I_{o5}$ (3.92×10^{-4} kg·m^2) were used for comparison with the original I_{o5}. With different values for I_{o5}, only the natural frequency f_5, as shown in Table 9, was affected, while the other natural frequencies remained nearly the same. For instance, when I_{o5} varied, natural frequencies f_1, f_2, f_3, and f_4 showed minimal variation. However, the natural frequency f_5 exhibited a clear change, with an increase (or decrease) in I_{o5} resulting in higher (or lower) values, such as 6040.3 (or 3845.5) Hz compared with the original value of 4485.7 Hz.

Table 9. Comparison of natural frequencies along with inertia values of the unloaded gear for the positive-definite system.

Natural Frequency (Hz)	0.5 I_{o5}	Original I_{o5} (5.23×10^{-4})	1.5 I_{o5}
f_1	7.6	7.6	7.6
f_2	61.5	60.6	59.8
f_3	274.7	272.8	271.1
f_4	1897.0	1846.3	1795.8
f_5	6040.3	4485.7	3845.5

This effect of I_{o5} is evident in Figure 19b, Tables 10 and 11. Upon examining the mode shapes, it becomes apparent that only the fifth mode shape, corresponding to natural frequency f_5, was significantly influenced by I_{o5} compared with the other sub-systems, as shown in Table 11. Furthermore, the comparison between Tables 10 and 11 reveals that the mode shape was primarily affected at the fifth mode, except for the second mode, providing valuable insights into the relationship between unloaded gar inertia and gear dynamic motions for various I_{o5} values. Figure 21 displays the steady-state responses of q_3 and q_4 for the unloaded and loaded gear pairs, respectively, with 0.5 I_{o5}, 1.5 I_{o5}, and the original I_{o5}. As observed from the modal analysis regarding the effect of I_{o5} mentioned earlier, the steady-state responses of q_3 in the time domain also exhibited significant changes with different I_{o5} values, as depicted in Figure 21a.

Table 10. Components of second modal vectors along with various I_{o5} values.

Second Mode Shape	0.5 I_{o5}	I_{o5}	1.5 I_{o5}
1. flywheel	−0.0878	−0.0903	−0.0929
2. clutch hub	0.9853	0.8930	0.8908
3. input shaft	0.9991	0.9992	0.9992
4. unloaded gear	0.9992	0.9994	0.9996
5. output shaft	1	1	1
6. vehicle + wheel	−0.0014	−0.0014	−0.0015

Table 11. Components of fifth modal vectors along with various I_{o5} values.

Fifth Mode Shape	0.5 I_{o5}	I_{o5}	1.5 I_{o5}
1. flywheel	0	0	0
2. clutch hub	0.0001	0.0005	0.0010
3. input shaft	−0.1018	−0.2150	−0.3392
4. unloaded gear	1	1	1
5. output shaft	0.0055	0.0218	0.0487
6. vehicle + wheel	0	0	0

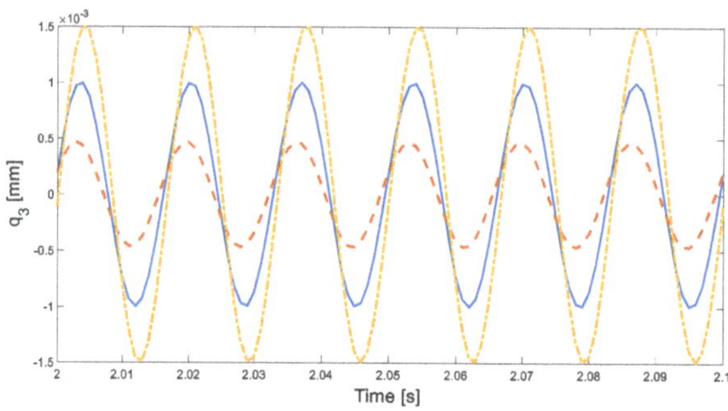

(**a**)

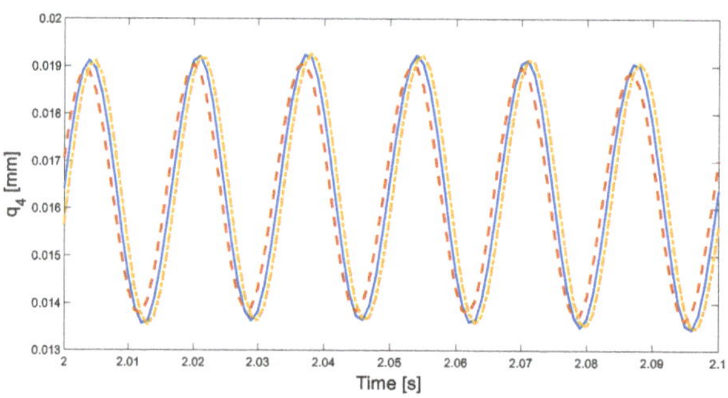

(**b**)

Figure 21. Comparisons of steady-state responses along with I_{o5} values for the positive−definite system: (**a**) steady-state responses of q_3; (**b**) steady-state responses of q_4. Key: ■■■■, 0.5 I_{o5}; ——, 1 I_{o5}; — · — · –, 1.5 I_{o5}.

For example, a reduction (or increase) in the inertia value of I_{o5} led to a corresponding reduction (or increase) in the amplitude of response. This was also evident when compared to the peak-to-peak values summarized in Table 12. On the other hand, q_4, associated with the loaded gear pair, demonstrated minimal variation across different I_{o5} values, as evident in Figure 21b and Table 12. These results underline the importance of I_{o5} in addressing gear rattle issues, even in the context of linear analysis. Generally, reducing the inertia of the unloaded gear aligns with prior studies' findings as an effective approach to improving gear rattling conditions [8]. In summary, it can be concluded that unloaded gears also have a significant influence on specific dynamic behaviors such as gear rattle. Therefore, altering the inertia values of unloaded gears relevant to the gear impact problems can improve the gear impact conditions when designing the clutch stiffness.

Table 12. Comparison of peak-to-peak value along with unloaded gear I_{o5} values.

	Peak-to-Peak Value q_3 (mm)	Peak-to-Peak Value q_4 (mm)
0.5 I_{o5}	9.2703×10^{-4}	52×10^{-4}
I_{o5}	20×10^{-4}	56×10^{-4}
1.5 I_{o5}	30×10^{-4}	56×10^{-4}

6.3. Case Study III: Gear Mesh Stiffness K_g

Table 13 presents a comparison of natural frequencies for various K_g values. Upon examining the calculated natural frequencies, it becomes evident that there were no discernible differences below 300 Hz. However, in the higher frequency range, above 300 Hz, substantial differences emerged, as illustrated in Table 13. Notably, the natural frequencies, associated with dominant gear dynamic modes were significantly influenced by variations in 0.1 K_g (2.7×10^7 N·m^{-1}), K_g, and 10 K_g (2.7×10^9 N·m^{-1}). Here, the natural frequencies corresponding to f_4 and f_5 exhibited pronounced changes, with f_4 and f_5 decreasing (or increasing) as K_g decreased (or increased).

Table 13. Comparison of natural frequencies along with gear mesh stiffnesses K_g for the positive-definite system.

Natural Frequency (Hz)	0.1 K_g	Original K_g (2.7×10^8)	10 K_g
f_1	7.6	7.6	7.6
f_2	60.6	60.6	60.6
f_3	264.5	272.8	273.5
f_4	603.5	1846.3	5820
f_5	1422.1	4485.7	14,170

Despite these variations in natural frequencies, the mode shapes for fourth and fifth modes remained nearly identical, as clearly shown in Figure 22, Tables 14 and 15. While the modal vectors themselves did not exhibit differences with various Kg, the mode shapes for both the fourth and fifth modes suggest that gear dynamic motions were indeed dominant, as evident in Figure 22. The fourth mode shapes depicted in Figure 22a indicate that both the unloaded and loaded gear pairs contributed significantly to the dominant motions. On the other hand, the fifth mode was influenced more by the unloaded gear pairs, as observed in Figure 22b. Overall, both the fourth and fifth modes are expected to play a crucial role in gear dynamic behaviors.

Table 14. Components of fourth modal vectors along with various K_g.

4th Mode Shape	0.1 K_g	K_g	10 K_g
1. flywheel	0.0001	0	0
2. clutch hub	−0.1100	−0.0104	−0.0010
3. input shaft	0.7801	0.7942	0.7954
4. unloaded gear	1	1	1
5. output shaft	−0.8204	−0.9503	−0.9625
6. vehicle + wheel	0	0	0

Table 15. Components of fifth modal vectors along with various K_g.

5th Mode Shape	0.1 K_g	K_g	10 K_g
1. flywheel	0	0	0
2. clutch hub	0.0049	0.0005	0
3. input shaft	−0.2212	−0.2150	−0.2144
4. unloaded gear	1	1	1
5. output shaft	0.0224	0.0218	0.0218
6. vehicle + wheel	0	0	0

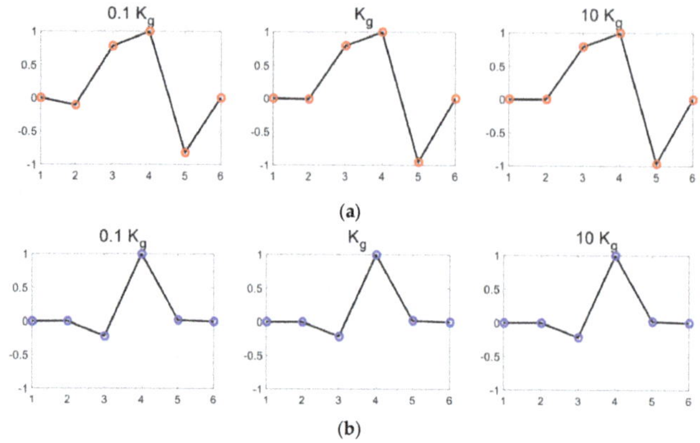

Figure 22. Mode shapes according to different K_g: (**a**) fourth mode shapes; (**b**) fifth mode shapes.

Figure 23 presents comparisons of steady-state responses in the time domain, occurring at the unloaded and loaded gear pairs. When evaluating q_3 along with different K_g values, it becomes apparent that the amplitudes of steady-state responses decreased as K_g increased, as shown in Figure 23a. These significant changes are further corroborated by the peak-to-peak values, as detailed in Table 16. Similarly, when comparing the time responses of the loaded gear pair, the same influences observed in the unloaded gear pair were evident, as depicted in Figure 23b and Table 16. Moreover, the mean values of time responses for the loaded gear pair were significantly affected. Specifically, as K_g increased, the mean values of steady-state responses approached zero. These results underline the impact of K_g on gear dynamic motions, particularly noticeable in the case of the loaded gear pairs, which are central to issues such as gear whine vibrations [38–40].

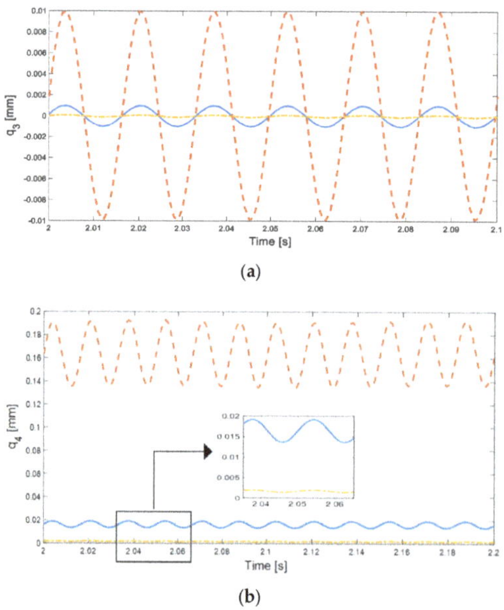

Figure 23. Comparisons steady-state responses along with K_g values using the positive definite system: (**a**) third steady-state responses based relative motion q_3; (**b**) fourth steady-state responses based relative motion q_4. Key: ■■■■, 0.1 K_g; ———,1 K_g; - - - -, 10 K_g.

Table 16. Comparison of peak-to-peak value along with various K_g values.

	Peak-to-Peak Value q_3 (mm)	Peak-to-Peak Value q_4 (mm)
0.1 K_g	199×10^{-4}	567×10^{-4}
1 K_g	20×10^{-4}	56×10^{-4}
10 K_g	1.9951×10^{-4}	5.5771×10^{-4}

In summary, it can be concluded that mesh stiffness significantly influences other specific gear dynamic behaviors, such as gear whine, which can occur in both loaded and unloaded gear pairs. Therefore, designing appropriate mesh stiffness values contributes to improving gear parametric excitations.

7. Conclusions

This study investigated the vehicle driveline under symmetric conditions using a linear analytical method to explore the dynamic characteristics of a specific manual transmission with a front-engine/front-wheel configuration. To examine these dynamics, a mathematical model based on a practical powertrain assembly was developed. The contributions of this study can be summarized as follows. First, a mathematical technique to reduce the number of DOFs for simulation was proposed, and its validity was established. Consequently, the dynamic characteristics, including eigensolutions and frequency response functions (FRFs) of the reduced system model, demonstrated a strong correlation with the simulated results of the original model. Second, this study extensively investigated a linear analysis model based on state variable equations. The linear results derived from the state variable equations were successfully validated against those from the FRF method. Furthermore, the findings based on the state variable equations effectively captured the relationships between key parameters and their associated vibratory issues, such as gear rattle and whine problems. Thus, this study successfully proposed the utilization of linear analytical methods to investigate and determine gear impact and excitation problems and their relevant key parameters in order to simplify the analysis compared to the complexity of nonlinear analysis.

The present study primarily emphasized the linear analytical method, specifically concentrating on the state variable equations, which can serve as the foundation for nonlinear analysis, incorporating features such as gear contact models. However, the current study has limitations in capturing the actual nonlinear dynamic behaviors, such as gear impact and gear time-varying motions. Therefore, the next stage of this research will implement a more advanced analytical model utilizing state variable forms while incorporating nonlinear properties, such as gear backlash and multi-staged clutch dampers.

Author Contributions: J.-M.B., S.-H.K. and J.-Y.Y. initiated and developed the ideas related to this research work. J.-M.B. and S.-H.K. developed parametric studies, analyzed relevant cases, and carried out performance analyses and numerical analyses. J.-M.B. wrote the paper draft under J.-Y.Y.'s guidance, and J.-Y.Y. finalized the paper. All authors have read and agreed to the published version of the manuscript.

Funding: This work was supported by an Incheon National University Research Grant in 2023 (20230059).

Data Availability Statement: No data were used to support this study.

Conflicts of Interest: The authors declare no conflict of interest.

References

1. Yoon, J.Y.; Singh, R. Effect of the multi-staged clutch damper characteristics on the transmission gear rattle under two engine conditions. *Proc. Inst. Mech. Eng. Part D J. Automob. Eng.* **2013**, *227*, 1273–1294. [CrossRef]
2. Guo, D.; Zhou, Y.; Zhou, Y.; Wang, Y.; Chen, F.; Shi, X. Numerical and experimental study of gear rattle based on a refined dynamic model. *Appl. Acoust.* **2022**, *185*, 108407. [CrossRef]

3. Zhou, Y.; Shi, X.; Guo, D.; Dazer, M.; Bertsche, B.; Mei, Z. Experimental investigation of gear rattle and nonlinear dynamic interaction in a dual-clutch transmission. *Appl. Acoust.* **2023**, *204*, 109233. [CrossRef]
4. Diez-Ibarbia, A.; Fernandez-del-Rincon, P.; Garcia, P.; Viadero, F. Gear rattle dynamics under non-stationary conditions: The lubricant role. *Mech. Mach. Theory* **2020**, *151*, 103929. [CrossRef]
5. Bozca, M. Transmission error model-based optimization of the geometric design parameters of an automotive transmission gearbox to reduce gear-rattle noise. *Appl. Acoust.* **2018**, *130*, 247–259. [CrossRef]
6. Rigaud, E.; Liaudet, J.P. Investigation of gear rattle noise including visualization of vibro-impact regimes. *J. Sound Vib.* **2020**, *467*, 115026. [CrossRef]
7. Donmez, A.; Kahraman, A. Influence of various manufacturing errors on gear rattle. *Mech. Mach. Theory* **2022**, *173*, 104868. [CrossRef]
8. Singh, R.; Xie, H.; Comparin, R.J. Analysis of automotive neutral gear rattle. *J. Sound Vib.* **1989**, *131*, 177–196. [CrossRef]
9. Pizzolante, F.; Battarra, M.; D'Elia, G.; Mucchi, E. A rattle index formulation for single and multiple branch geartrains. *Mech. Mach. Theory* **2021**, *158*, 104246. [CrossRef]
10. Guo, D.; Ning, Q.; Ge, S.; Wang, Y.; Zhou, Y.; Zhou, Y.; Shi, X. Nonlinear characteristic analysis of gear rattle based on refined dynamic model. *Nonlinear Dyn.* **2022**, *110*, 3109–3133. [CrossRef]
11. Shangguan, W.B.; Liu, X.L.; Yin, Y.; Rakheja, S. Modeling of automotive driveline system for reducing gear rattles. *J. Sound Vib.* **2018**, *416*, 136–153. [CrossRef]
12. Trochon, E.P. Analytical Formulation of Automotive Drivetrain Rattle Problems. Master's Thesis, The Ohio State University, Columbus, OH, USA, 1997.
13. Idehara, S.J.; Flach, F.L.; Lemes, D. Modeling of nonlinear torsional vibration of the automotive powertrain. *J. Vib. Control.* **2018**, *24*, 1774–1786. [CrossRef]
14. Beinstingel, A.; Parker, R.G.; Marburg, S. Experimental measurement and numerical computation of parametric instabilities in a planetary gearbox. *J. Sound Vib.* **2022**, *536*, 117160. [CrossRef]
15. Pizzolante, F.; Battarra, M.; Mucchi, E. The role of gear layout and meshing phase for whine noise reduction in ordinary geartrains. *Mech. Mach. Theory* **2023**, *181*, 105209. [CrossRef]
16. Palermo, A.; Britte, L.; Janssens, K.; Mundo, D.; Desmet, W. The measurement of Gear Transmission Error as an NVH indicator: Theoretical discussion and industrial application via low-cost digital encoders to an all-electric vehicle gearbox. *Mech. Syst. Signal Process.* **2018**, *110*, 368–389. [CrossRef]
17. Mughal, H.; Sivayogan, G.; Nader, D.; Ramin, R. An efficient analytical approach to assess root cause of nonlinear electric vehicle gear whine. *Nonlinear Dyn.* **2022**, *110*, 3167–3186. [CrossRef]
18. Yoo, H.G.; Chung, W.J.; Kim, B.S.; Park, Y.J.; Kang, M.R.; Lee, H.K.; Kim, M.S.; Lee, K.S. Effect of hybrid metal-composite gear on the reduction of dynamic transmission error. *J. Mech. Sci.* **2023**, *37*, 3445–3457. [CrossRef]
19. Barthod, M.; Hayne, B.; Tébec, J.L.; Pin, J.C. Experimental study of gear rattle excited by a multi-harmonic excitation. *Appl. Acoust.* **2007**, *68*, 1003–1025. [CrossRef]
20. Cui, L.; Liu, T.; Huang, J.; Wang, H. Improvement on meshing stiffness algorithms of gear with peeling. *Symmetry* **2019**, *11*, 609. [CrossRef]
21. Chen, L.; Zhang, X.; Yan, Z.; Zeng, R. Matching model of dual mess flywheel and power transmission based on the structural sensitivity analysis method. *Symmetry* **2019**, *11*, 187. [CrossRef]
22. Ren, Z.; Xin, X.; Sun, G.; Wei, X. The effect of gear meshing on the high-speed vehicle dynamics. *Veh. Syst. Dyn.* **2021**, *59*, 743–764. [CrossRef]
23. Garvey, S.D.; Penny, J.E.T.; Friswell, M.I. The relationship between the real and imaginary parts of complex modes. *J. Sound Vib.* **1998**, *212*, 75–83. [CrossRef]
24. Rubens, G.S.J. Advanced in the Theory of Linear Dynamical Systems through Coordinate Decoupling. Ph.D. Thesis, University of California, Berkeley, CA, USA, 2019.
25. Tisseur, F.; Meerbergen, K. The quadratic eigenvalue problem. *SIAM Rev.* **2006**, *43*, 235–286. [CrossRef]
26. Afolabi, D. Linearization of the quadratic eigenvalue problem. *Comput. Struct.* **1987**, *26*, 1039–1040. [CrossRef]
27. Chen, H.C.; Taylor, R.L. Solution of eigenproblems for damped structural systems by the Lanczos algorithm. *Comput. Struct.* **1988**, *30*, 151–161. [CrossRef]
28. Datta, B.N.; Elhay, S.; Ram, Y.M. Orthogonality and partial pole assignment for the symmetric definite quadratic pencil. *Linear Algebra Appl.* **1997**, *257*, 29–48. [CrossRef]
29. Golub, G.H. Some modified matrix eigenvalue problems. *SIAM Rev.* **1973**, *15*, 318–334. [CrossRef]
30. Arnoldi, W.E. The principle of minimized iterations in the solution of the matrix eigenvalue problem. *Q. Appl. Math.* **1951**, *9*, 17–29. [CrossRef]
31. Den Hartog, J.P. *Mechanical Vibrations*; Dover Publications: New York, NY, USA, 1985.
32. Inman, D.J. *Engineering Vibration*; Pearson Education Asia: Seoul, Korea, 2013.
33. Luis, J.; Zabala, M.; All, M.; Connor, J.J.; Sussman, J.M. State-Space Formulation for Structural Dynamics. Master's Thesis, Massachusetts Institute of Technology, Cambridge, MA, USA, 1996.

34. Koch, U.; Wiedemann, D.; Sundqvist, N.; Ulbrich, H. State-space modelling and decoupling control of electromagnetic actuators for car vibration excitation. In Proceedings of the IEEE 2009 International Conference on Mechatronics, ICM 2009, Málaga, Spain, 14–17 April 2009.
35. Leonard, I.E. The matrix exponential. *SIAM Rev.* **1996**, *38*, 507–512. [CrossRef]
36. Bay, J. *Fundamentals of Linear State Space Systems*; WCB McGraw-Hill: Boston, MA, USA, 1999.
37. Macías, J.A.R.; Expósito, A.G.; Soler, A.B. A comparison of techniques for state-space transient analysis of transmission lines. *IEEE Trans. Power Deliv.* **2005**, *20*, 894–903. [CrossRef]
38. Rook, T.E.; Singh, R. Modeling of automotive gear rattle phenomenon: State of the art. *J. Passeng. Cars* **1995**, *104*, 2332–2343.
39. Padmanabhan, C.; Barlow, R.C.; Rook, R.C.; Singh, R. Computational issues associated with gear rattle analysis. *J. Mech. Des.* **1995**, *117*, 185–192. [CrossRef]
40. Curtis, S.; Pears, J.; Palmer, D.; Eccles, M.; Poon, A.; Kim, M.G.; Jeon, G.Y.; Kim, J.K.; Joo, S.H. An analytical method to reduce gear whine noise, including validation with test data. *J. Passeng. Car Mech. Syst. J.* **2005**, *114*, 2195–2202.

symmetry MDPI

Article

Analysis of High-Speed Rotor Vibration Failure Due to Sudden Angular Deformation of Bolt Joints

Fayong Wu [1,2], Jie Hong [1,3], Xueqi Chen [3,*] and Yanhong Ma [3]

1 School of Energy and Power Engineering, Beihang University, Beijing 102206, China; wufayong@buaa.edu.cn (F.W.); hongjie@buaa.edu.cn (J.H.)
2 Shenyang Engine Design and Research Institute, Aero Engine Corporation of China, Shenyang 110015, China
3 Research Institute of Aero-Engine, Beihang University, Beijing 102206, China; mayanhong@buaa.edu.cn
* Correspondence: chenxueqi@buaa.edu.cn

Abstract: As the efficiency of advanced aero engines improves, the operational speed of their rotors increases. This heightened operational speed makes the rotor dynamics highly sensitive to changes in the rotor's mass asymmetry state, or unbalance state. During the use of a dual-spool turbofan engine, when its supercritical high-pressure rotor (HPR) exceeds a certain operational speed, the rotor's vibration spikes and continues to increase with the operational speed until it drops sharply near the maximum operational speed. Analysis of the bolt joints in the faulty rotor reveals various phenomena such as joint interface damage, changes in bolt loosening torque distribution, and alterations in rotor initial unbalance. This paper proposes that at high operational speeds, the bolt joint of the HPR undergoes sudden angular deformation, resulting in the slanting of the principal axis of inertia of the turbine disk. This slant leads to changes in the unbalanced state of the HPR. The additional unbalance causes a sudden rotational inertia load excitation, triggering the rotor vibration failure. Subsequently, a rotor dynamic model that incorporates the angular deformation of the joints is established to simulate how this joint deformation influences the dynamic response of the rotor. The simulation results align well with the observed failure phenomenon and validate the proposed failure mechanism. Finally, troubleshooting measures are proposed and implemented in the faulty engine, effectively mitigating the vibration fault.

Keywords: vibration spike; rotational inertia load; joint; aero engine; rotor dynamic

1. Introduction

As advanced aero engines develop toward high performance and structural efficiency, their structural systems have become more complex. A rotor with a poor unbalance state can exert excessive rotational inertia load during high-speed rotation, resulting in rotor vibration failure.

Existing rotor dynamic models, such as the Jeffcott model [1] or the four-degree-of-freedom (DOF) rotor model [2–5], often equate the rotor's unbalance state to the center of mass (CM) offsets on one or more rotor cross sections and simplify the rotational inertia load to a lateral harmonic excitation force in these CM offset planes. According to this assumption, research has been conducted on rotor dynamics [6–8]. In these rotor dynamic models, when the operational speed of a CM-offset rotor exceeds its first and second critical speeds, the rotational inertia force caused by the CM offset will decrease as the operational speed increases owing to the inward turn of the CM. Therefore, when the operational speed of a supercritical rotor is largely far away from its critical speed, the CM offset is unlikely to cause an excessive dynamic response. However, for a thin disk (i.e., disks with a ratio of polar moment of inertia I_p to diameter moment of inertia I_d greater than 1) with a slanted principal axis of inertia (PAI), its rotational inertia load can increase with operational speed under supercritical conditions [9,10] and cause rotor dynamic response to increase continuously, which has been found in several rotor vibration failures [11–13]. Therefore, considering the influence of disk slant in the dynamic response analysis of a supercritical rotor is critical.

Several studies have also delved into the impact of the mechanical characteristics of joints on rotor dynamics. Qin [14,15] and Luan [16] modeled the bending stiffness of the rotor's joints as a function of operational loads, examining its influence on rotor dynamics. Zhuo [17] and Chen [18,19] discovered that bending moments can induce non-axisymmetric changes in the contact stress distribution at the joint interface, leading to reductions in joint bending stiffness and bending mode frequency. Beyond the stiffness considerations, the deformation of these joints has also been a point of focus. Yu [20] explored the bolt flange's deformation under bending moments, suggesting that micro-slips between interfaces can introduce damping nonlinearity in the rotor, potentially leading to self-excited vibrations in the rotor system under supercritical conditions. Shi [21] and Jaszak [22] studied the macro deformation of the flange-bolt joints in pipes under axial force and bending moment, highlighting that increasing bending moments can induce an abrupt angular deformation due to macro slips at the interface. In an experimental investigation, Hong [23] subjected a bolt-jointed rotor to repeated bending, concluding that such joint deformations can degrade the rotor's dynamic response. Liu [24] also pinpointed the detrimental effects of bolt joint deformations on rotor dynamic response, which originate from interfacial cutting faults. Neither of the two investigations adequately addressed the root causes of these abnormal experimental observations. To elucidate how the deformations of rotor joints impact rotor dynamic response, Sun [25] and Chen [26] developed rotor unbalance models considering joint deformation. They observed that joint deformation can change the rotor's unbalanced state, subsequently affecting its dynamic response. Yet, they equated the influence of joint deformation on rotor unbalance with an added CM offset, but this equation cannot explain the supercritical rotor's dynamic response. Therefore, understanding rotor joint deformation and its negative implications for the thin disk's PAI slant is essential to unveiling the mechanistic progression of how joint deformation influences the rotor's supercritical dynamic response.

In summary, this paper analyzes the failure of a turbofan engine. A rotor dynamic model that accounts for abrupt angular deformation of the joint and its impact on the disk's PAI slant is established. The simulation of rotor dynamic response is carried out to reveal the failure process. Finally, corresponding vibration troubleshooting measures are proposed, and their effectiveness is verified in another engine test.

2. Phenomenon of Vibration Failure

Figure 1a shows a dual-spool low-bypass-ratio turbofan engine. To monitor the engine's vibration, accelerometers are arranged at the engine's intermediate frame and rear turbine frame (points A and B in the figure). The operational speed of the high-pressure rotor (HPR) of the engine is relatively high, far exceeding the first/second critical speed. However, owing to the high bending stiffness of the rotor, its operational speed is much lower than the third critical speed. Therefore, at its maximum operational speed, the rotor will not undergo modal vibration, and its dynamic response amplitude depends on rotor unbalance.

The HPR consists of multiple components that are connected by bolt joints. The deformation of the joints during operation can affect the rotor's unbalance and dynamic response, particularly for bolt joints located before the turbine, which operates in a high-temperature environment with high linear velocity and significant axial force. As shown in Figure 1b, the bolt joint before the turbine connects the rear flange of the drum shaft, the labyrinth disk, and the front flange of the turbine disk. During operation, the different geometries and operational loads of these components can cause them to experience varying radial deformations. This can negatively impact the contact characteristics of the joint interfaces, leading to interfacial slippage and joint deformation.

During usage in real scenarios, the engine mounted on aircraft underwent severe vibrations that exceeded the allowable limit. Following the failure, an analysis of the vibration signals achieved by the accelerometer at Point A and a comprehensive inspection of the faulty HPR were conducted.

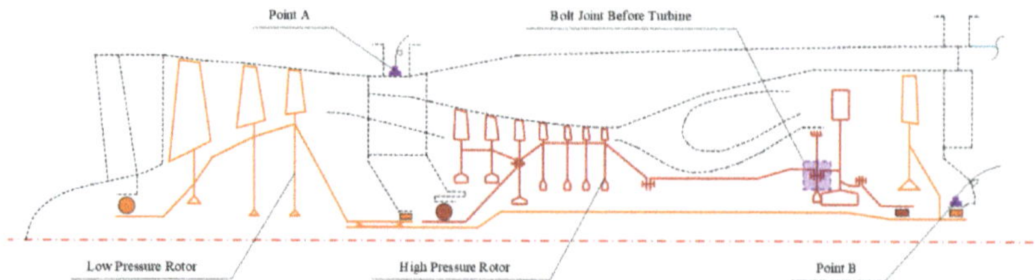

(a)

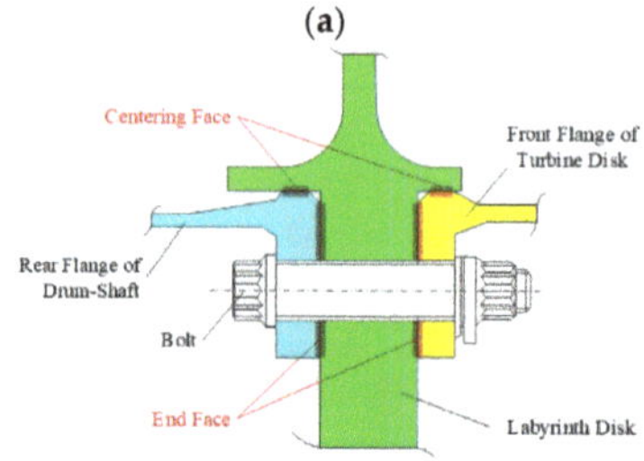

(b)

Figure 1. Typical dual-spool low-bypass-ratio turbofan engine: (**a**) structure of the engine and the position of vibration measurement; (**b**) bolt joint located before high-pressure turbine disk.

2.1. Signal Analysis of Engine Vibration

As shown in Figure 2a, the engine's vibration amplitude changes with the operational speed of its HPR and low-pressure rotor. When the HPR's operational speed is increased to 16,000 RPM (~75% of its maximum operational speed and significantly exceeding its first and second critical speeds), the engine's vibration suddenly increases (S.I.) with minor changes in operational speed. As the operational speed continues to increase, the vibration rapidly rises, which can be described as vibration following (V.F.). When the HPR's operational speed reaches ~21,000 RPM (close to its maximum operational speed), the vibration suddenly decreases (S.D.). At this point, the engine's vibration reaches 80–90 mm/s, far exceeding the limit. It is important to emphasize that the operational speed of the HPR was deliberately varied, oscillating at different rates. As depicted in the figure, once the operational speed of the HPR remains at 16,000 RPM for several seconds, S.I. occurs; once the speed escalates to 21,000 RPM, S.D. occurs. Consequently, the repeatability of this unusual experimental phenomenon can be validated.

Taking time interval T in Figure 2a as an example, Figure 2b shows the frequency composition of the engine vibration. The energy of the vibration is primarily attributed to the rotation of the HPR (as indicated by N2 in the figure), with virtually no other frequency components present. This suggests that the sudden increase in vibration is attributable to a sudden change in the HPR's unbalance.

However, the HPRs of aeroengines typically operate well below the critical speed of bending modes. As a result, they are often characterized as rotors exhibiting rigid behavior [27]. In engineering analyses, the 4-DOF rotor dynamic model, excited by CM offset, is frequently employed to describe their dynamics. Based on this 4-DOF rotor dynamic model, when the rotor's operational speed surpasses both the first and second critical speeds, the rotor's dynamic response, which is generated by the excitation of CM offset, will decrease with an increase in operational speed. This is due to the inward turn of the rotor's CM, regardless of the magnitude of this CM offset. This differs significantly from the observed experimental phenomenon, where the vibration of the rotor increased rapidly with operational speed.

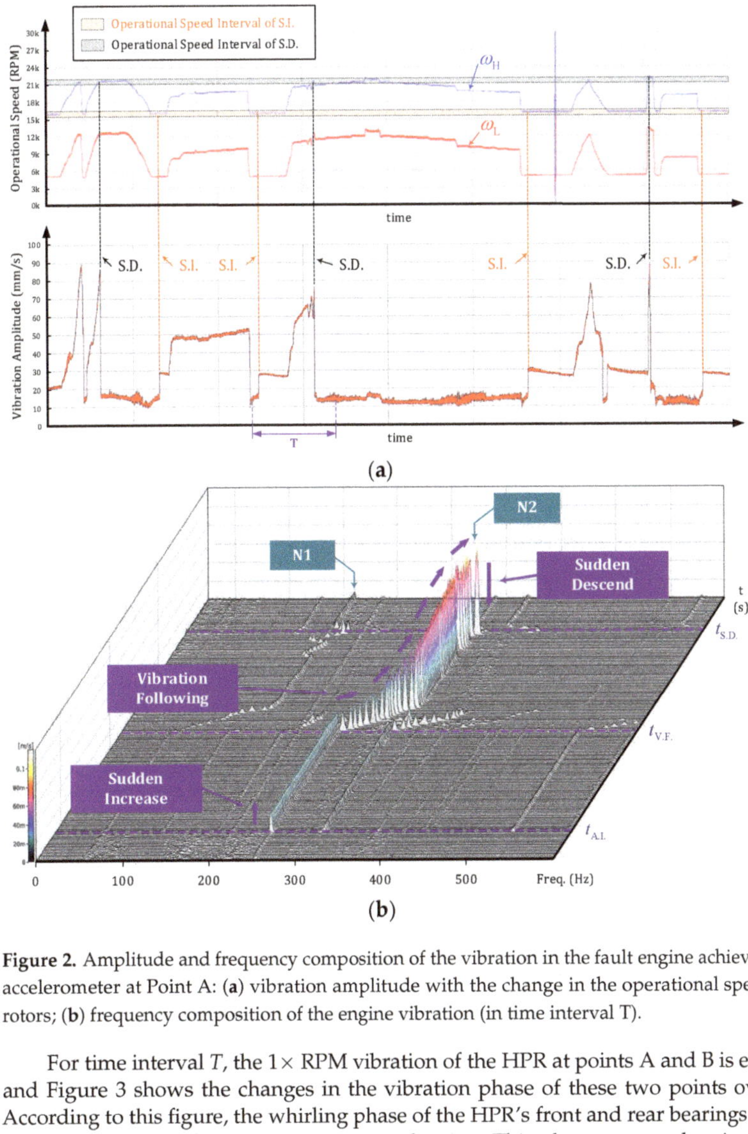

(a)

(b)

Figure 2. Amplitude and frequency composition of the vibration in the fault engine achieved by the accelerometer at Point A: (**a**) vibration amplitude with the change in the operational speed of the rotors; (**b**) frequency composition of the engine vibration (in time interval T).

For time interval *T*, the 1× RPM vibration of the HPR at points A and B is extracted, and Figure 3 shows the changes in the vibration phase of these two points over time. According to this figure, the whirling phase of the HPR's front and rear bearings changes when there is a sudden change in engine vibration. This phenomenon also significantly differs from the whirling phase change caused by the inward turn of the CM. When a 4-DOF rotor with a CM offset passes through its first and second critical speeds, the change in the whirling phase of its front/rear bearing should be 0° or 180°. However, during this engine vibration failure, throughout the S.I. and S.D., the changes in the whirling phase of the front bearing are 170° and 210°, respectively, while those of the rear bearing are 120° and 150°. Therefore, this phenomenon cannot be explained by the traditional 4-DOF rotor dynamics model excited by the CM offset.

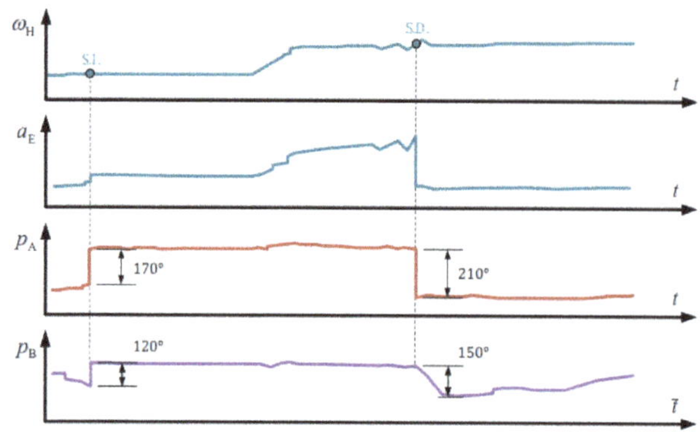

Figure 3. Phase of the 1x RPM vibration of HPR at points A and B: ω_H—operational speed of the HPR; a_E—vibration amplitude of the engine; p_A, p_B—vibration amplitude of point A and point B.

2.2. *Interface Damage and Deformation of Rotor Joints*

During the thorough inspection of the faulty engine, all components were found to be within acceptable limits as specified by the maintenance manual, except for the unusual deformation and the interfacial damage of the bolt joints in the HPR.

2.2.1. Loosening Torque Distribution of Bolt Joints

Owing to the irreversible slip of joint interfaces under operational loads, the pre-strain of the bolts changes relative to their initial state. Consequently, a significant difference can exist between the loosening torque and the tightening torque of the bolts [28]. The interfacial-slip-induced deformation of joints can be evaluated by measuring the loosening torque of the bolts and analyzing their distribution during rotor disassembly.

Figure 4 shows the distribution of loosening torque in the HPR joints. All loosening torques show a significant increase (>200%), indicating that the joint interfaces experience irreversible slip owing to the harsh operational load. However, a comparison of the joints reveals that the loosening torques of the bolt joints before and after the turbine are distributed in a "spindle shape". This indicates that the two joints were subjected to a large bending moment during operation and underwent angular deformation, resulting in a systematic orientation of deformation in their bolts [22,28].

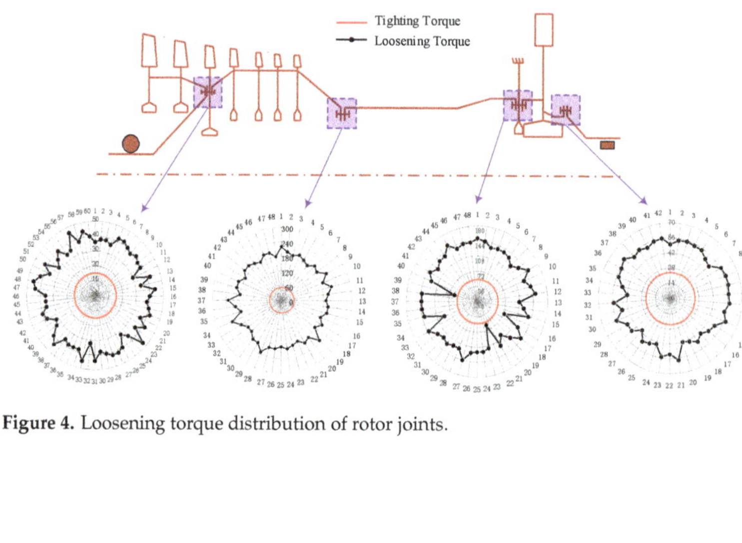

Figure 4. Loosening torque distribution of rotor joints.

2.2.2. Interface Damage of Bolt Joint End Faces

As depicted in Figure 5, triangular marks are present on the turbine disk's front-end face, distributed outside the bolt hole. Microscopic inspection and surface topography scanning reveal that these marks are protrusions of material from the substrate surface, and they are referred to as "stains" in this paper. Physicochemical analysis of the stain compositions (Table 1) indicates that they are identical to the composition of the substrate materials, except for the presence of molybdenum. Considering the location of the stains, the most likely source of the additional molybdenum is the bolt lubrication containing MoS_2. It can be inferred that during operation, the end faces of the bolt joint did not make proper contact with each other, allowing some lubrication to leak into the gap between the end faces and corrode the surface. Owing to the high temperatures, the corroded surface experienced material growth, resulting in the formation of stains.

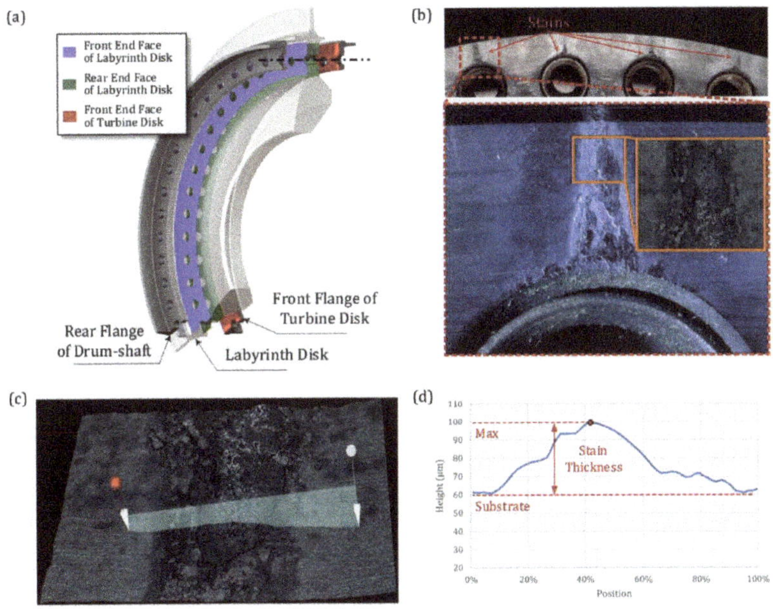

Figure 5. Damage at the interface of the bolt joint located before the turbine: (**a**) Bolt joint located before the high-pressure turbine; (**b**) microscopic examination of interface damage on the turbine disk's front-end face, using bolt 48# as an example; (**c**) 3D modeling of stains, the red and white circle represnet the start point and the end point of the cross-section of the stains; (**d**) Stain profile at the cross section shown in subfigure (**c**), allowing for determination of stain thickness.

Table 1. Composition of the Stains on Turbine Disk's Front-End Face.

Element	Mo	Cr	Co	W	Zr	Al	Nb
Bolt 18#	5.17	10.11	9.25	6.89	0.073	1.16	0.93
Bolt 22#	4.92	10.05	9.51	6.77	0.072	1.38	0.97
Bolt 40#	4.57	10.39	9.25	6.94	0.077	1.28	0.91
Bolt 48#	5.39	10.35	9.12	6.73	0.072	1.19	0.98
Substrate	4.02	10.77	9.58	6.99	0.076	1.23	0.88
Design Value	3.8–4.05	10.5–11.5	8.5–10.5	6.8–10.2	0.07–0.09	1.15–1.60	0.85–1.05

Considering that the size of the gap between the end faces determines the amount of lubrication leakage, the poorer the contact between the end faces, the thicker and larger the stains. The distribution of the stain thickness along the circumferential direction (Figure 6) indicates that the gap in the joint's top half is larger than that in the lower half, and the joint underwent angular deformation during operation. Furthermore, as the gap between the end faces can cause the bolt to stretch and increase the preload of the bolt, the sectors with larger bolt loosening torque are also located in the top half of this bolt joint, overlapping with severely stained sectors (Figure 4). This further confirms that angular deformation occurred in this joint during the operation.

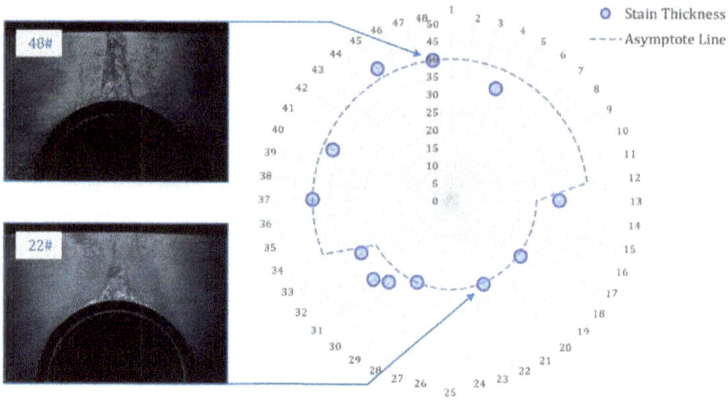

Figure 6. Thickness of the stains varies across different bolt sectors.

2.2.3. Indentations on Bolt Screws

In addition to damage on the end face, some indentations also occur on the surface of the screws of the bolt joint before the turbine (Figure 7). These indentations result from the extrusion of the bolt hole on the screw due to uncoordinated radial deformation between the labyrinth disk and the turbine disk under centrifugal force and thermal deformation. Analysis of the position of these indentations and their correlation with the bolt's loosening torque reveals a typical distribution pattern, which is illustrated in Figure 8. This figure shows that the abnormally indentated screws are relatively concentrated, with most of them found in sectors with relatively small loosening torque. This is related to the lower frictional constraint in sectors with smaller loosening torque, which cannot effectively restrain the relative displacement between the end faces, resulting in the squeezing of the screw by the screw hole.

The phenomenon of joint deformation and interface damage resulting from the combined action of centrifugal force, thermal deformation, and axial force reveals that the contact between the components of the bolt joint before the turbine is not effective, leading to irreversible interfacial slip. Consequently, changes in bolt loosening torque, end face damage, and indentations on the screw surfaces occur. These changes are unevenly distributed across the circumference of the bolt joint. This suggests that the bolt joint before the turbine is subjected to a large bending moment during operation and experiences angular deformation.

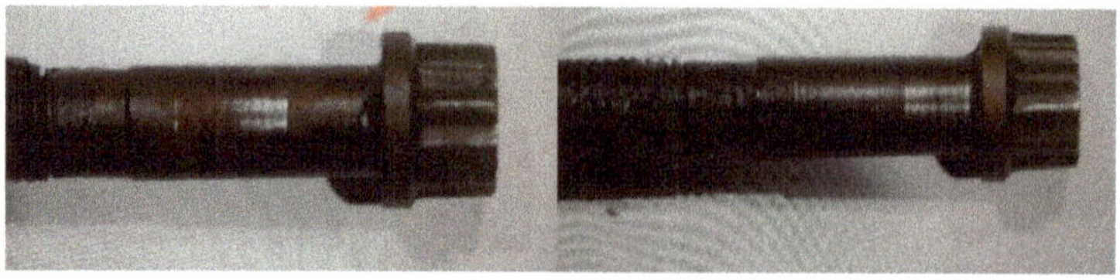

Figure 7. Patterns of the indentations on the surface of bolt screws.

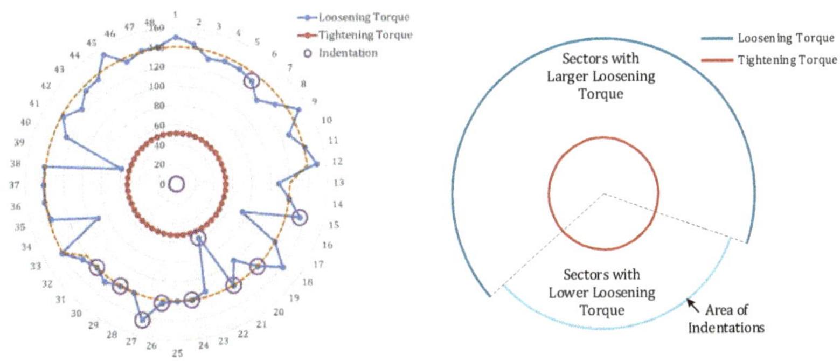

(a) (b)

Figure 8. Distribution of indented bolts: (**a**) experiment results, where the orange line represents the fitted curve of the loosening torque of the bolts; (**b**) distribution of bolt loosening torque and indentations.

2.3. Change in Rotor Unbalance

Table 2 presents the changes in the HPR's unbalanced state before and after engine usage. It is important to emphasize that, when measuring its unbalance, the rotor lacks blades and balancing mass. Consequently, the measured rotor unbalance stems exclusively from the rotor's asymmetrical deformation. The unbalanced state of the compressor module undergoes relatively minor changes before and after the test. However, the unbalance state of the turbine module experiences significant changes in its front balance plane (B.P.), with the amplitude of unbalance increasing from 540 to 1190 g/mm and the phase of unbalance shifting from 343° to 210°. Additionally, a substantial change in unbalance exists in the rear B.P. of the HPR, with the phase of unbalance changing from 23° to 330°. These changes align with the unbalanced changes in the turbine module.

Table 2. Rotor unbalance before and after engine experiment.

Module	Before Engine Test (gmm)			After Engine Test (gmm)		
	Front B.P.	Rear B.P.	Static Unbalance	Front B.P.	Rear B.P.	Static Unbalance
Compressor Module	3440∠330°	2170∠158°	1340∠316°	3180∠314°	2180∠162°	1600∠275°
Turbine Module	540∠343°	1580∠201°	1201∠217°	1190∠210°	1240∠180°	2350∠195°
HPR	560∠60°	370∠269°	300∠23°	730∠47°	1890∠308°	1920∠330°

Considering the circumferential nonuniform distribution of bolt loosening torque, interface damage, screw indentation, and changes in HPR distribution, it can be concluded that during HPR operation, the harsh environment leads to interfacial slip and angular deformation of the bolt joint before the turbine. This, in turn, results in a change in the unbalance of the turbine module by slanting the thin turbine disk. The slant of the turbine disk generates significant rotational inertia moment excitation in the rotor's supercritical state, which likely contributes to the vibration failure of the HPR.

3. Dynamic Analysis of the Jointed High-Speed Rotor System

Analysis of the failure phenomena reveals that the angular deformation of the bolt joint triggers the vibration failure of the HPR by altering the rotor's unbalance. In this section, a rotor dynamics model that considers the sudden angular deformation of the joints is established. Through simulation of sudden changes in rotor dynamics, the mechanism underlying the rotor vibration failure is revealed.

3.1. *Joint Interfacial Slip and Angular Deformation*

3.1.1. Bolt Joint Deformation Model

Figure 9 shows the operational loads applied to the bolt joint before the turbine disk. In its initial state, owing to the axial compression of the bolt and the interference fit of the centering faces, the flange is subjected to axial compression force from the bolt F_P, contact stress from the end face $P_E(y)$, and contact stress from the centering face $P_R(x)$.

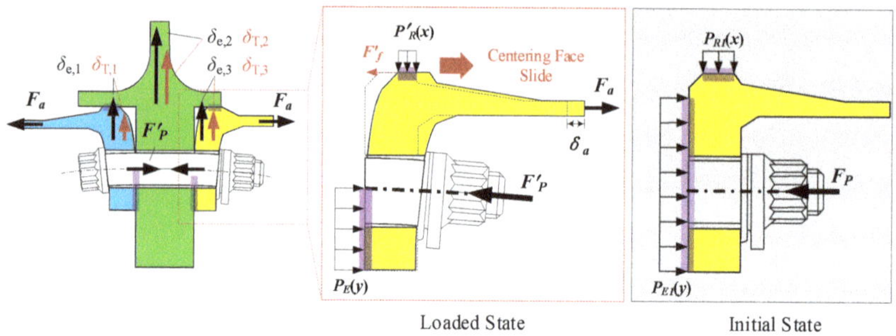

Figure 9. Operational loads applied to the bolt joint before the turbine disk: purple areas represent contacted area; blue areas represent the rear flange of the drum shaft; green areas represent the labyrinth disk; yellow areas represent the front flange of the turbine disk; F_P represents axial compression force applied to the flange from the bolt; $P_E(y)$ represents contact stress from the end face; $P_R(x)$ represents contact stress from the centering face; F_P represents bolt preload; F_a, δ_a represent axial force applied to the bolt joint and corresponding axial slid; F_f represents frictional force of the centering face.

When an operational load is applied to the joint, under the action of axial force F_a, centrifugal force, and thermal load, the joint's centering face undergoes axial relative sliding. The direction of deformation δ_a and the direction of frictional force F_f on the flange's centering face are shown in this figure. The axial deformation of the flange causes axial deformation of the bolt, increasing residual preload inside the screw. Additionally, under the action of external loads, interfacial contact stress distribution $P_E(y)$ and $P_R(x)$ will also change. According to force balancing,

$$F_a + F_c' = F_f' + k_\theta \delta_a + F_P', \tag{1}$$

where k_θ is the axial stiffness of the flange, F_c' is the reaction force applied to the end face, F_f' is the frictional force applied to the centering face, and can be expressed as

$$F_c' = \int P_E(y)\mathrm{d}y, F_f' = \mu \int P_R'(x)\mathrm{d}x. \tag{2}$$

From Equation (2), the axial slip of the centering face under operational force δ_a can be obtained. This slip depends on the axial force F_a, the residual preload of the bolt F_P', and the contact stress distribution of the centering face $P_R'(x)$.

As shown in Figure 10, when the joint is subjected to a bending moment load M_J, its upper sectors experience tension while its lower sectors undergo compression. This indicates that the operational load applied to the joint is non-axisymmetric. Consequently, the interfacial slip and axial deformation of each sector can vary, resulting in the angular deformation of the joint. Similarly, owing to variations in bolt preload and the contact status of the centering faces, the interfacial constraint of the joint also exhibits a non-axisymmetric characteristic, which can contribute to angular deformation of the bolt joint [29]. Additionally, considering that interfacial slip often occurs abruptly when the applied load surpasses a certain threshold, changes in the joint's angular deformation can also exhibit a sudden characteristic.

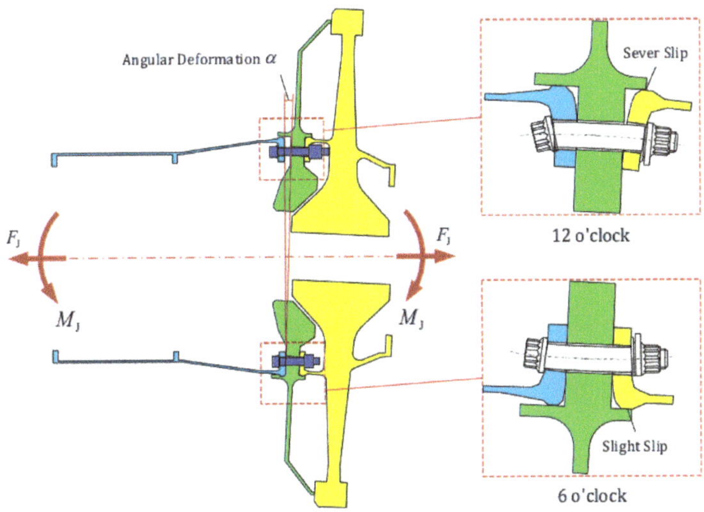

Figure 10. Angular deformation of the bolt joint before the turbine disk: blue areas represent drum shaft; green areas represent labyrinth disk; yellow areas represent turbine disk; F_J, M_J represent axial force and bending moment applied to the bolt joint before the turbine, α represents angular deformation of the bolt joint.

3.1.2. Simulation on Bolt Joint Angular Deformation

A simulation model of the bolt joint before the turbine disk is established to simulate the angular deformation of the joint (Figure 11). Both the FEM modeling, the boundary condition setup, and the solving of this model are realized in the commercial FEM software ANSYS 2021. Two sets of simulations are conducted to investigate the mechanism through which the unevenness of the bolt preload and moment load affects the deformation of the bolt joint. In Set A, let ω, F_a, and $F_{pre}(i)$ satisfy the relationship as

$$\omega/\omega_{max} = F_a/F_{a,max} = \beta, \quad (3)$$

$$F_{pre}(i)/F_{pre,max} = 1 + (|2i-2-N|/N-1)\gamma. \quad (4)$$

Equation (3) indicates that the rotor's operational speed and the axial force applied to the bolt joint will increase linearly. Equation (4) indicates that bolt preload varies linearly with its circumferential position, where γ is a factor that determines the difference between the maximum and minimum preloads of the bolts, and γ can be expressed as

$$\gamma = \frac{F_{pre,max} - F_{pre,min}}{F_{pre,max}}. \quad (5)$$

Figures 12 and 13 show the FEM simulation results of the joint's angular deformation. In Figure 12, when the operational speed reaches $72\%\omega_{max}$, the top half of the bolts, characterized by lower preload, experience larger axial deformation, resulting in an abrupt angular deformation of the joint (~10^{-4} rad). In Figure 13, when M_J is low and no macroscopic slip occurs on the centering face, the angular deformation of the bolt joint is relatively small. However, once M_J exceeds a certain threshold, the joint also undergoes sudden angular deformation (~10^{-4} rad) owing to the slip of the upper bolt sectors.

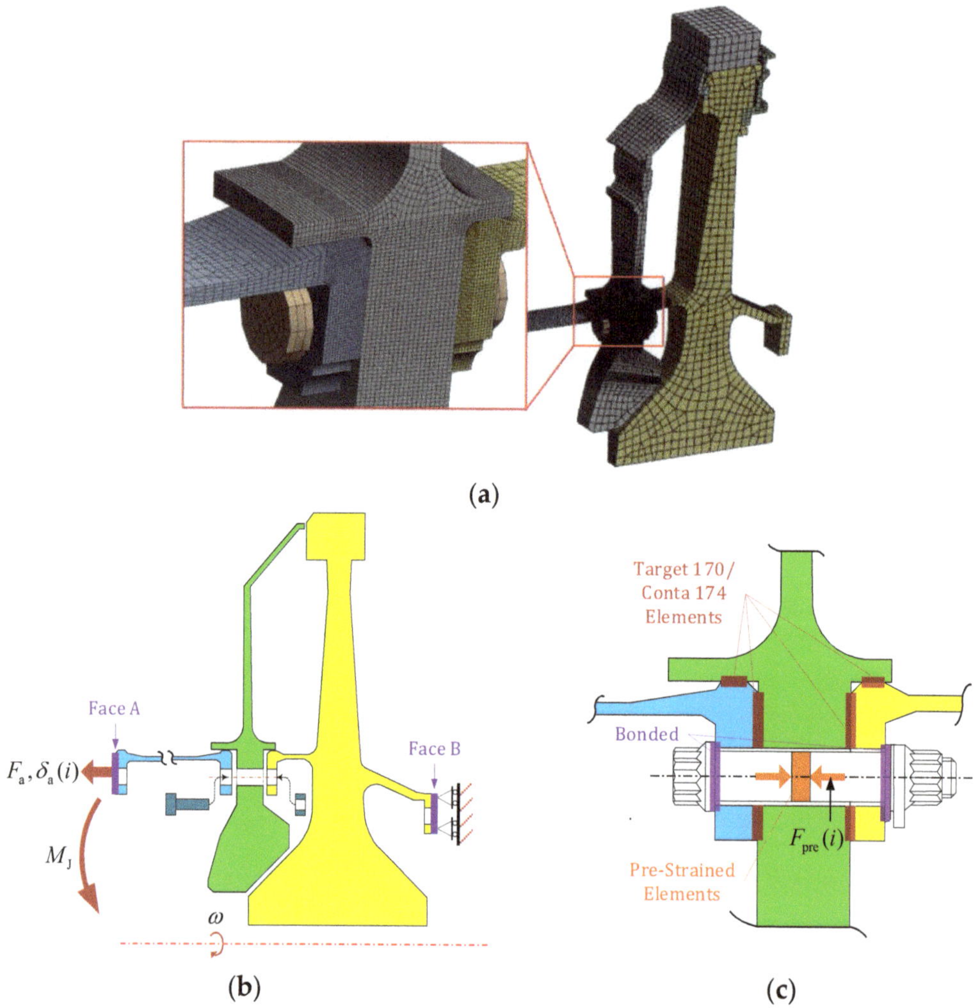

Figure 11. Simulation model of the bolt joint before the turbine disk: (**a**) a finite element model (FEM) of this bolt joint, which consists of Solid 185 elements, and the elements near the interface have been refined; (**b**) the boundary condition of the simulation, where blue areas represent the rear flange of the drum shaft; green areas represent the labyrinth disk; yellow areas represent the front flange of the turbine disk; F_a represents axial force applied at the front end face of the drum shaft (Face A), M_J represents moment load applied at Face A, ω represents the operational speed of the joint, $\delta_a(i)$ represents axial deformation of Face A at the #*i* bolt caused by operational loads. Meanwhile, axial displacement constraints are applied at the rear end face of the turbine disk (Face B); (**c**) contact elements and bolt preload of the joint before the turbine, where $F_{pre}(i)$ represents preload in bolt #*i* obtained by screw elements' pre-strain; the contact characteristics of the centering faces and end faces are obtained by Target 170/Conta 174 elements located at these surfaces.

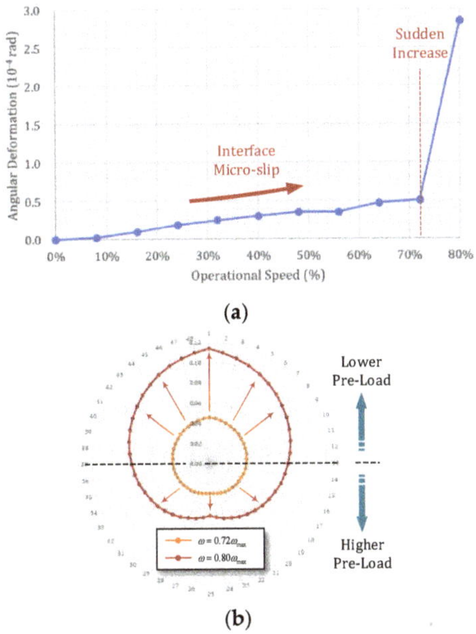

Figure 12. Non-axisymmetric axial deformation of sectors with increasing operational speed and its influence on joint angular deformation: (**a**) angular deformation of the joint with rotor operational speed; (**b**) axial deformation of sectors.

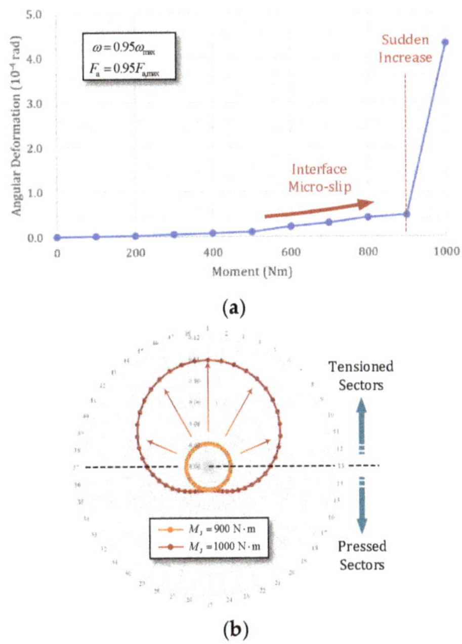

Figure 13. Non-axisymmetric axial deformation of sectors under moment load and its influence on joint angular deformation: (**a**) angular deformation of the joint with the applied moment, when $\omega = 0.95\omega_{\text{max}}$, $F_a = 0.95F_{a,\text{max}}$; (**b**) axial deformation of sectors.

According to these simulation results, both the non-axisymmetric bolt preloads and the bending moment can induce a sudden angular deformation of the bolt joint before the turbine disk. This angular deformation of the joint can result in a change in the PAI slant of the turbine disk, resulting in a significant rotational inertia moment excitation at high operational speeds, thereby affecting the rotor's dynamic response.

3.2. Rotor Dynamic Analysis Considering Joint Angular Deformation

3.2.1. Rotor Dynamic Model with Distributed Rotational Inertia Load Excitation

As shown in Figure 14a, when the HPR undergoes bending deformation while rotating, the relative position relationship between its components changes, particularly in terms of the angular relative position. Each component exhibits a different whirling state owing to variations in materials and geometric configurations. Consequently, the influence of the components on rotor dynamics, including rotational inertia load excitation and gyro-effect, also varies. At this stage, considering the rotor as a whole is no longer appropriate, and it becomes necessary to decompose the HPR into the five elements illustrated in the figure.

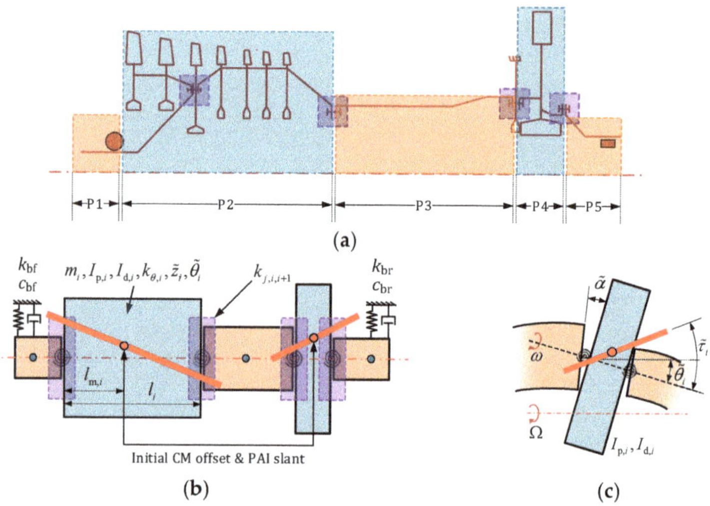

Figure 14. Rotor dynamic model with distributed excitation arising from rotational inertia load: (**a**) decomposition of the HPR into several elements; the high-pressure compressor with multi-disks can be treated as a single thick disk; (**b**) The mass, stiffness, moment of inertia, and displacement of the rotor, where m_i—the mass of element i, $I_{p,i}$, $I_{d,i}$—the polar moment of inertia and diameter moment of inertia of element i, $k_{\theta,i}$—bending stiffness of element i, $\tilde{z}_i$, $\tilde{\theta}_i$—lateral and angular displacement of element i described by the complex vector, $k_{b,f}$, $k_{b,r}$, $c_{b,f}$, $c_{b,r}$—stiffness and damping ratio of front and rear bearings, $k_{j,j+1}$—joint bending stiffness between element j and j + 1; (**c**) description of the rotational inertia load applied to rotor system.

According to the mass and stiffness of the elements, they can be categorized into two groups: elastic elements and mass elements. Elastic elements are characterized by relatively small masses and large deformations. The front shaft element (P1), drum shaft element (P3), and rear shaft element (P5) are elastic elements. Mass elements are characterized by relatively large mass and moment of inertia, and their deformation can be neglected. The compressor element (P2) and the turbine element (P4) are mass elements.

As illustrated in Figure 14b, m_i, $I_{p,i}$, $I_{d,i}$, $k_{\theta,i}$ represent the mass, moment of inertia, and stiffness of the five elements. The lateral and angular displacements of each element's CM are represented by complex vectors $\tilde{z}_i$ and $\tilde{\theta}_i$, respectively. The bending stiffness of joints between two elements is represented by $k_{j,i,i+1}$, and front and rear support constraint stiffness and damping effects are also considered.

Although both the compressor and turbine elements are mass elements, the compressor element has an inertia ratio $I_{p,2}/I_{d,2} < 1$, which should be treated as a thick disk, while the turbine element has an inertia ratio $I_{p,4}/I_{d,4} > 1$, which should be treated as a thin disk. Therefore, the rotational inertia load generated by these two elements during rotation exhibits distinct changes with operational speed [9]. As shown in Figure 14b, it is necessary to separately consider the unbalance of both compressor and turbine elements and the distributed excitation generated by these two elements. The unbalance of the compressor element is described by the CM offset $\widetilde{e}_2$ and the PAI slant $\widetilde{\tau}_2$, while that of the turbine element is described by $\widetilde{e}_4, \widetilde{\tau}_4$.

The motion differential equation of the rotor can be obtained via the Lagrange method as

$$\frac{d}{dt}\frac{\partial L}{\partial \dot{u}_i} - \frac{\partial L}{\partial u_i} = Q_i, \tag{6}$$

$$L = \sum_{i=1}^{n} T_i - \left(\sum_{i=1}^{n-1} V_{i,i+1} + V_{\mathrm{bf}} + V_{\mathrm{br}}\right), \tag{7}$$

where T_i represents the kinetic energy of the ith element; $V_{i,i+1}$ represents the elastic energy between the element i and $i+1$; $V_{\mathrm{bf}}, V_{\mathrm{br}}$ represents the elastic energies of the front and rear bearings, respectively; u_i represents the lateral or angular displacement of the DOF i; and Q_i represents the generalized force applied to the DOF i.

Substituting Equation (7) into Equation (6), the motion of the rotor system can be expressed as

$$\mathbf{M}\ddot{\mathbf{u}} + (\mathbf{C} + \mathbf{G}(\omega))\dot{\mathbf{u}} + \mathbf{K}\mathbf{u} = \mathbf{Q}, \tag{8}$$

where, $\mathbf{M}$ represents the mass matrix, $\mathbf{G}(\omega)$ represents the gyroscopic matrix, $\mathbf{C}$ represents the damping model, $\mathbf{K}$ represents stiffness matrix, $\mathbf{u}$ represents the displacement of the rotor system, and $\mathbf{Q}$ represents the rotational inertia load applied to the rotor system arising from rotor unbalance. $\mathbf{u}$ and $\mathbf{Q}$ can be expressed as

$$\mathbf{u} = \left(\cdots, \widetilde{z}_{i-1}, \widetilde{\theta}_{i-1}, \widetilde{z}_i, \widetilde{\theta}_i, \widetilde{z}_{i+1}, \widetilde{\theta}_{i+1}, \cdots\right)^{\mathrm{T}} \tag{9}$$

$$\mathbf{Q} = \left(\cdots, \widetilde{F}_2, \widetilde{M}_2, 0, 0, \widetilde{F}_4, \widetilde{M}_4, \cdots\right)^{\mathrm{T}} \tag{10}$$

Only excitation arising from the unbalance of mass elements is considered, which can be expressed as

$$\begin{cases} \widetilde{F}_2 = m_2\widetilde{e}_2 \cdot \omega^2 \mathrm{e}^{i\omega t} \\ \widetilde{M}_2 = (I_{d,2} - I_{p,2}) \cdot \widetilde{\tau}_2 \cdot \omega^2 \mathrm{e}^{i\omega t} \\ \widetilde{F}_4 = m_4\widetilde{e}_4 \cdot \omega^2 \mathrm{e}^{i\omega t} \\ \widetilde{M}_4 = (I_{d,4} - I_{p,4})(\widetilde{\tau}_4 + \widetilde{\alpha}) \cdot \omega^2 \mathrm{e}^{i\omega t} \end{cases} \tag{11}$$

Substituting Equation (11) to Equation (10), $\mathbf{Q}$ can be rewritten as $\mathbf{Q} = \mathbf{Q}_0 e^{i\omega t}$.

Owing to the angular deformation of the joint before the turbine, the rotational inertia moment of the turbine element considers the initial slant of the PAI and the angular deformation $\widetilde{a}$ of the joint (if the additional slant caused by the joint is equal to its angular deformation, as illustrated in Figure 14c).

According to Equation (8), the displacement $\mathbf{u}$ of the rotor can be solved to reveal its dynamics. The detailed solution methodology can be found in Appendix A.

3.2.2. Sudden Angular Deformation of Joints in Rotor Dynamic Model

According to the analysis in Section 3.1, as the operational speed increases, the centrifugal load, axial force, thermal deformation, and bending moment applied to the joint gradually increase. When the speed reaches a certain threshold ω_t, the joint interfaces undergo circumferentially uneven slip, causing sudden angular deformation $\widetilde{\alpha}_0 = a_0\angle\varphi_a$. This causes the turbine element to slant and generate additional rotational inertia moment

$\widetilde{M}_g$ acts on the joints and subjects them to bending moment $\widetilde{M}_J$. Finally, this bending moment induces a reaction load at the rear bearing. According to force balance,

$$\widetilde{M}_J = \widetilde{M}_g + \widetilde{l}_s \times \widetilde{F}_b, \tag{12}$$

where, $\widetilde{l}_s$ represents the position vector of the rear bearing relative to the CM of the turbine element. As the rotor's operational speed increases, the rotational inertia moment generated by the slant of the turbine element's PAI increases rapidly; consequently, the angular deformation of the joint is restored, and the turbine disk tends to return to its original position. Once the amplitude of the bending moment $\widetilde{M}_J$ reaches the threshold M_t, the angular deformation of the joint reduces (though there is still residual angular deformation $\widetilde{\alpha}_r$). This process is shown in Figure 15, and the angular deformation of the joint can be expressed as

$$\widetilde{\alpha}(\omega, \widetilde{M}_J(\omega)) = \begin{cases} 0 & ,\omega < \omega_t \\ \alpha_0\angle\varphi_\alpha & ,\omega \geq \omega_t \;\&\; \left|\widetilde{M}_J(\omega)\right| < M_t \\ \alpha_r\angle\varphi_\alpha & ,\left|\widetilde{M}_J(\omega)\right| \geq M_t \end{cases}. \tag{13}$$

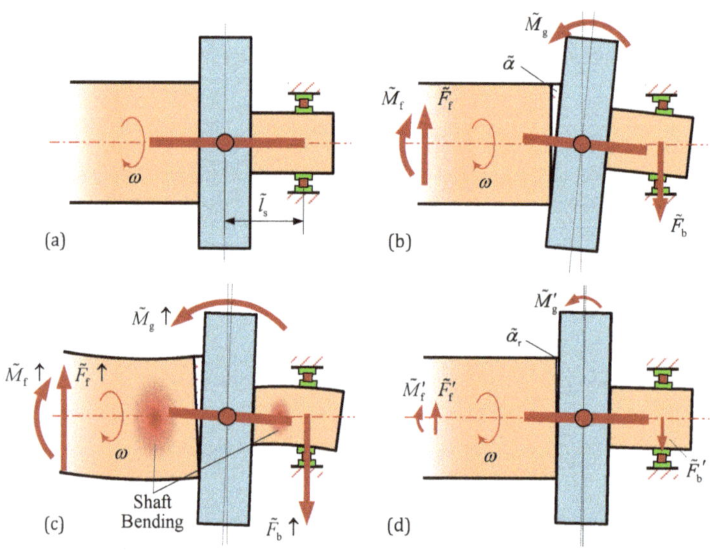

Figure 15. Deformation of the joint before the turbine under different operational speeds: (**a**) $\omega < \omega_t$; (**b**) $\omega = \omega_t$; (**c**) $\omega \geq \omega_t$ & $\left|\widetilde{M}_J(\omega)\right| < M_t$; (**d**) $\left|\widetilde{M}_J(\omega)\right| = M_t$.

By substituting Equation (13) into Equation (11), the influence of additional angular deformation of the joint can be considered on rotor dynamics.

3.3. *Rotor Dynamic Response to Increase in Operational Speed*

Let $\widetilde{e}_2 m_2 = \widetilde{e}_3 m_3 = 2500\ \text{g}\cdot\text{mm}\angle 0°$, $\widetilde{\tau}_2 = \widetilde{\tau}_3 = 0\angle 0°$, $\omega_t = 75\%\omega_{max}$, $M_t = 950\ \text{N}\cdot\text{m}$, $\alpha_0 = 5 \times 10^{-4}$ rad, $\varphi_\alpha = 75°$, $\widetilde{\alpha}_r = 0\angle 0°$. The rotor dynamic response to operational speed is analyzed.

3.3.1. Amplitude and Phase of Rotor Bearing Load

Figure 16 illustrates the vibration amplitude and the phase of the rotor's front and rear bearings. According to Figure 16a, when the rotor's operational speed is below the first and second critical speeds, the dynamic response increases with speed until a local maximum value is reached at the critical speed position. The dynamic response at the second critical speed is quite minimal. This can be attributed to the modal shape at the second critical speed, characterized as a rotor-angular-pitch shape. Such a mode shape renders the rotor's dynamic response at the second critical speed highly sensitive to the rotor's couple unbalance. And due to the unbalance setup ($\tilde{e}_2 m_2 = \tilde{e}_3 m_3 = 2500\ \text{g} \cdot \text{mm}\angle 0°$), the couple unbalance of the HPR is minimal. On all accounts, once the operational speed exceeds the first and second critical speeds and the rotor's CM turns inward, the dynamic loads of the front and rear bearings gradually decrease with increasing operational speed. This phenomenon aligns with the predictions of a 4-DOF rotor dynamic model.

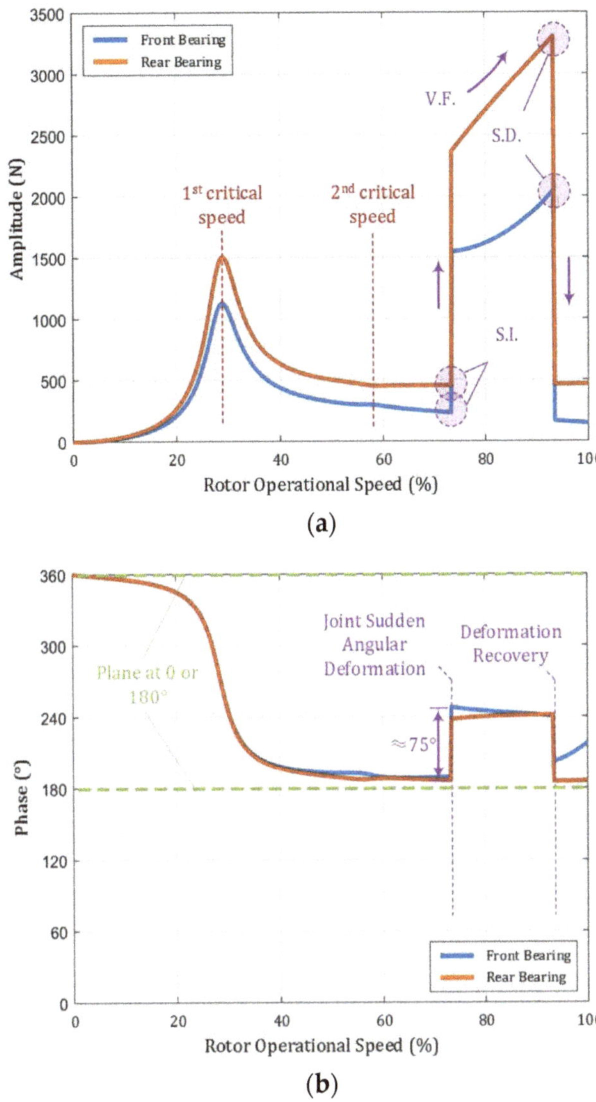

Figure 16. Vibration amplitude and phase of the front and rear bearing of the rotor: (**a**) vibration amplitude; (**b**) vibration phase.

However, when the joint suddenly undergoes angular deformation $\widetilde{\alpha}_0$ at $\omega = \omega_t$ and causes slanting of the turbine element, a spike in the rotational inertia moment arising from the turbine element results in a sudden increase in bearing load. Owing to the turbine element being a thin disk, its rotational inertia moment increases rapidly with operational speed, even in the supercritical state. Consequently, the bearing load amplitude exhibits speed-following characteristics until the rotational inertia moment generated by the turbine element becomes sufficiently large to satisfy the condition $\left|\widetilde{M}_J(\omega)\right| = M_t$. At that point, the angular deformation of the joint decreases, and the turbine element is immediately "straightened", causing the bearing load to also decrease suddenly. As the speed continues to increase, the dynamic loads of the front and rear bearings hardly change. This bearing load phenomenon is consistent with the rotor vibration failure mentioned in Section 2.1.

As shown in Figure 16b, in the low-speed region (except near-critical speed), the vibration phases of both the front and rear bearings are in the plane where the rotor's CM is located (0/180° plane). This dynamic behavior also aligns with a 4-DOF rotor dynamic model. However, when the joint undergoes sudden angular deformation, the vibration phases of the front and rear bearings simultaneously experience a sudden change of ~75°. As the operational speed increases, the vibration phases of the front and rear bearings gradually change and tend to align with this 75° plane. The reasons for this phenomenon are as follows: Before the joint's angular deformation, the rotor deformation is primarily influenced by the rotational inertia force generated by the rotor's CM offset (static imbalance). Consequently, the displacements of both the front and rear bearings lie in the same plane where the rotor's CM offset is located. However, once the joint's angular deformation leads to additional slanting of the turbine element, the rotor deformation is dominated by the rotational inertia moment generated by the slanting of the turbine element's PAI. Consequently, the vibration phases of both the front and rear bearings abruptly shift toward the plane where the turbine disk is slanted ($\varphi_\alpha = 75°$). This moment gradually increases with operational speed, causing the vibration phases of the rotor's front and rear bearings to approach the plane $\varphi_\alpha = 75°$. This phenomenon is also consistent with that of the faulty engine.

3.3.2. Bending Moment Applied to Rotor Joints

Figure 17 illustrates the changes in the bending moment applied to each joint as the rotor's operational speed increases. The following can be observed from the figure:

(a) Prior to the sudden angular deformation of the joint, the bending moment applied to the joints follows a pattern: it gradually increases with operational speed, reaches a maximum at the position of the first critical speed, and then decreases. This pattern is a result of the bending moment primarily originating from the rotational inertia force generated by the rotor's CM offset.

(b) After the joint before the turbine element undergoes sudden angular deformation, it generates a large rotational inertia moment at high speeds owing to the additional slant of the PAI of the turbine element. This results in a significant increase in the bending moment applied to the rotor joint. Among the four joints, the joint before the turbine disk experiences the highest bending moment because it is closest to the turbine element.

(c) As the operational speed continues to increase, the bending moment applied to the joints also increases because the rotational inertia moment generated by the slanted thin disk continues to rise with operational speed. Once the bending moment is applied to the joint before the turbine element reaches a certain threshold M_t, the joint interfaces experience a slip, leading to a sudden reduction in joint angular deformation and the disappearance of the slant in the turbine element. Consequently, the rotational inertia moment from the turbine element also decreases significantly, resulting in a sudden decrease in the bending moment load applied to the rotor joints.

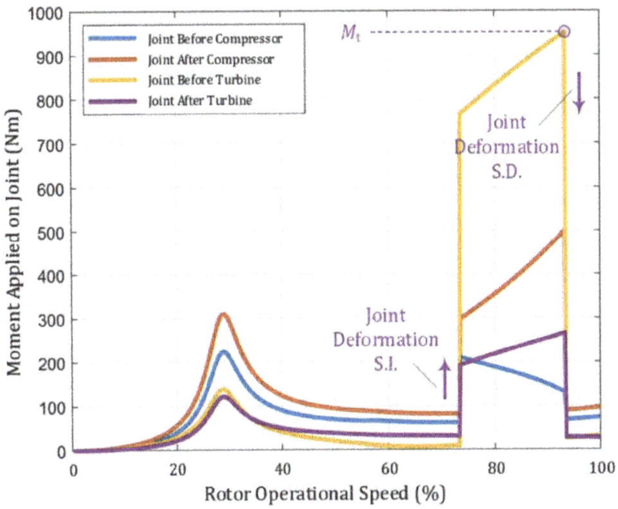

Figure 17. Bending moment applied to rotor joints in response to rotor operational speed.

4. Failure Mechanism and Troubleshooting Measures

4.1. Failure Process of Rotor Vibration Failure

According to the observed failure phenomena and simulations of the dynamic response of the jointed high-speed rotor system, the rotor vibration process can be described as illustrated in Figure 18:

(1) When the operational speed of the HPR has just passed the first and second critical speeds, the bearing load is small and changes minimally with the operational speed owing to the inward turn of the rotor's CM.

(2) Once the operational speed exceeds a certain threshold, the non-axisymmetric bolt preload of the joint before the turbine disk causes a sudden angular deformation of this joint. This results in a slanting of the turbine's PAI and an increase in the additional rotational inertia moment that grows rapidly with the rotor's operational speed. Consequently, the bearing load and the bending moment applied to the joint before the turbine increase.

(3) As the speed continues to increase, the rotational inertia moment generated by the turbine disks keeps growing until the bending moment is applied to the joint before the turbine surpasses a certain threshold. This triggers non-axisymmetric slip at the joint interfaces and leads to the redistribution of unbalance in the HPR. The PAI of the turbine disk momentarily coincides with the rotor's rotation axis again, causing a drop in the rear bearing load.

Therefore, the rotor vibration failure mainly originates from two aspects: (a) improper control of the slant of the turbine disk, resulting in the generation of a rotational inertia moment at high speeds, which causes bending deformation of the rotor and affects the axial distribution of the rotational inertia moment excitation load. (b) Poor contact between the rotor joints leads to interfacial slip when the joints experience the bending moment. This causes the turbine disk, which is a thin disk, to suddenly slant, thereby amplifying the impact of rotor bending deformation on the distribution of rotational inertia moment excitation and rotor dynamic response.

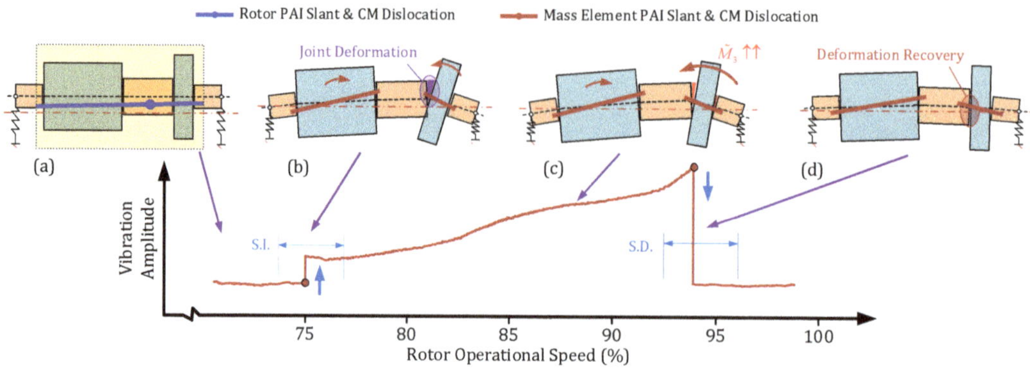

Figure 18. Process of the S.I. and S.D. of rotor vibration: (**a**) Rotor CM turns inward, and vibration caused by rotor CM offset decreases with operational speed. (**b**) The bolt joint before the turbine undergoes sudden angular deformation. (**c**) The rotational inertia moment arising from the slant of the turbine's PAI increases with operational speed. (**d**) The rotational inertia moment reaches a threshold, and the turbine disk returns to its original angular position.

4.2. Troubleshooting Measures toward Rotor Vibration Failure

Considering the identified failure process, the troubleshooting measures shown in Table 3 are implemented to control the rotational inertia moment and suppress joint deformation. Engine tests confirm the effectiveness of these measures in eliminating vibration failure. This provides evidence for the accuracy of the above-described failure mechanism and the successful implementation of the troubleshooting measures.

Table 3. Troubleshooting measures toward rotor vibration failure.

Type	Measures	How It Works
Rotational inertia moment control	Control initial static unbalance of turbine disk and labyrinth disk, and control initial static unbalance and initial couple unbalance of the turbine module.	Control initial mass asymmetry of the components of the turbine module.
	After the engine is assembled, measure and control the runout of the turbine disk rim relative to the rotor's rotation axis.	Prevent initial turbine disk slant caused by engine assembly.
Joint deformation control	While tightening the bolts before the turbine, apply axial compressive force to assembled components.	Ensure end faces are tightly attached and their contact states are uniform circumferentially during joint assembly.
	Evenly select eight bolt sectors of the bolt joint before the turbine disk, and select screws with relatively large diameters to make the screw and the bolt hole nearly contact.	Constrain relative displacement between flanges through screw to prevent angular deformation of the joint.
	Measure and control the uniformity of bolts' preload.	Optimize circumferential evenness of the constraint of the bolts before the turbine disk.

5. Conclusions

This paper investigates the vibration failure of the HPR in a dual-spool low-bypass-ratio turbofan engine. The failure phenomenon is analyzed, and a dynamic model of the jointed high-speed rotor system is established to simulate the dynamic response of the HPR and reveal the mechanism behind the rotor vibration failure. Troubleshooting measures are proposed, and their effectiveness is demonstrated through engine tests. The conclusions of the research are as follows:

(1) As the HPR operates at high operational speeds, it undergoes bending deformation, inducing angular displacement between the turbine and compressor elements. At low operational speeds, the dynamics are predominantly influenced by the rotor's CM offset. Conversely, at higher speeds, the slant of the turbine disk's PAI and the associated rotational inertia moment become dominant excitation factors. This causes the rotor dynamic responses to increase with the operational speed in the supercritical domain.
(2) The non-axisymmetric contact at the bolt joint before the turbine induces an asymmetrical slip within the joint interfaces. This leads to angular deformation in the joint, resulting in a slant in the turbine disk's PAI. This condition then triggers a sudden increase in the rotor dynamic response. As the speed increases, the rotational inertia moment also amplifies. Once this moment reaches a critical threshold that induces an opposite angular deformation in the joint, the turbine reverts to its initial angular position, causing a sudden decrease in the rotor dynamic response. Notably, these joint deformations are consistent with interface damages identified in faulty engines.
(3) Troubleshooting measures are proposed to control the rotational inertia moment induced by the turbine disk's PAI slant and the deformation of the bolt joint before the turbine disk. These measures effectively prevent the recurrence of rotor vibration failure, validating the accuracy of the mechanism analysis and the efficacy of the troubleshooting measures.
(4) To prevent similar aero-engine vibration issues fundamentally from the design phase, a robust structural design should be emphasized to mitigate joint deformations from operational loads, including axial force, bending moment, and operational speed. Following this, simulations and tests should evaluate the design's effectiveness.

Author Contributions: Conceptualization, F.W.; Software, X.C.; Validation, F.W.; Investigation, Y.M.; Resources, F.W.; Data curation, X.C.; Writing—original draft, F.W.; Writing—review & editing, J.H.; Supervision, Y.M.; Project administration, X.C. All authors have read and agreed to the published version of the manuscript.

Funding: The authors gratefully acknowledge the financial support provided by the National Science and Technology Major Project (Y2019-VIII-0011-0172) and the Science Center for Gas Turbine Project (P2021-A-I-002-002).

Data Availability Statement: Data is unavailable due to privacy.

Conflicts of Interest: The authors declare that they have no known competing financial interest or personal relationship that could have appeared to influence the work reported in this paper.

Appendix A. Solution Methodology to Rotor Dynamics

Based on Equation (8), the damping matrix of the rotor system is obtained by the Rayleigh damping model, which means

$$\mathbf{C} = \mu\mathbf{M} + \nu\mathbf{K}, \tag{A1}$$

where, factor μ and ν can be obtained by

$$\begin{cases} \frac{\mu}{2(2\pi f_1)} + \frac{\nu(2\pi f_1)}{2} = \xi_n \\ \frac{\mu}{2(2\pi f)} + \frac{\nu(2\pi f_2)}{2} = \xi_n \end{cases}, \tag{A2}$$

where, f_1, f_2 represents the first and second critical speed of the rotor system, ξ_n represents the damping ratio of the rotor system. According to Reference [30], ξ_n is set to be 0.05 in this paper.

Assuming the displacement vector $\mathbf{u}$ has the form of $\mathbf{u} = \mathbf{u}_0 e^{i\omega t}$, then Equation (8) can be rewritten as

$$\left(-\omega^2\mathbf{M} + i\omega(\mathbf{C} + \mathbf{G}(\omega)) + \mathbf{K}\right) \cdot \mathbf{u}_0 = \mathbf{Q}_0, \tag{A3}$$

Finally, the displacement of the rotor system at operational speed ω can be obtained by

$$\mathbf{u}_0 = \left(-\omega^2\mathbf{M} + i\omega(\mathbf{C} + \mathbf{G}(\omega)) + \mathbf{K}\right)^{-1} \cdot \mathbf{Q}_0 \tag{A4}$$

References

1. Jeffcott, H.H. The lateral vibration of loaded shafts in the neighbourhood of a whirling speed—The effect of want of balance. *Lond. Edinb. Dublin Philos. Mag. J. Sci.* **1919**, *37*, 304–314. [CrossRef]
2. Smith, D.M. The motion of a rotor carried by a flexible shaft in flexible bearings. *Proc. Roy. Soc.* **1933**, *142*, 92–118.
3. Den Hartog, J.P. *Mechanical Vibration*; McGraw-Hill Book Company: New York, NY, USA, 1940.
4. Timoshenko, S.P. *Vibration Problems in Engineering*; D. Van Nostrand: New York, NY, USA, 1955.
5. Carnegie, W. Rotary inertia and gyroscopic effects in overhung shaft systems. *Bull. Mech. Eng. Educ.* **1964**, *3*, 191.
6. Kikuchi, K. Analysis of unbalance vibration of rotating shaft system with many bearings and disks. *Bull. JSME* **1970**, *13*, 864–872. [CrossRef]
7. Cruz, W.; Arzola, N.; Araque, O. Modeling, and experimental validation of the vibration in an unbalance multi-stage rotor. *J. Mech. Eng. Sci.* **2019**, *13*, 5703–5716. [CrossRef]
8. Datz, J.; Karimi, M.; Marburg, S. Effect of Uncertainty in the Balancing Weights on the Vibration Response of a High-Speed Rotor. *J. Vib. Acoust.* **2021**, *143*, 061002. [CrossRef]
9. Hong, J.; Yan, Q.; Sun, B.; Ma, Y. Vibration Failure Analysis of Multi-Disk High-Speed Rotor Based on Rotary Inertia Load Model. In *Turbo Expo: Power for Land, Sea, and Air*; American Society of Mechanical Engineers: New York, NY, USA, 2022; Volume 86076, p. V08BT26A014.
10. Ishida, Y.; Yamamoto, T. *Linear and Nonlinear Rotordynamics: A Modern Treatment with Applications*; John Wiley & Sons: Hoboken, NJ, USA, 2013.
11. Liu, Y.; Liu, H.; Wang, N. Effects of typical machining errors on the nonlinear dynamic characteristics of rod-fastened rotor bearing system. *J. Vib. Acoust.* **2017**, *139*, 011004. [CrossRef]
12. Han, Z.; Wang, D.; Wang, Y.; Hong, J.; Cheng, R. Dynamic effects of the initial skewness of the inertial principal axis on the rotor with bolted joint. *Mech. Syst. Signal Process.* **2023**, *200*, 110564. [CrossRef]
13. Barragan, J.M. *Engine Vibration Monitoring and Diagnosis Based on On-Board Captured Data*; MTU Aero Engines Gmbh Munich: Munich, Germany, 2003.
14. Qin, Z.Y.; Han, Q.K.; Chu, F.L. Analytical model of bolted disk-drum joints and its application to dynamic analysis of jointed rotor. *Proc. Inst. Mech. Eng. Part C J. Mech. Eng. Sci.* **2013**, *228*, 55808444. [CrossRef]
15. Qin, Z.; Han, Q.; Chu, F. Bolt loosening at rotating joint interface and its influence on rotor dynamics. *Eng. Fail. Anal.* **2016**, *59*, 456–466. [CrossRef]
16. Luan, Y.; Guan, Z.Q.; Cheng, G.D.; Liu, S. A simplified nonlinear dynamic model for the analysis of pipe structures with bolted flange joints. *J. Sound Vib.* **2012**, *331*, 325–344. [CrossRef]
17. Zhuo, M.; Yang, L.H.; Yu, L. Contact Stiffness Calculation and Effects on Rotordynamic of Rod Fastened Rotor. In Proceedings of the ASME 2016 International Mechanical Engineering Congress and Exposition, Phoenix, AZ, USA, 11–17 November 2016; p. V009T12A014.
18. Hong, J.; Chen, X.; Wang, Y.; Ma, Y. Optimization of dynamics of non-continuous rotor based on model of rotor stiffness. *Mech. Syst. Signal Process.* **2019**, *131*, 166–182. [CrossRef]
19. Ma, Y.; Ni, Y.; Chen, X. Bending stiffness loss of rod-rabbet joints and its effect on vibration response of rotor systems. *Acta Aeronaut. Astronaut. Sin.* **2021**, *42*, 223861.
20. Yu, P.; Li, L.; Chen, G.; Yang, M. Dynamic modelling and vibration characteristics analysis for the bolted joint with spigot in the rotor system. *Appl. Math. Model.* **2021**, *94*, 306–331. [CrossRef]
21. Shi, W.; Zhang, Z. Elastostatic properties for flange-bolted joints. *Int. J. Press. Vessel. Pip.* **2023**, *205*, 104966. [CrossRef]
22. Jaszak, P. Prediction of the durability of a gasket operating in a bolted-flange-joint subjected to cyclic bending. *Eng. Fail. Anal.* **2021**, *120*, 105027. [CrossRef]
23. Liu, S.; Ma, Y.; Zhang, D.; Hong, J. Studies on dynamic characteristics of the joint in the aero-engine rotor system. *Mech. Syst. Signal Process.* **2012**, *29*, 120–136.
24. Liu, Y.; Liu, H.; Fan, B.W. Nonlinear dynamic properties of disk-bolt rotor with interfacial cutting faults on assembly surfaces. *J. Vib. Control.* **2018**, *24*, 4369–4382. [CrossRef]
25. Zhao, B.; Sun, Q.; Zhao, R.; Yang, Y.; Sun, K.; Mu, X. Mass-eccentricity nonlinear evolution mechanism of combined rotor in fretting slip process. *Tribol. Int.* **2023**, *179*, 108195. [CrossRef]

26. Chen, X.; Ma, Y.; Hong, J. Vibration suppression of additional unbalance caused by the non-continuous characteristics of a typical aero-engine rotor. In Proceedings of the 10th International Conference on Rotor Dynamics–IFToMM, Rio de Janeiro, Brazil, 23–27 September 2018; Springer: New York, NY, USA, 2018; pp. 34–48.
27. *ISO 21940-11:2016*; Mechanical Vibration—Rotor Balancing—Part 11: Procedures and Tolerances for Rotors with Rigid Behaviour. Available online: https://www.iso.org/standard/54074.html (accessed on 16 October 2023).
28. Hong, J.; Yang, Z.; Wang, Y.; Cheng, R.; Ma, Y. Combination resonances of rotor systems with asymmetric residual preloads in bolted joints. *Mech. Syst. Signal Process.* **2020**, *183*, 109626. [CrossRef]
29. Zhao, B.; Sun, Q.; Yang, Y.; Sun, K.; Liu, Z. Study on interface non-uniform slip of combined rotor considering real preload distribution. *Tribol. Int.* **2022**, *169*, 107482. [CrossRef]
30. Chandra, N.H.; Sekhar, A.S. Swept sine testing of rotor-bearing system for damping estimation. *J. Sound Vib.* **2014**, *333*, 604–620. [CrossRef]

symmetry

MDPI

Article

Distributed Rotational Inertia Load Excitation Model and Its Impact on High-Speed Jointed Rotor Dynamic Response

Fayong Wu [1,2], Jie Hong [1,3] and Xueqi Chen [3,*]

1 School of Energy and Power Engineering, Beihang University, Beijing 102206, China
2 Shenyang Engine Design and Research Institute, Aero Engine Corporation of China, Shenyang 110015, China
3 Research Institute of Aero-Engine, Beihang University, Beijing 102206, China
* Correspondence: chenxueqi@buaa.edu.cn

Abstract: Contemporary aero-engines aim for enhanced efficiency and weight reduction. They are designed to increase rotor operational speed while reducing rotor bending stiffness. This may result in bending deformation in rotor systems within the operational speed range. Such deformation can change the relative positions of rotor components, potentially causing increased mass asymmetry or unbalance. Traditional rotor dynamic models typically assume a constant rotor state. They approximate unbalance using constant mass eccentricities at certain rotor cross-sections. However, this approach has its limitations. This paper focuses on a high-speed jointed rotor system. A distributed rotational inertia load excitation model is proposed. This model explicitly considers the rotor's variable unbalance state at different operational speeds. The study involves both simulations and experimental investigations. The results show that at high speeds, bending deformation causes the unbalance and rotational inertia load to shift from a concentrated to a distributed state. Notably, the localized rotational inertia moment from thin-disk components like turbine disks becomes significant at high speeds. This results in a rapid increase in bearing load with rotational speed. It also profoundly affects the rotor's joints, causing interfacial slip and sudden changes in rotor vibration characteristics.

Keywords: rotor dynamics; rotational inertia load; dynamic response; unbalance state; joint interface

1. Introduction

As aero-engines continuously pursue performance improvement and structural efficiency, their structural systems confront progressively intricate dynamic challenges. Notably, the rotor systems of these aero-engines may cause vibrations that exceed acceptable limits if they exhibit poor unbalance states. This concern becomes particularly pronounced for jointed rotors, where joints may undergo deformation due to interfacial slip damage, resulting in alterations in the rotor's unbalance state. Therefore, gaining insights into jointed rotors with mass asymmetry and understanding the influence of joints on the rotor's dynamic response become pivotal in vibration control for aero-engines.

In 1919, Jeffcott [1] introduced the "Jeffcott rotor" model, which initially identified the deviation between the rotor's center of mass (CM) and center of shape as the source of rotational inertia load, leading to the primary cause of rotor whirling [2]. Smith and others [3–6] expanded on the "Jeffcott rotor" model by developing the "four degrees-of-freedom rotor" model. In both models, the rotor's mass asymmetry is represented by the offsets of the rotor's CM, and the rotational inertia loads arising from rotating unbalance are approximated as transverse harmonic force excitations within the specified plane. Hereafter, this model of rotational inertia load excitation is described as the "concentrated rotational inertia load excitation model" or the "concentrated excitation model". As per this excitation model, the rotor's dynamic response induced by unbalance excitation may decrease with the operational speed once the rotor's CM turns inward, unless the operational speed is in proximity to the critical speeds, causing the rotor's modal vibration. Based on this concentrated excitation model, researchers have extensively analyzed rotor dynamics, leading to significant research findings [7–9].

However, as the investigation into rotor dynamics advanced, certain limitations of the concentrated excitation model came to light. Firstly, the approach of representing the rotor's mass asymmetry solely through the CM offset, while neglecting the rotor's axial length, can introduce significant errors when the rotor undergoes bending deformation, rendering it no longer in a constant state [10–13]. Secondly, the rotational inertia load excitations resulting from the slant of the principal axis of inertia (PAI) of rotor components can significantly influence rotor vibrations. Particularly, for a thin-disk, when a slant of its PAI occurs, it induces a continuously increasing rotational inertia moment of excitation with the operational speed in even a supercritical state, leading to a rapid escalation in rotor dynamic response [14,15]. This supercritical vibration has been observed in heavy gas turbine engines and aero-engines, where the slant of the turbine disk was identified as the underlying cause of elevated rotor vibration under supercritical conditions [16,17]. Therefore, it is critical to consider the effect of the slant of a thin-disk's PAI and its axial distribution on the rotor's dynamic response at high speeds.

In study of jointed rotor dynamics, Qin [18] and Luan [19] represented the bending stiffness of the rotor's joints as a function of operational loads and investigated its impact on the rotor's modal characteristics. Zhuo [20] and Chen [21,22] conducted studies in which they observed that under the influence of bending moments, the contact stress distribution at the joint interface undergoes non-axisymmetric changes, resulting in a significant reduction in the joint's bending stiffness. Consequently, the bending mode frequency of the rotor becomes highly sensitive to the rotor's bending deformation. These studies primarily focused on the influence of joint bending stiffness on the rotor's modal characteristics. However, in the rotor designs of aero-engines, critical-speed-avoidance principles are commonly adopted, and variations in joint bending stiffness are unlikely to induce rotor modal vibrations within the operating speed range of the rotor.

Nevertheless, the phenomenon of joint deformation affecting the rotor's unbalance state and subsequently influencing the rotor's dynamic response has gained considerable attention. Chen et al. [23] conducted a study on the impact of joint deformation on the distribution of the rotor's CM offset, confirming that rotor unbalance primarily originates from components' slanting relative to the rotation axis caused by joint deformation. Hong et al. [24] observed a deformation in a rotor's bolt joint after cumulative operating cycles, resulting in an increase in the rotor's unbalance and vibration. Liu et al. [25,26] investigated cases involving interface defects and parallelism errors, revealing that a rotor's CM offset exhibits variation with the operational speed, leading to increased rotational inertia excitation and vibration. In engine experiments, engineers have also discovered that as the high-pressure rotor speed increases from 80% to 100% of the maximum operational speed, the turbine disk's slant caused by centrifugal force and thermal deformation at the joint leads to continuous growth in rotor vibration [27]. Furthermore, Sun et al. [28,29] found that if the pre-forces of a rotor's bolt joints are uneven circumferentially, the radial slip distance in various sectors of the bolt's end face under centrifugal loads can be asymmetric with respect to the rotation axis, resulting in changes in the concentricity of the joint and exacerbating the rotor's unbalance state.

Previous studies have identified that joint deformation can induce changes in the rotor's unbalance state leading to an impact on the rotor's vibration response. However, these investigations still simplify the unbalance change caused by joint deformation as merely an increase in rotor CM offset. This oversimplification fails to provide a comprehensive explanation for the observed increase in rotor vibration with the operational speed in the supercritical state. Consequently, there is an urgent need to reveal the underlying mechanisms by which joint deformation affects a rotor's dynamic response at high operational speeds.

In conclusion, the primary objective of this paper is to explore the dynamic behavior of high-speed, jointed rotor systems, taking into consideration the distributed mass asymmetries of rotor components, including both the CM offset and PAI slant, and their variations resulting from joint deformation. To achieve this, a distributed rotational inertia load excitation model, referred to as the "distributed excitation model", is proposed. Through simulation analyses of rotor dynamic responses at varying operational speeds, the underlying mechanisms behind the changes in the distribution of rotational inertia loads within the rotor and their influence on rotor dynamics can be revealed. Additionally, experimental investigations will be carried out to validate the correctness of the proposed model.

2. Distributed Rotational Excitation Model for High-Speed Rotor System

In the context of engineering, the initial mass distribution of a rotor in its static state may not exhibit complete axisymmetry due to manufacturing and assembly errors. This initial mass asymmetry is commonly referred to as the "initial unbalance" of the rotor. As the rotor undergoes rotation, this mass asymmetry induces rotational excitation at the rotor's rotational frequency, which is known as "rotational inertia load excitation". It is important to note that different types and locations of mass asymmetry can have varying effects on the dynamic response of the rotor [14]. Furthermore, as the operational speed of the rotor increases, the rotor experiences bending deformation and localized deformation at the joints, leading to changes in its mass asymmetry. This variated mass asymmetry during operation is referred to as the "operational unbalance" of the rotor. Consequently, relying solely on a constant and unchanging CM offset within a specific cross-section to represent the mass asymmetry of the rotor across its entire speed range is inadequate. Instead, a comprehensive analysis should consider the axial distribution of its mass asymmetry and how these distributions evolve with the operational speed.

2.1. Configuration of a Typical High-Speed Jointed Rotor

Figure 1 illustrates a typical high-speed jointed rotor system consisting of multiple components interconnected axially through joints. The inherent variations in materials and geometric configurations among these components lead to inevitable changes in their relative angular positions during high operational speeds, resulting in distinct motions of these components. Consequently, the rotational inertia load excitations arising from these motions also undergo corresponding changes. Therefore, the rotor no longer complies with the assumption of being in constant state.

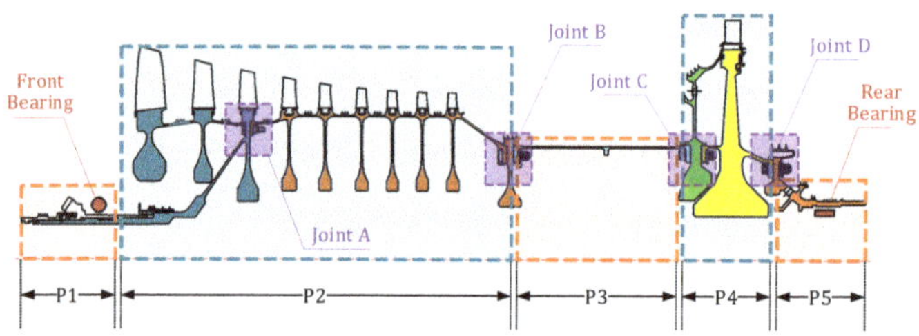

Figure 1. A typical high-speed jointed rotor: P1—front shaft element; P2—compressor element; P3—drum shaft element; P4—turbine element; P5—rear shaft element.

Based on the structural characteristics of the rotor and its deformations during operation, certain connecting components that share similar characteristics and motion states can be aggregated and represented as unified entities, denoted as "structural elements" or simply "elements" for conciseness. These elements can be classified into two categories based on their mass and stiffness: "mass elements" and "elastic elements". "Mass elements" refer to elements with large mass and moments of inertia, whose own deformations can be neglected during motion. In the rotor system, the compressor element (P2) and turbine element (P4) are both considered as mass elements. Meanwhile, "elastic elements" refer to elements with a smaller mass but significant deformations, which determine the stiffness of the rotor system. In this rotor, the front shaft element (P1), drum shaft element (P3), and rear shaft element (P5) are all classified as elastic elements.

2.2. Description of the Rotational Inertia Load Distributed in the Rotor

2.2.1. Description Method Based on PAI Slant and CM Offset

In a rotating rotor, its elements experience non-inertial motion, leading to continuous changes in their linear and angular momentum due to the constraints imposed by bearings. Consequently, the rotor generates reaction forces on the constrained boundaries, known as rotational inertia loads, which can be expressed as

$$\begin{cases} \mathbf{F}_\mathrm{I} = \oiiint \rho(-\mathbf{a})\mathrm{d}V = -\mathrm{d}\mathbf{P}/\mathrm{d}t \\ \mathbf{M}_\mathrm{I} = \oiiint \rho\mathbf{r}' \times (-\mathbf{a})\mathrm{d}V = -\mathrm{d}\mathbf{L}/\mathrm{d}t \end{cases} \tag{1}$$

where, $\mathbf{F}_\mathrm{I}$, $\mathbf{M}_\mathrm{I}$—rotational inertial force and rotational inertia moment generated by the element, respectively, $\mathbf{a}$—acceleration of the element, $\mathbf{r}'$—displacement vector of the element relative to its centroid, and $\mathbf{P}$, $\mathbf{L}$—linear and angular momentum of the element.

Obviously, an element's linear momentum is determined by its mass and velocity, while its angular momentum is determined by its moment of inertia and angular velocity. As shown in Figure 2, it is assumed that the rotor's elements are geometrically symmetrical, but due to manufacturing and assembly errors, its CM, referred as G, does not coincide exactly with its centroid, referred as M. Similarly, the polar PAI, referred as I_p, does not coincide exactly with the rotor's rotation axis. Therefore, it is necessary to establish two coordinate systems, one of which is a body-fixed translation coordinate system x_1My_1, based on M, while the other is based on G, seen in Figure 2b as x_0Gy_0. These two coordinate systems are used to describe the mass asymmetry of the element and its linear momentum and angular momentum during rotation. It is worth noting that this paper assumes that the rotor motion is always synchronized and in the forward direction.

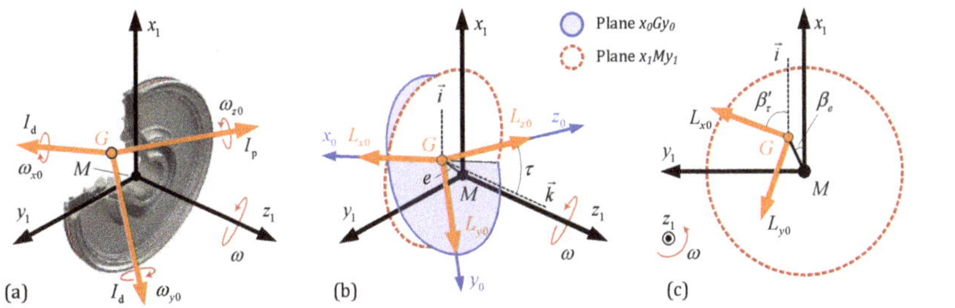

Figure 2. Mass asymmetry of an element and its description method: (**a**) body-fixed translation coordinate system of the element, referred to as x_1My_1. This coordinate system has the element's centroid M as its origin, and its coordinate axes align with the absolute coordinate system. In this coordinate system, G—the element's CM, I_p, I_d—the element's PAI, $\omega_{x0}, \omega_{y0}, \omega_{z0}$—components of rotor's rotation ω in the direction of the element's PAI; (**b**) mass center coordinate system of the element, referred to as x_0Gy_0. This coordinate system has the element's CM G as its origin, and the line of intersection between planes x_1My_1 and x_0Gy_0 as x-axis. In this coordinate system, $\vec{i}$, $\vec{k}$—directions vector of x-axis and z-axis, respectively. e—distance between the element's CM and centroid, τ—angle between the PAI and the rotor's rotation axis, L_{x0}, L_{y0}, L_{z0}—components of the element's angular momentum along the x, y, z axes of coordinate x_1My_1. (**c**) The projection of the element's PAI slant and CM offset onto plane x_1My_1, where β_e, β'_τ—phase of the CM offset and PAI slant.

Among them, the distance between G and M is called the eccentricity distance, denoted as e; the angle between $\overline{GM}$ and Mx_1 is called the eccentric phase angle, denoted as β_e.

During the rotation of the element, the position of its CM in coordinate x_1My_1 can be represented as

$$\begin{cases} x_1 = e\cos(\omega t + \beta_e) \\ y_1 = e\sin(\omega t + \beta_e) \\ z_1 = \text{const} \end{cases} \tag{2}$$

Substituting Equation (2) into Equation (1), the expression for the rotational inertia force can be obtained as

$$\begin{cases} F_{\mathrm{I},x} = -\mathrm{d}P_x/\mathrm{d}t = me\omega^2\cos(\omega t + \beta_e) \\ F_{\mathrm{I},y} = -\mathrm{d}P_y/\mathrm{d}t = me\omega^2\sin(\omega t + \beta_e) \end{cases} \tag{3}$$

Based on the geometric relationship, the expressions for the components of the rotor's self-rotation ω in the coordinate x_0Gy_0 can be obtained as

$$\begin{pmatrix} \omega_{x0} \\ \omega_{y0} \\ \omega_{z0} \end{pmatrix} = \mathbf{A}^{\mathrm{T}} \begin{pmatrix} 0 \\ 0 \\ \omega \end{pmatrix} \tag{4}$$

where, $\mathbf{A}$ is a transformation matrix to transfer vectors in coordinate x_0Gy_0 to x_1My_1. Assuming the initial phase of the motion of coordinate x_0Gy_0 relative to coordinate x_1My_1 is 0, and $\mathbf{A}$ can be expressed as

$$\mathbf{A} = \begin{pmatrix} \cos(\omega t) & -\sin(\omega t) & 0 \\ \sin(\omega t) & \cos(\omega t) & 0 \\ 0 & 0 & 1 \end{pmatrix} \begin{pmatrix} \cos\beta'_\tau & -\sin\beta'_\tau & \tau\sin\beta'_\tau \\ \sin\beta'_\tau & \cos\beta'_\tau & -\tau\cos\beta'_\tau \\ 0 & \tau & 1 \end{pmatrix} \tag{5}$$

by which the angular momentum vector $\mathbf{L}_0 = (L_{x0}, L_{y0}, L_{z0})^{\mathrm{T}}$ of this element in coordinate x_0Gy_0 can be obtained as

$$\begin{pmatrix} L_{x0} \\ L_{y0} \\ L_{z0} \end{pmatrix} = \mathbf{I} \begin{pmatrix} \omega_{x0} \\ \omega_{y0} \\ \omega_{z0} \end{pmatrix} = \mathbf{I}\mathbf{A}^{\mathrm{T}} \begin{pmatrix} 0 \\ 0 \\ \omega \end{pmatrix} \tag{6}$$

where, $\mathbf{I}$ represents the matrix of the moment of inertia, and it can be expressed as

$$\mathbf{I} = diag(I_{\mathrm{d}}, I_{\mathrm{d}}, I_{\mathrm{p}}) \tag{7}$$

Transforming $\mathbf{L}_0$ into coordinate x_iMy_i, which can be expressed as

$$\begin{pmatrix} L_{x1} \\ L_{y1} \\ L_{z1} \end{pmatrix} = \mathbf{A}\mathbf{I}\mathbf{A}^{\mathrm{T}} \begin{pmatrix} 0 \\ 0 \\ \omega \end{pmatrix} \tag{8}$$

Neglecting the term containing higher-order small quantities τ^2, the angular momentum of the element can finally be obtained as

$$\begin{pmatrix} L_{x1} \\ L_{y1} \\ L_{z1} \end{pmatrix} = \begin{pmatrix} -(I_{\mathrm{d}} - I_{\mathrm{p}})\tau\omega\sin(\omega t + \beta'_\tau) \\ (I_{\mathrm{d}} - I_{\mathrm{p}})\tau\omega\cos(\omega t + \beta'_\tau) \\ I_{\mathrm{p}}\omega \end{pmatrix} \tag{9}$$

Substituting Equation (9) into Equation (1), the rotational inertia moment can be obtained as

$$\begin{cases} M_{\mathrm{I},x} = -\mathrm{d}L_{x1}/\mathrm{d}t = (I_{\mathrm{d}} - I_{\mathrm{p}})\tau\omega^2\cos(\omega t + \beta'_\tau) \\ M_{\mathrm{I},y} = -\mathrm{d}L_{y1}/\mathrm{d}t = (I_{\mathrm{d}} - I_{\mathrm{p}})\tau\omega^2\sin(\omega t + \beta'_\tau) \end{cases} \tag{10}$$

Combining Equations (3) and (10), then expressing them in terms of complex notation, the expression for rotational inertia load excitation can be obtained as

$$\begin{cases} \widetilde{F}_{\mathrm{I}} = m\widetilde{e}\omega^2 \cdot e^{\mathrm{i}\omega \mathrm{t}} \\ \widetilde{M}_{\mathrm{I}} = (I_{\mathrm{d}} - I_{\mathrm{p}})\widetilde{\tau}\omega^2 \cdot e^{\mathrm{i}\omega \mathrm{t}} \end{cases} \tag{11}$$

where, $\widetilde{e} = e\angle\beta_e$, $\widetilde{\tau} = \tau\angle\beta_\tau$, $\beta_\tau = \beta'_\tau - \pi/2$, and from a geometric perspective, β_τ represents the angle between the plane containing the PAI slant and plane x_1My_1.

From Equation (11), it can be seen that the rotational inertia force $\widetilde{F}_{\mathrm{I}}$ depends on the mass and CM offset of the element, while the rotational inertia moment $\widetilde{M}_{\mathrm{I}}$ not only depends on the slant of the element's PAI $\widetilde{\tau}$ but also on the relative size of the polar and diametral moments of inertia. Therefore, the elements that satisfy $I_{\mathrm{p}} < I_{\mathrm{d}}$ are referred to as thick-disks, while those that satisfy $I_{\mathrm{p}} > I_{\mathrm{d}}$ are referred to as thin-disks. These two types of rotors exhibit significant differences in the direction and variation of the rotational inertia moment with the operational speed [14].

2.2.2. Axial Distribution of the Rotational Inertia Load of a Rotor

In high-speed rotors, because most of the mass and the moment of inertia are concentrated in the compressor and turbine elements, only the mass asymmetry of these two mass elements should be considered. However, the compressor element is typically a thick-disk, while the turbine element is a thin-disk, resulting in significant differences in the direction and variation patterns of the rotational inertia loads generated by their rotations at varying operational speeds. If the compressor and turbine element are treated as a single entity and combine their unbalances, it would not accurately represent the actual state of the rotational inertia loads acting on the rotor.

Therefore, as shown in Figure 3, it is necessary to describe the unbalance of the compressor and turbine elements separately by using the CM offset $\widetilde{e}_2, \widetilde{e}_4$ and the PAI slant $\widetilde{\tau}_2, \widetilde{\tau}_4$, based on which the rotational inertia loads applied to the rotor can be described as

$$\begin{cases} \widetilde{F}_2 = m_2\widetilde{e}_2 \cdot \omega^2 e^{\mathrm{i}\omega t}, \widetilde{M}_2 = (I_{\mathrm{d},2} - I_{\mathrm{p},2}) \cdot \widetilde{\tau}_2 \cdot \omega^2 e^{\mathrm{i}\omega t} \\ \widetilde{F}_3 = m_3\widetilde{e}_3 \cdot \omega^2 e^{\mathrm{i}\omega t}, \widetilde{M}_3 = (I_{\mathrm{d},3} - I_{\mathrm{p},3}) \cdot \widetilde{\tau}_3 \cdot \omega^2 e^{\mathrm{i}\omega t} \end{cases} \tag{12}$$

where $\widetilde{F}_2, \widetilde{M}_2$ and $\widetilde{F}_4, \widetilde{M}_4$ represent the rotational inertia force and rotational inertia moment generated by the compressor element and the turbine element. These rotational inertia load excitations act on elements' CM.

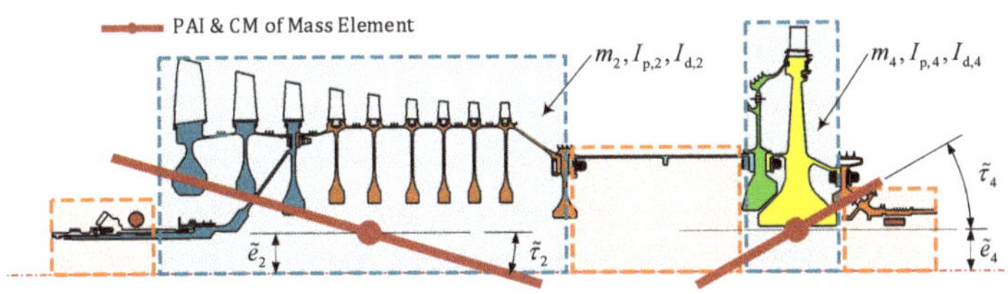

Figure 3. Axial distribution and representation of mass asymmetry in high-speed rotors: m_i—mass of the ith element; $I_{\mathrm{p},i}$, $I_{\mathrm{d},i}$—polar and diametral moment of inertia of the ith element; $\widetilde{\tau}_i$—angles between the ith element's PAI and the centroid line of this element, $\widetilde{e}_i$—distance between the ith element's CM and the centroid line of this element.

2.3. *Influence of the Motion State of the Rotor*

As illustrated in Figure 4, the motion of the rotor in the low-speed range is depicted with respect to the operational speed. In this low-speed region, the effect of rotor bending deformation on the relative positions between elements can be negligible. Therefore, the rotor can be treated as a thick-disk rotor due to $I_{\mathrm{p,r}} < I_{\mathrm{d,r}}$. The mass asymmetry of the compressor and the turbine elements are aggregated and represented as the rotor's CM offset $\tilde{e}_r$ and PAI slant $\tilde{\tau}_r$. The rotor's motion state variation with the operational speed ω can be described using a four degrees-of-freedom rotor model [15]. Specifically, before reaching the critical speed, the amplitude of the rotor dynamic response increases with ω. However, after surpassing the critical speed, the dynamic response decreases with ω due to a turning inwards of the rotor's CM or PAI.

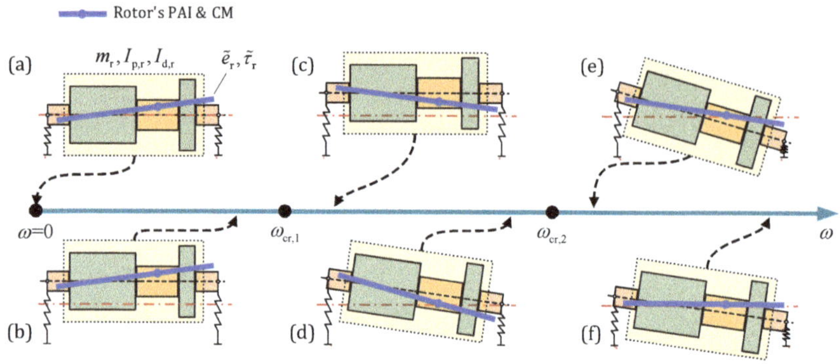

Figure 4. Variation of a rotor's motion with the operational speed (low-speed range): (**a**) while rotating at low speed, the rotor can be regarded as a thick-disk, where, m_r, $I_{p,r}$, $I_{d,r}$ —mass and moment of inertia of the rotor, $\tilde{e}_r$, $\tilde{\tau}_r$—CM offset and PAI slant of the rotor, ω—operational speed; (**b**) the rotational inertia force generated by the CM offset increases with the rotational speed, enhancing the rotor's lateral displacement; (**c**) when ω exceeds the first critical speed $\omega_{cr,1}$, the rotor undergoes the turning inward of CM; (**d**) the rotor's CM gradually approaches the rotor's rotation axis, resulting in a gradual decrease in the rotational inertia force. However, the rotational inertia moment produced by the rotor's PAI slant increases rapidly with the operational speed, enhancing the rotor's angular displacement; (**e**) as ω surpasses the second critical speed $\omega_{cr,2}$, the rotor's PAI changes direction; (**f**) the rotor's PAI gradually approaches the rotor's rotation axis, leading to a gradual reduction in the rotational inertia moment and a consequent decrease in the angular displacement of the rotor.

As illustrated in Figure 5, the motion of the rotor in the high-speed range shows significant changes due to rotor bending deformation and the corresponding alterations in the relative positions between its mass elements. Consequently, the rotor can no longer be considered as a constant-state rotor, and it becomes imperative to separately consider the mass asymmetry and rotational inertia load excitations for the compressor element and the turbine element, respectively. Notably, the turbine element is classified as a thin-disk, where the effect of rotational inertia moment is relatively strong, by which the turbine element undergoes angular displacement, gradually approaching and aligning its PAI with the rotor's rotation axis. Subsequently, the angular deformation of the turbine element ceases to exhibit significant increments with further increases in ω. Moreover, with a further increase in ω, the rotational inertia moment generated by the compressor element also becomes significant, causing its PAI to align with the rotor's rotation axis.

Hence, from the comprehensive analysis presented, it can be deduced that at high-speed range, the rotor experiences bending deformation, resulting in alterations in the relative angular positions between the compressor element and the turbine element. This, in turn, impacts the rotor's mass asymmetry, leading to the manifestation of localized rotational inertia load excitations caused by elements' mass asymmetry.

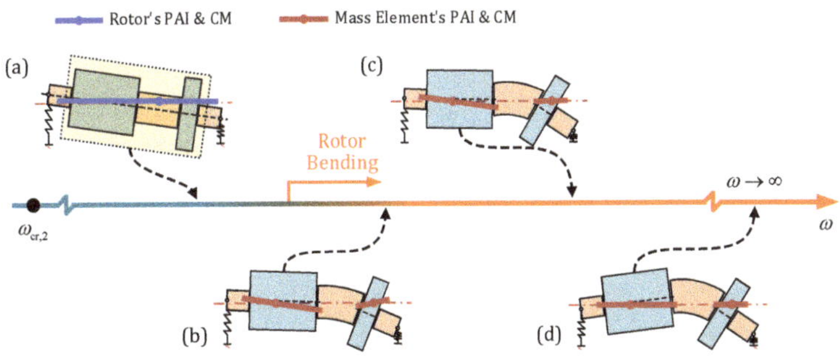

Figure 5. Variation of a rotor's motion with the operational speed (high-speed range): (**a**) the CM and PAI of the rotor are nearly aligned with the rotation axis; (**b**) with increasing ω, the rotor undergoes bending deformation, leading to the disruption of the structural integrity of the rotor. At this stage, it becomes necessary to separately consider the mass asymmetry and motion states of the compressor and turbine element; (**c**) due to the thin-disk characteristic of the turbine element, the rotational inertia moment generated by this element becomes more pronounced compared to the compressor element as ω increases. Consequently, the PAI of the turbine element aligns with the rotation axis first, and the angular displacement of the turbine element no longer increases with further operational speed increase; (**d**) upon reaching sufficiently high operational speed, the rotational inertia load induced by the compressor element also increases to the point where its PAI aligns with the rotation axis. After this threshold, the angular displacement of the compressor element ceases to increase with further operational speed increments.

2.4. *Influence of the Mechanical Characteristic of the Joints*

2.4.1. Bending Stiffness Loss of the Joints

The impact of operational loads, such as centrifugal forces, thermal deformations, bending moments, and aerodynamic loads, can lead to changes in the contacting area and redistribution of contact stresses at the joints' interfaces, resulting in a reduction in their bending stiffness [30–32]. Consequently, in the high-speed range, the bending deformation of the rotor can become more pronounced due to the weakened bending stiffness of its joints. This alteration in the rotor's bending behavior affects the distribution of rotational inertia load excitations and subsequently influences the rotor's dynamic response.

As shown in Figure 6, taking the bending stiffness of Joints C/D ($k_{J,C}$, $k_{J,D}$) as examples, the influence of joints' bending stiffness loss on the distribution of the rotational inertia loads in the high-speed range ($\omega > \omega_{cr,2}$) can be concluded as follows:

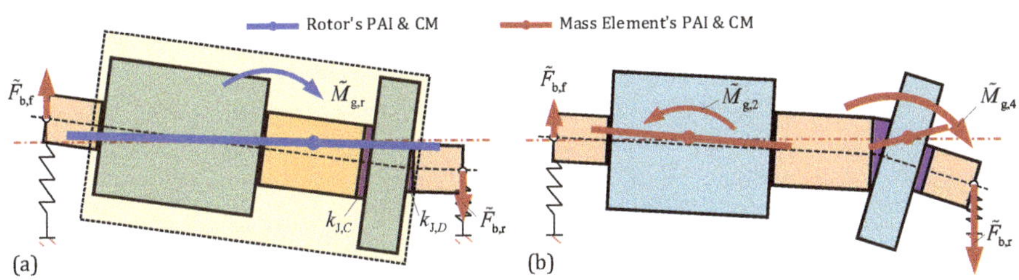

Figure 6. Influence of a joint's bend stiffness loss on the distribution of rotational inertia load in a rotor at high operational speed: (**a**) while the bending stiffness of Joint C, denoted as $k_{J,C}$ is greater than that of Joint D, denoted as $k_{J,D}$. In the diagram, $\tilde{F}_{b,f}$, $\tilde{F}_{b,r}$—bearing load at the front and rear bearing, respectively, $\tilde{M}_{g,r}$—rotational inertia moment generated by the rotor; (**b**) when $k_{J,C} < k_{J,D}$. In the diagram $\tilde{M}_{g,2}$, $\tilde{M}_{g,4}$—local rotational inertia moment generated by the compressor element and turbine element, respectively.

(a) When $k_{J,C} > k_{J,D}$: As $k_{J,C}$ determines the bending stiffness between the compressor element and the turbine element, a higher $k_{J,C}$ can effectively restrains the angular dislocation between the turbine element and the compressor element, by which the integrity of the rotor is enhanced. Therefore, the rotor can still be regarded as a thick-disk, whose rotational inertia moment excitation $\widetilde{M}_{g,r}$ decreases with the operational speed. Simultaneously, the decrease in $k_{J,D}$ weakens the bearing constraint exerted on the rotor motion, and this weakening in constraint reduces the reaction forces at the rear bearing.

(b) When $k_{J,C} < k_{J,D}$: A lower value of $k_{J,C}$ exacerbates the change in the relative position between the compressor and turbine elements, leading to the appearance of local rotational inertia moment excitation, denoted as $\widetilde{M}_{g,4}$, caused by the slant of the turbine element's PAI. This rotational inertia moment excitation can rapidly increase with the operational speed. Simultaneously, a higher $k_{J,D}$ imposes stronger constraints on the rotor motion, inducing a significant increase in the bearing load in the high-speed range.

2.4.2. Sudden Angular Deformation

During the variation of operational conditions, the joints may experience non-uniform interface constraints or be subjected to bending moments, leading to changes in the distribution of interface contact stresses or macroscopic interfacial slide, resulting in angular deformation of the joints [28,29]. This, in turn, causes additional slant of the PAI of nearby mass elements, as depicted in Figure 7. To accurately account for the effect of the angular deformation of the joints on the slant of the PAI of the mass element and the resulting rotational inertia moment, it is necessary to modify Equation (12) as

$$\begin{cases} \widetilde{F}_2 = m_2\widetilde{e}_2 \cdot \omega^2 e^{i\omega t}, \widetilde{M}_2 = (I_{d,2} - I_{p,2}) \cdot \widetilde{\tau}_2' \cdot \omega^2 e^{i\omega t} \\ \widetilde{F}_4 = m_4\widetilde{e}_4 \cdot \omega^2 e^{i\omega t}, \widetilde{M}_4 = (I_{d,4} - I_{p,4}) \cdot \widetilde{\tau}_4' \cdot \omega^2 e^{i\omega t} \end{cases} \tag{13}$$

where, $\widetilde{\tau}_2'$, $\widetilde{\tau}_4'$ represents the slant angles of the compressor element and the turbine element, respectively, considering the influence of the angular deformation of the joints, and they are denoted as $\widetilde{\tau}_2' = \widetilde{\tau}_2 + \widetilde{\alpha}_2$ and $\widetilde{\tau}_4' = \widetilde{\tau}_4 + \widetilde{\alpha}_4$.

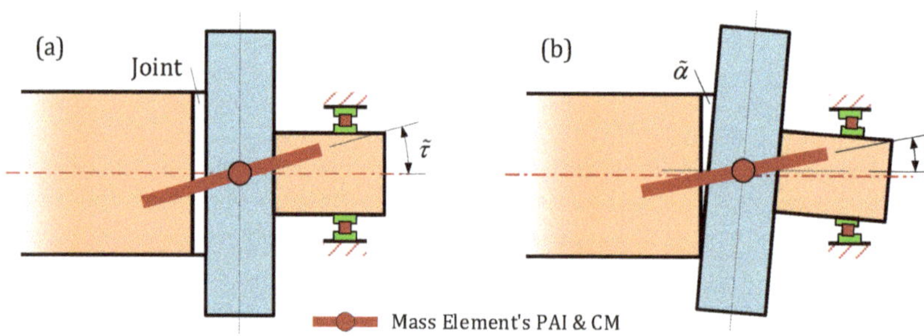

Figure 7. Influence of angular deformation of the joints on the slant of an adjacent mass element's PAI: (**a**) in the initial state, the slant of the PAI of the mass element is $\widetilde{\tau}$; (**b**) due to the local angular deformation $\widetilde{\alpha}$ of the joint near this mass element, its PAI undergoes an additional slant, resulting in an equivalent slant angle $\widetilde{\tau}' = \widetilde{\tau} + \widetilde{\alpha}$.

It is critical to note that the mechanical characteristics of the joints' interfaces are discontinuous, resulting in sudden changes in interface contact stress and sliding damage with varying operational loads. Therefore, the impact of angular deformation of the joint on the rotational inertia load distribution of the elements also exhibits abrupt changes.

3. Dynamic Model for High-Speed Jointed Rotor System

3.1. Parameters of Rotor System

As shown in Figure 8, the rotor's motion state is achieved by the lateral displacement $\widetilde{r}_i$ and angular displacement $\widetilde{\theta}_i$ of each element's CM, denoted as

$$\mathbf{u} = (\cdots, \widetilde{r}_i, \widetilde{\theta}_i, \cdots)^{\mathrm{T}} \tag{14}$$

where, $\mathbf{u}$ represents the displacement vector of the rotor system. To analyze the dynamic response of the high-speed jointed rotor, it is necessary to obtain the mass, moment of inertia, and bending stiffness of its constituent element, i.e., m_i, $I_{\mathrm{p},i}$, $I_{\mathrm{d},i}$, $k_{\theta,i}$ shown in Figure 8, as well as the bending stiffness of the joints $k_{\mathrm{J},i}$. The boundary conditions of the rotor, namely the stiffness k_{bf}, k_{br} and damping coefficients c_{bf}, c_{br} of the front and rear bearings also need to be considered. Additionally, the CM offsets and PAI slants in the mass elements, denoted as $\widetilde{e}_2$, $\widetilde{\tau}_2$, $\widetilde{e}_4$, $\widetilde{\tau}_4$, must be considered.

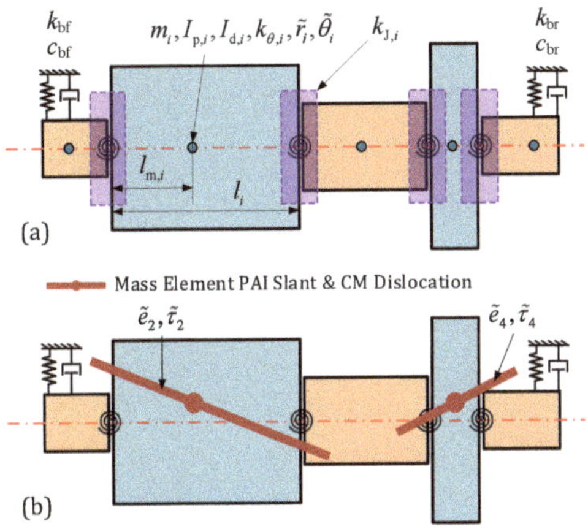

Figure 8. Structural and mechanical parameters of the rotor and its elements: (**a**) for the ith element, m_i-mass, $I_{\mathrm{p},i}$, $I_{\mathrm{d},i}$—polar moment of inertia and diametral moment of inertia, $k_{\theta,i}$—bending stiffness, $\widetilde{z}_i$, $\widetilde{\theta}_i$—lateral displacement and angular displacement of the element's CM, l_i—axial length, $l_{\mathrm{m},i}$—axial distance between the front end face of the element and its CM. k_{bf}, k_{br}, c_{bf}, c_{br}—stiffness and damping coefficient of the front and rear bearing, $k_{\mathrm{J},i}$—bending stiffness of the ith joint; (**b**) during rotor dynamic analysis, only the rotational inertia load excitations caused by the CM offset and the PAI slant of the mass elements are considered.

3.2. Rotor Motion Differential Equations Based on Lagrange's Method

Based on the above-mentioned description method, the motion differential equations of this rotor system can be obtained using Lagrange's method as

$$\frac{d}{dt}\frac{\partial L}{\partial \dot{u}_i} - \frac{\partial L}{\partial u_i} = Q_i \tag{15}$$

where, L can be expressed as

$$L = \sum_{i=1}^{n} T_i - (\sum_{i=1}^{n-1} V_{i,i+1} + V_{\mathrm{bf}} + V_{\mathrm{br}}) \tag{16}$$

where, T_i—kinetic energy of the ith degree of freedom (DoF), $V_{i,i+1}$—elastic energy between the ith and $(i+1)$th DoF, V_{bf}, V_{br}—elastic energy of front and rear bearings, u_i—displacement of the ith DoF, Q_i —generalized force applied to the ith DoF.

Based on Equation (15), the mass matrix **M**, damping matrix **C**, and stiffness matrix **K** of the rotor system can be obtained, and simultaneously, the motion differential equations of this rotor system can be expressed as

$$\mathbf{M}\ddot{\mathbf{u}} + \mathbf{C}\dot{\mathbf{u}} + \mathbf{K}\mathbf{u} = \mathbf{Q} \tag{17}$$

where the vector of generalized force can be obtained by Equation (13) as

$$\mathbf{Q} = \left(\cdots, \widetilde{F}_2, \widetilde{M}_2, \cdots, \widetilde{F}_4, \widetilde{M}_4, \cdots\right)^{\mathrm{T}} \tag{18}$$

3.3. *Mechanical Characteristics of Joints in the Dynamic Rotor Model*

3.3.1. Bending Stiffness Loss of the Joints

In order to characterize the influence of the joints' bending stiffness loss on the rotor dynamic model, the stiffness matrix **K** in Equation (17) needs to be modified. As shown in Figure 9, the bending stiffness between the ith and $(i+1)$th elements' angular DoF, denoted as $k_{\mathrm{t},i,i+1}$, can be considered as a series connection of the bending stiffness of the ith element's rear section $k^{\mathrm{r}}_{\theta,i}$, the bending stiffness of the $(i+1)$th element's front section $k^{\mathrm{f}}_{\theta,i+1}$, and the bending stiffness of the joint between the two elements $k_{\mathrm{J},n}$. Therefore, $k_{\mathrm{t},i,i+1}$ can be expressed as

$$\frac{1}{k_{\mathrm{t},i,i+1}} = \frac{1}{k^{\mathrm{r}}_{\theta,i}} + \frac{1}{k^{\mathrm{f}}_{\theta,i+1}} + \frac{1}{k_{\mathrm{J},n}} \tag{19}$$

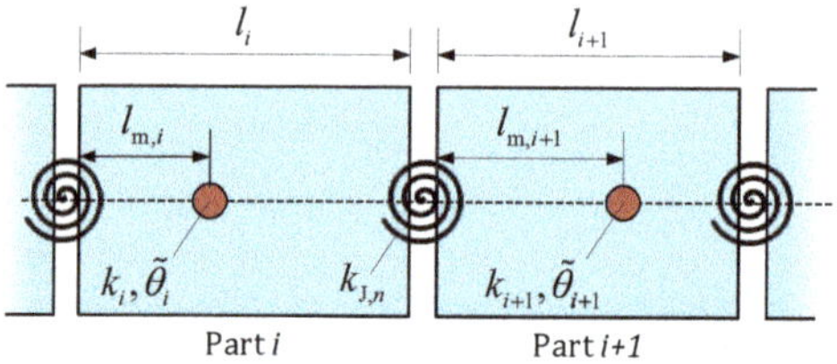

Figure 9. Equivalent method for the influence of joints' bending stiffness loss on rotor stiffness. In this method, based on the assumption of constant section beams, the bending stiffness between the ith and $(i+1)$th elements' angular DoF is obtained by serially connecting the joint and its adjacent components.

Assuming that the elements are constant section beams, it is easy to determine the bending stiffness of the front and rear sections of the ith element by Equation (20)

$$k^{\mathrm{f}}_{\theta,i} = \frac{l_i}{l_{\mathrm{m},i}} k_{\theta,i}, k^{\mathrm{r}}_{\theta,i} = \frac{l_i}{l_i - l_{\mathrm{m},i}} k_{\theta,i} \tag{20}$$

Meanwhile, the stiffness loss coefficient β_n is used to represent the degree of stiffness loss at the nth joint by

$$\beta_n = 1 - k_{\mathrm{J},n} / k^0_{\mathrm{J},n} \tag{21}$$

where, $k^0_{\mathrm{J},n}$—bending stiffness of the joint when its interfaces are perfected merged, $k_{\mathrm{J},n}$—actual bending stiffness of the joint.

3.3.2. Sudden Angular Deformation of the Joint

Due to the turbine element having a configuration of a thin-disk, the impact of its rotational inertia moment is more pronounced. Furthermore, Joint C is subjected to harsh loading conditions, making it more susceptible to deformation during operation. As a result, this paper specifically investigates the influence of the angular deformation of Joint C on the rotational inertia moment excitation caused by the turbine element.

As shown in Figure 10, for a turbine element with an initial PAI slant $\tilde{\tau}$, when the initial slant is small and the operational speed ω is low, the rotational inertia moment generated by the turbine element and the resulting bending deformation of the rotor can almost be neglected. However, as ω increases, various loads such as centrifugal force, axial tension force, and thermal deformation acting on the joints gradually increase with ω, until ω reaches a certain threshold ω_t. At this key point, the contact stress distribution or interface slip damage at the joint's interfaces experiences a sudden change, leading to a sudden angular deformation in this joint, denoted as $\widetilde{\alpha}$. As a result, the PAI slant of the turbine element suddenly increases to $\widetilde{\tau}'$. It can be assumed that the additional slant of the turbine element's PAI equals the joint's angular deformation, which can be expressed as

$$\widetilde{\tau}' = \widetilde{\tau} + \widetilde{\alpha} \tag{22}$$

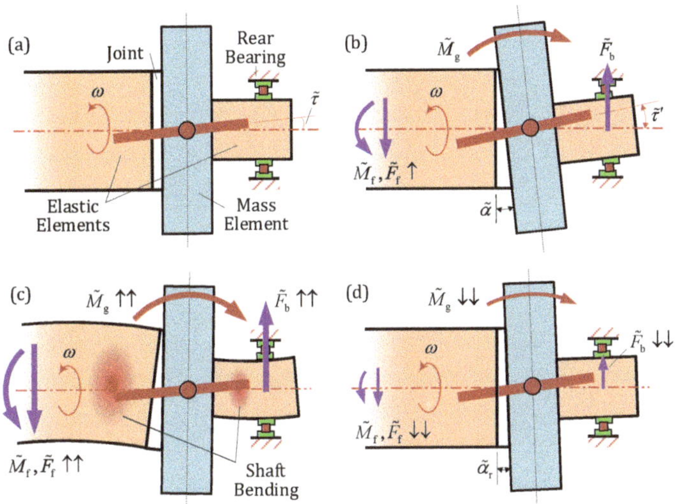

Figure 10. Process of the effect of joints' angular deformation on the dynamic characteristics of a high-speed rotor: (**a**) in low-speed ranges, since the initial PAI slant $\widetilde{\tau}$ is relatively small and the operational speed ω is low, the rotational inertia load generated by the elements during operation and the resulting rotor bending deformation can be negligible; (**b**) when ω increases to ω_t, the loads applied to the joint cause their interfaces to slip, resulting in sudden angular deformation $\widetilde{\alpha}$ of the joint. This leads to the turbine element's slant and corresponding rotational inertia moment $\widetilde{M}_g$. This $\widetilde{M}_g$ not only acts back on the joint, causing it to experience bending moment $\widetilde{M}_J$, but also induces bearing loads $\widetilde{F}_b$ at the rear bearing; (**c**) as ω continues to increase, $\widetilde{M}_g$ increases rapidly, and the bending deformation of the adjacent elastic elements intensifies accordingly. Consequently, as the rear bearing acts as a constraint boundary, in order to balance the bending deformation within the elastic elements, the bearing loads $\widetilde{F}_b$ also increase rapidly with ω; (**d**) when $\widetilde{M}_J$ exceeds a certain threshold, the joint undergoes a sudden angular deformation change, with residual angular deformation $\widetilde{\alpha}_r$, causing a sudden decrease in the angle between the element's PAI and the rotor's rotation axis. Simultaneously, as the rotational inertia moment generated by the turbine element decreases, the bending deformation of the elastic elements and the resulting bearing reaction force also experience a sudden drop.

In response, the rotational inertia moment excitation generated by the turbine element also suddenly increases to $\widetilde{M}_g$. This rotational inertia moment acts on the two adjacent elastic elements, causing bending deformations in these elements. As the rear bearing acts as a constraint boundary, it can generate a significant reaction force, denoted as $\widetilde{F}_b$, that balances the bending deformations in the elastic elements. Additionally, apart from inducing elastic elements' bending deformations and bearing loads, the rotational inertia moment also applies a bending moment, denoted as $\widetilde{M}_J$, to the joint. Force equilibrium can be formulated as

$$\widetilde{M}_J = \widetilde{M}_g + \widetilde{l}_s \times \widetilde{F}_b \tag{23}$$

where, $\widetilde{l}_s$—position vector of the rear bearing relative to the CM of the turbine element.

As ω increases, the rotational inertia moment rapidly increases and the turbine element's PAI tends to align with the rotor's rotation axis. However, when $\widetilde{M}_J$ acting on the joint reaches the threshold, denoted as M_t, the joint can undergo a sudden angular deformation in the same direction as $\widetilde{M}_J$, resulting in a residual angular deformation $\widetilde{\alpha}_r$. As a result, the angle between the turbine element's PAI and the rotor's rotation axis suddenly decreases, leading to a sharp reduction in the rotational inertia moment generated by the turbine element. Consequently, the rotor's bending deformation and the corresponding bearing loads also experience a sudden decrease.

Therefore, the angular deformation of the turbine front joint, denoted as $\widetilde{\alpha}_C$, is expressed as a piecewise function of ω and the bending moment applied to Joint C (denoted as $\widetilde{M}_{J,C}$) given by

$$\widetilde{\alpha}_C(\omega, \widetilde{M}_{J,C}) = \begin{cases} 0 & ,\omega < \omega_t \\ \widetilde{\alpha} & ,\omega \geq \omega_t \& \left|\widetilde{M}_{J,C}\right| < M_{t,C} \\ \widetilde{\alpha}_r & ,\left|\widetilde{M}_{J,C}\right| \geq M_{t,C} \end{cases} \tag{24}$$

where ω_t—threshold operational speed at which Joint C undergoes a sudden increase in angular deformation, and $M_{t,C}$—threshold bending moment at which Joint C undergoes a sudden angular deformation decrease. These threshold values can be obtained through static analysis of the bolt joint. By substituting Equation (24) into Equation (13), the impact of the sudden angular deformation caused by the bolt joint on the rotor's dynamics can be analyzed.

4. Simulation of High-Speed Jointed Rotor System

Table 1 presents the mass, moment of inertia, and bending stiffness values of the rotor and its elements. The stiffnesses of the front and rear bearings are set as $k_{bf} = 5 \times 10^7$ N/m and $k_{br} = 1 \times 10^8$ N/m. Moreover, the local bending stiffnesses of the bolt joints are obtained through static mechanical analysis.

Table 1. Mass/moment of inertia/bending stiffness of the rotor and its elements.

Element ID	P1	P2	P3	P4	P5	Rotor
m (kg)	10	100	10	100	10	230
I_p (kg·mm^2)	7.0×10^4	3.5×10^6	1.5×10^5	5.5×10^6	1.5×10^5	9.4×10^6
I_d (kg·mm^2)	7.8×10^5	4.7×10^6	1.2×10^5	2.8×10^6	8.3×10^4	2.8×10^7
I_p/I_d	0.09	0.74	1.26	1.94	1.80	0.33
k_θ (N·m/rad)	7.3×10^9	10.0×10^{10}	2.6×10^9	10.0×10^{10}	0.4×10^9	-

4.1. Rotor Dynamic Response with Different Excitation Model

The influence of sudden angular deformation of the rotor joints is temporarily ignored and the distributed rotor unbalance state can be set as $\widetilde{e}_2 m_2 = \widetilde{e}_4 m_4 = 300$ g·mm$\angle 0°$, $\widetilde{\tau}_2 = 0\angle 0°$, and $\widetilde{\tau}_4 = 1 \times 10^{-4} \angle 90°$ rad, referred as Group A.

Meanwhile, in order to compare the effects of different excitation models on rotor dynamics, this section also designs a second group of rotor unbalance states, referred to as Group B. In this Group B, the CM offset of the mass element remains unchanged, and the mass elements' PAI slants are equivalent to the CM offset, denoted as $\Delta\widetilde{e}_2 m_2, \Delta\widetilde{e}_4 m_4$, in the planes of the compressor and turbine elements' CM. Based on force balance, $\Delta\widetilde{e}_2 m_2, \Delta\widetilde{e}_4 m_4$ can be obtained by Equations (25) and (26), where $\widetilde{L}_{24}$ represents the vector between the turbine element's CM and the compressor element's CM.

$$\widetilde{L}_{24} \times \Delta\widetilde{e}_4 m_4 = (I_{d,2} - I_{p,2})\widetilde{\tau}_2 + (I_{d,4} - I_{p,4})\widetilde{\tau}_4 \tag{25}$$

$$\Delta\widetilde{e}_2 m_2 + \Delta\widetilde{e}_4 m_4 = 0 \tag{26}$$

Clearly, in Group B, the rotor's mass asymmetries, regardless of their types or locations, are represented as the rotor's dynamic unbalance, denoted as $(\widetilde{e}_2 + \Delta\widetilde{e}_2)m_2$ and $(\widetilde{e}_4 + \Delta\widetilde{e}_4)m_4$. This unbalance equivalent method considers the rotor as a four degrees-of-freedom rotor, and the distributed rotational inertia load excitations caused by mass elements' CM offset and PAI slant are consolidated into two planes.

The results of comparing the rotor's dynamic response under the two excitation models are depicted in Figure 11. It is evident that when using the concentrated excitation model, the front and rear bearing loads of the rotor follow a trend where the bearing load amplitude increases with the operational speed, reaches a maximum at the critical speed, and then decreases with further increase in operational speed. As the rotor approaches its maximum operating speed, the bearing load remains relatively constant. Additionally, upon observing the phase of the bearing load, it can be inferred that within the non-critical speed region, the vibration phase of both the front and rear bearing remains nearly constant with the operational speed, and the phase difference between the front and rear bearing remains consistent, which aligns with the predictions of the four degrees-of-freedom rotor model.

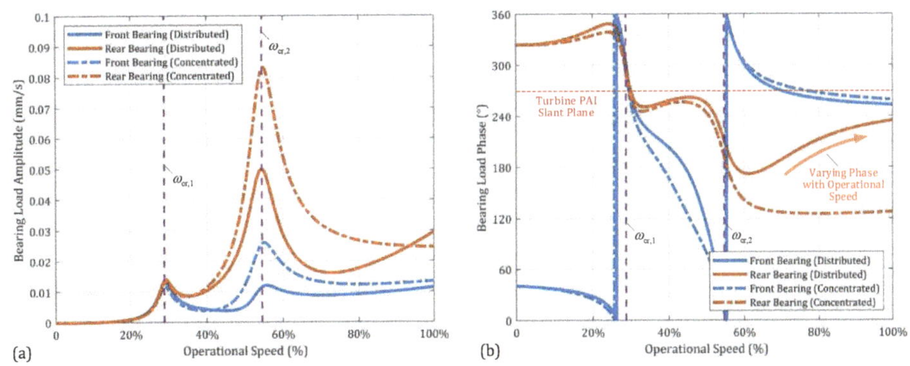

Figure 11. Rotor's dynamic response under different rotational inertia load excitation models: (**a**) amplitude of front and rear bearing loads. $\omega_{cr,1}$, $\omega_{cr,2}$—first/second critical speed of the rotor; (**b**) phase of front and rear bearing loads.

When utilizing the distributed excitation model, in the low-speed range ($\omega < \omega_{cr,2}$), the rotor bearing loads, both in terms of amplitude and phase, exhibit a pattern nearly identical to that of the concentrated excitation model group. This similarity arises because, at this low-speed range, the rotor undergoes no significant bending deformation, allowing it to be treated as having integrity, and thus, the dynamic response of the rotor remains similar for different rotational inertia load excitation models. However, as the rotational speed enters the high-speed range ($\omega > \omega_{cr,2}$), the rotational inertia moment generated by the turbine element increases continuously with the operational speed, leading to a rapid increase in the amplitude of the bearing load with the operational speed (referred to as vibration following, or VF). Furthermore, the phase of the rear bearing load also rapidly approaches the plane of the turbine element's PAI slant (90/270° plane) with increasing operational speed.

The angular displacements of the mass elements under different excitation models are illustrated in Figure 12. In the low-speed range ($\omega < \omega_{cr,2}$), it can be observed from Figure 12a that the excitation model has little influence on the amplitude of the angular displacements of the mass elements. However, as the operational speed enters the high-speed range ($\omega > \omega_{cr,2}$), using the concentrated excitation model results in both the compressor element and turbine element experiencing a decrease in their angular displacements with increasing operational speed, while their difference in angular displacements remains almost unchanged. This indicates that the rotor undergoes little bending deformation in this speed range. However, when the distributed excitation model is used, the turbine element experiences a significant local rotational inertia moment, leading to an increase in its angular displacement with the operational speed. Furthermore, the difference in angular displacements between the turbine element and the compressor element increases rapidly with the operational speed, suggesting that the bending deformation of the rotor is progressively intensifying.

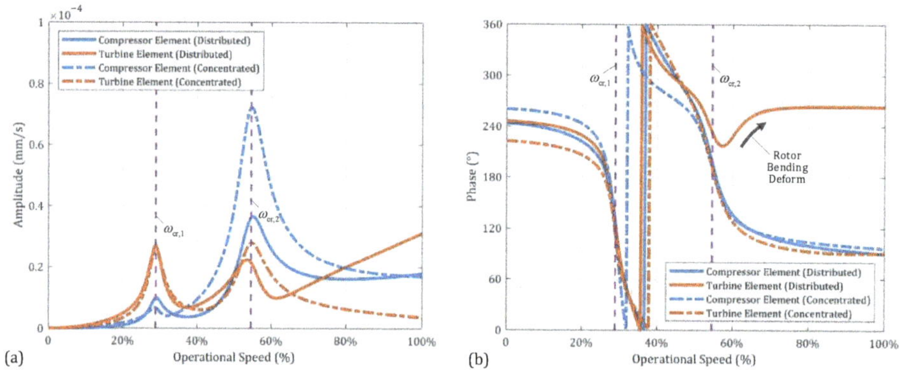

Figure 12. Rotor mass elements' angular displacement under different rotational inertia load excitation models: (**a**) amplitude of mass element's angular displacement; (**b**) phase of mass element's angular displacement.

From Figure 12b, it can be observed that when using the concentrated excitation model, the angular displacements of the compressor element and turbine element occur in the same plane throughout the entire speed range, which confirms the rotor's integrity and minimal bending deformation. However, when the distributed excitation model is used, the phase difference of the turbine element's angular displacement relative to that of the compressor element rapidly changes by 180° as the operational speed exceeds the second critical speed. This confirms the conclusion that the bending deformation of the rotor significantly intensifies under the distributed excitation model, leading to substantial changes in the relative positions of the mass elements.

4.2. Impact of Joint Bending Stiffness on Rotor Dynamics

Since the turbine element is a thin-disk, the rotational inertia load it generates is more sensitive to the angular deformation of nearby joints. Therefore, this section primarily focuses on investigating the impact of the bending stiffness loss in the bolt joints before and after the turbine element.

To make the analysis results more general, parameter λ is defined as the ratio of the bending stiffnesses of the front and rear bolt joints of the turbine element., i.e., $\lambda = k_{J,C}/k_{J,D}$, where $k_{J,C}$ represents the bending stiffnesses of the joints before the turbine element, and $k_{J,D}$ represents the bending stiffnesses of the joints after the turbine element. While keeping $k_{J,D}$ constant, the rotor's bearing loads as $\lambda = 0.5$, 1.0, 2.0, 5.0 are solved. The results are illustrated in Figure 13. Based on this figure, conclusions can be made:

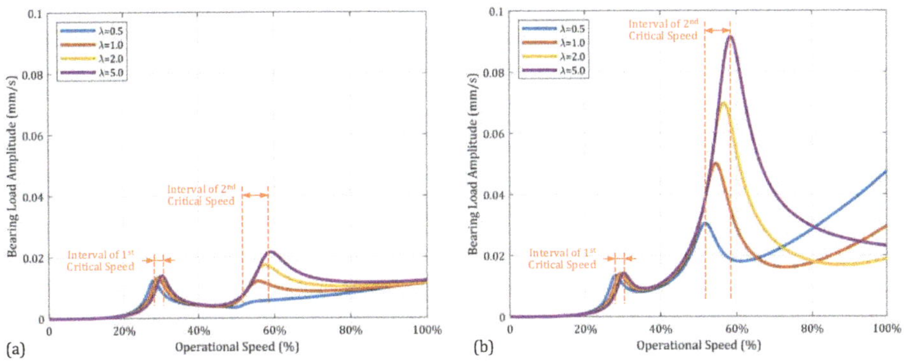

Figure 13. Impact of joints' bending stiffness on bearing load: (**a**) amplitude of front bearing load at different joints' bending stiffness, where λ represents the ratio of the bending stiffnesses of the front and rear bolt joints of the turbine element; (**b**) amplitude of rear bearing load at different joints' bending stiffness.

(a) As the bending stiffnesses of the rotor's joints change, the first two critical speeds of the rotor also change, but the variation is relatively small (less than 10%). This is because the modal shapes corresponding to the first two critical speeds are rigid body modes, which makes them less sensitive to changes in rotor bending stiffness;

(b) In the low-speed range ($\omega < \omega_{cr,2}$), the rotor's bearing loads are not sensitive to changes in joints' bending stiffness. This is because the rotor does not undergo significant bending deformation within this speed range. However, it should be noted that near the second critical speed, with an increase in $k_{J,C}$, the bearing loads at the second critical speed rapidly increase. This is because the second critical speed corresponds to a rotor's pitching modal shape, and an increase in bending stiffness improves the integrity of the rotor. This enables the rotational inertia moment caused by the turbine element to act on the entire rotor, leading to a significant increase in this modal vibration.

(c) In the high-speed range ($\omega > \omega_{cr,2}$), the amplitude of the rear bearing loads is extremely sensitive to the joints' bending stiffness. When $k_{J,C}$ is low, the rear bearing loads can continuously increase with the operational speed. The reason behind this phenomenon is that as the $k_{J,C}$ decreases, the integrity of the rotor is also weakened, resulting in the manifestation of the rotational inertia moment excitation caused by the turbine element's PAI slant. This rotational moment rapidly increases with the operational speed, leading to an increase in the rear bearing loads, which is consistent with the analysis in Section 2.4.1.

In addition, both the concentrated excitation model and the distributed excitation model were employed to solve the rotor's dynamic characteristics. To assess the influence of different excitation models on the simulation results, the error coefficient $\eta_E(\omega)$ was introduced, which is defined as

$$\eta_E(\omega) = (F_{br,con}(\omega) - F_{br,dis}(\omega))/F_{br,dis}(\omega) \tag{27}$$

where $F_{br,con}(\omega)$, $F_{br,dis}(\omega)$ represent the amplitudes of rear bearing loads obtained by employing the concentrated excitation model and the distributed excitation model, respectively. The curves of the error coefficient $\eta_E(\omega)$ are illustrated in Figure 14.

It can be observed that when the operational speed is low or close to the first critical speed, employing either the concentrated excitation model or the distributed excitation model has little impact on the prediction of the rotor's dynamic response. However, in the high-speed range, there is a significant difference in the rotor's dynamic response characteristics obtained from the two excitation models. If the concentrated excitation model is employed instead of the distributed excitation model in this speed range, the rotor's dynamic response characteristics will be greatly underestimated. Moreover, when the bending stiffness of the joints between the compressor element and the turbine element is low, the poor integrity of the rotor can result in a more severe underestimation of the rotor's dynamic response in this high-speed range. Therefore, it is essential to consider the influence of joint bending stiffness and use the distributed excitation model for accurate prediction of the rotor's dynamic response in the high-speed range.

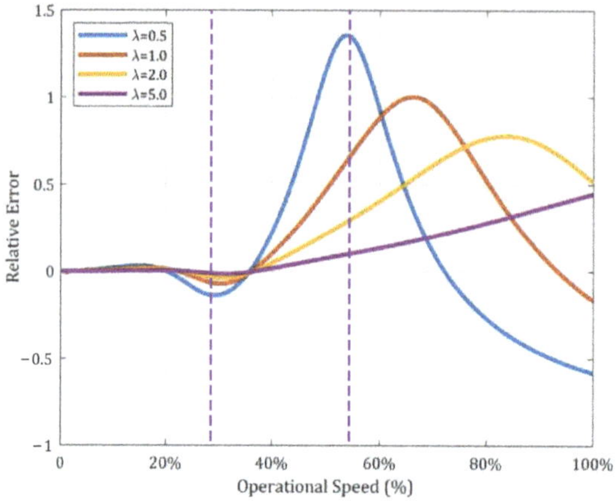

Figure 14. Impact of rotational inertia load excitation model and joints' bending stiffness on rotor dynamics.

4.3. Impact of Joints' Angular Deformation on Rotor Dynamics

Employing Group A as the rotor's unbalance state, let $\omega_t = 72\%\omega_{max}$, $M_t = 1400\ \mathrm{N \cdot m}$, $\widetilde{\alpha} = 3 \times 10^{-4} \angle 165^\circ$ rad, $\widetilde{\alpha}_r = 0 \angle 0^\circ$. The simulation results of the rotor dynamic response are illustrated in Figure 15.

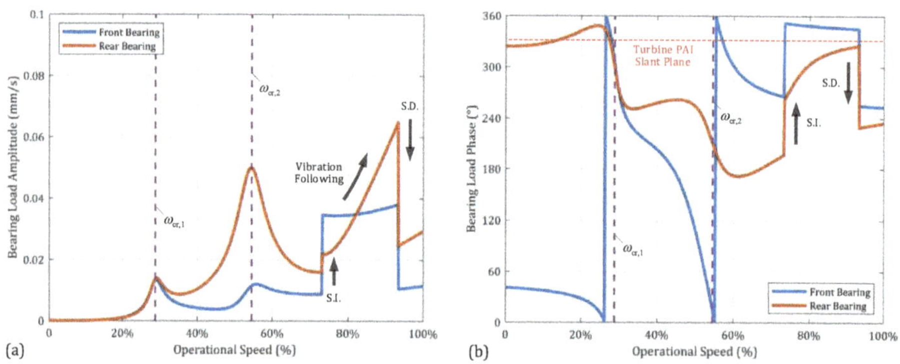

Figure 15. Rotor dynamic response characteristics considering the joint's sudden angular deformation: (**a**) amplitude of bearing load; (**b**) phase of bearing load.

From Figure 15a, it can be observed that the rotor's dynamic response increases with the operational speed until it reaches its maximum value at the first/second critical speed, then decreases with the operational speed. However, when the rotor's operational speed satisfies $\omega = \omega_t$, there is a sudden angular deformation in Joint C, causing the turbine element to slant. The resulting rotational inertia moment suddenly increases the bearing loads, especially for the rear bearing, which is closer to the turbine element. As the turbine element is a thin-disk, the rotational inertia moment it generates can increase with the operational speed even under this supercritical state, and this increasing rotational inertia moment leads to a speed-dependent increase in the bearing load. The vibration following continues until the rotational inertia moment generated by the turbine element becomes large enough that it makes $\widetilde{M}_J(\omega)$ satisfy $\left|\widetilde{M}_J(\omega)\right| = M_t$. At this point, the angular deformation of Joint C suddenly recovers, causing the turbine element's PAI to coincide with the rotor's rotation axis, i.e., "straightened", leading to a sudden decrease in bearing loads.

Furthermore, from Figure 15b, it can be noticed that at the moment when the rotor's dynamic response experiences a sudden increase, the phase of the front and rear bearing loads abruptly shifts towards the plane where the additional PAI slant of the turbine element occurs (165/345° plane), and with increasing operational speed, it gradually approaches this plane. This is because after the joint angular deformation induces a significant PAI slant of the turbine element, the dominant factor influencing the rotor dynamics becomes the rotational inertia moment generated by the turbine element. As a result, the phase of the front and rear bearing loads abruptly shifts towards the plane where this rotational inertia moment or the additional PAI slant is located. Meanwhile, the amplitude of this moment can increase with the operational speed, causing the phase of the front and rear bearing loads to come closer and closer to this plane.

As the rotor's operational speed increases, the variation of bending moments applied to different joints is illustrated in Figure 16. It can be observed that when Joint C undergoes a sudden angular deformation, the resulting spiked rotational inertia moment is caused by the turbine element, leading to a sudden increase in the bending moment applied to the rotor's joints. Among them, Joint C experiences the highest bending moment. This is because this joint is the closest joint to the turbine element and has a higher bending stiffness. Furthermore, as the operational speed causes the rotational inertia moment generated by the turbine element to increase, the resulting bending moment applied to Joint C also increases until it reaches the threshold value M_t, triggering this joint to angularly deform again. Consequently, the turbine element's PAI slant disappears, leading to a sudden decrease in the bending moment applied to the rotor's joints.

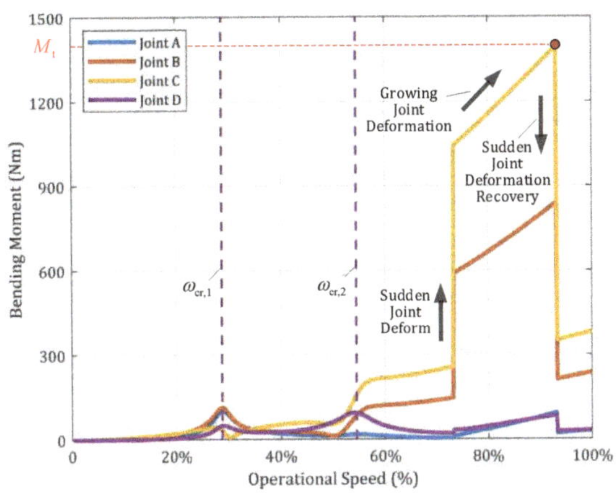

Figure 16. Variation of the bending moment applied to the joints of a high-speed rotor with the operational speed.

The angular displacement amplitudes and phases of the compressor and turbine elements are shown in Figure 17. By comparing with Figure 12, it can be observed that when $\omega = \omega_t$, the deformation of Joint C causes a sudden increase in the angular displacement of the turbine element. However, the rotational inertia moment generated by the turbine element tends to restore the angular displacement of the turbine element. Therefore, as the operational speed increases, the angular displacement of the turbine element gradually decreases until the angular deformation of the joint suddenly recovers, causing the angular displacement of the turbine element to almost return to 0. In addition, by observing the phase of the angular displacement of the mass elements, it can be seen that at the moment of the sudden increase in the angular deformation of Joint C, the angular displacement of the mass element jumps to the plane where the additional slant of the turbine element's PAI occurs (165/345° plane). This confirms that the deformation of the rotor in the high-speed range is dominated by the rotational inertia moment caused by the sudden additional slant of the turbine element's PAI.

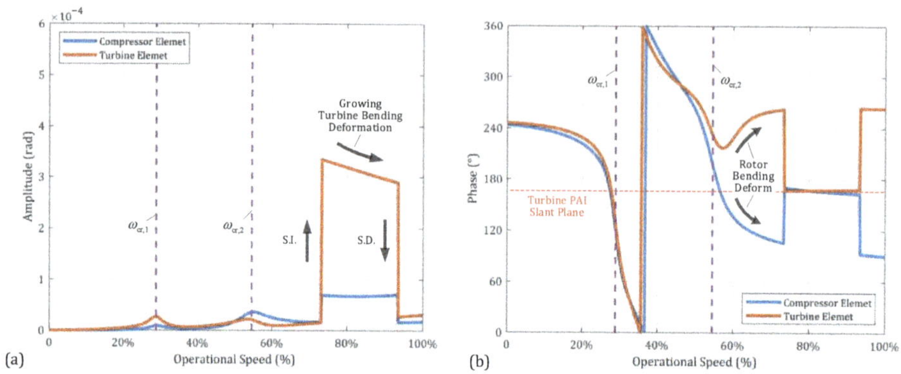

Figure 17. Variation of rotor mass elements' angular displacement with the operational speed considering joints' sudden angular deformation: (**a**) angular displacement amplitude of the mass elements; (**b**) angular displacement phase of the mass elements.

5. Dynamic Experiment of High-Speed Jointed Rotor System

5.1. Experimental Facility and Test Method

The experimental rotor used in this research is a full-scale rotor, constructed with materials and bearings identical to those utilized in real-world rotors. The rotor is driven by a high-speed motor at the front end. Acceleration sensors are installed at the front and rear bearings to measure the rotor bearing loads accurately. To capture the angular displacement of each element and the angular deformation of the joints as the operational speed changes, multiple sets of displacement sensors are placed at the turbine module. The arrangement of these displacement sensors at the turbine module is shown in Figure 18.

With the above-mentioned experimental setup, the lateral displacements of the drum shaft element, and the axial displacement of the turbine element and rear shaft element can be measured. The angular displacements of these elements can be obtained by

$$\begin{cases} \theta_{3,x} = \left(D^2_{3,x} - D^1_{3,x}\right)/L_3 \\ \theta_{3,y} = \left(D^1_{3,y} - D^2_{3,y}\right)/L_3 \end{cases} \tag{28}$$

$$\begin{cases} \theta_{4,i} = D_{4,i}/L_4 \\ \theta_{5,i} = D_{5,i}/L_5 \end{cases}, i = x, y \tag{29}$$

where, $\theta_{3,i}$, $\theta_{4,i}$, and $\theta_{5,i}$ represent the angular displacements of the drum shaft element, turbine element, and rear shaft element, respectively, in plane i ($i = x$ or y). It can be assumed that the bending deformation of each element is relatively small, thus the angular deformation difference between adjacent element mainly comes from the bending deformation of the joints, which can be expressed as

$$\begin{cases} \alpha_{C,i} = \theta_{4,i} - \theta_{3,i} \\ \alpha_{D,i} = \theta_{5,i} - \theta_{4,i} \end{cases}, i = x, y \tag{30}$$

where, $\alpha_{C,i}$, $\alpha_{D,i}$ represent the angular deformations of Joint C and D, respectively, in plane i ($i = x$ or y).

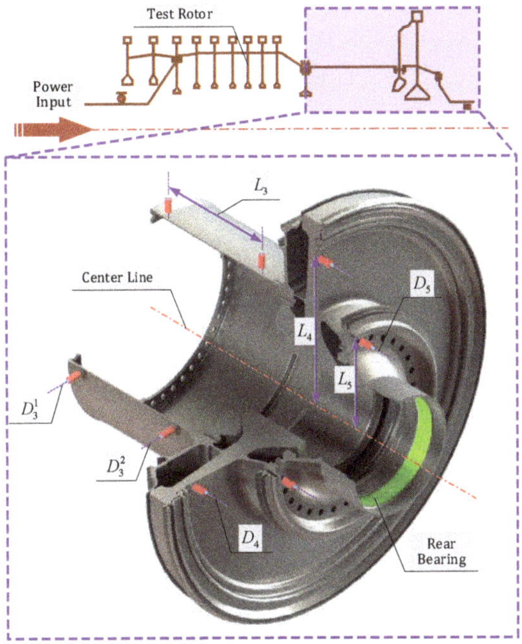

Figure 18. High-speed jointed rotor dynamic characteristics test rig: displacement sensors D_3^1, D_3^2 are used to measure the lateral displacement at two sections of the drum shaft element, with an axial distance of L_3 between the two sections. Each section is equipped with two displacement sensors to measure the horizontal/vertical displacement of the rotor at that position. Displacement sensors D_4, D_5 are used to measure the axial displacement of the turbine element and the rear shaft element, at distances L_4, L_5 from their respective centroid axes. Each element is equipped with two sensors to measure its axial displacement in the horizontal and vertical planes.

5.2. Rotor Dynamic Response and Joint Angular Deformation

The measurements presented in Figure 19 illustrate the time-varying amplitude and phase of the rotor's rear bearing load. It is important to note that these measurements were taken when the rotor's operational speed had significantly exceeded both the first and second critical speeds, and the operational speed continued to rise during the entire data collection interval.

From Figure 19, it is evident that a sudden significant increase in the bearing load occurs at time t_s, corresponding to $\omega = 76\%\omega_{\max}$. As the operational speed continues to increase, the amplitude of the rear bearing load also increases. Notably, despite the rotor speed increasing at a nearly constant rate, with only occasional speed increment pauses, the rate of the bearing load increase exhibits considerable variability. As the operational speed approaches $\omega = 90\%\omega_{\max}$, a sudden drop in the rear bearing load is observed.

Moreover, an analysis of the rear bearing load phase reveals interesting findings. Prior to the sudden increase in vibration response, the phase of the rear bearing load remains relatively stable around 90°, showing negligible variation with respect to the operational speed. However, during the moment of the sudden increase in vibration response, the phase of the rear bearing load undergoes a rapid shift of approximately 110°. Throughout the vibration-following stage, the phase of the rear bearing load continues to vary, and regions with a higher change rate of the phase align with areas of notable change rate of vibration amplitude.

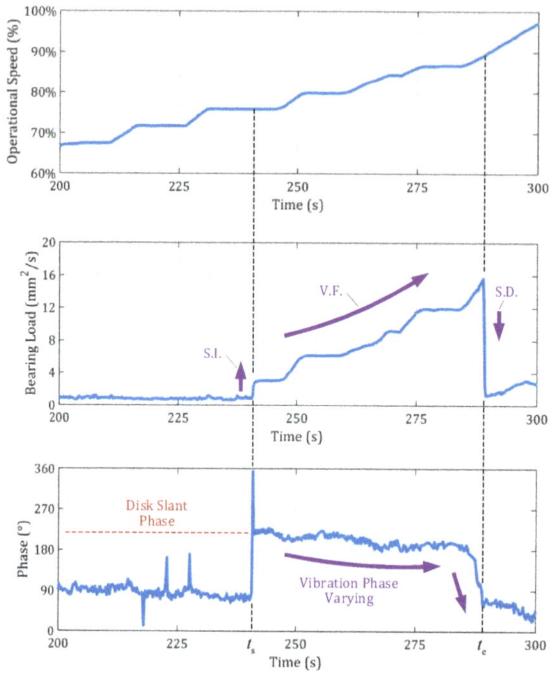

Figure 19. 1× RPM component of the rotor rear bearing load: the top graph shows the rotor operational speed, the middle graph shows the amplitude of the rear bearing load, and the bottom graph shows the phase of the rear bearing load. Here, t_s, t_e—the starting and ending moments of the rotor's abnormal vibration.

Time-varying angular deformations for Joint C and Joint D are depicted in Figure 20. It is evident that prior to time t_s, the angular deformations at both joints are negligible, and their corresponding phases remain nearly constant. However, at time t_s, a small increase in angular deformation is observed at both bolt joints, accompanied by significant changes in the deformation phases. Specifically, the phase of the angular deformation at Joint C aligns closely with the phase change observed in the rear bearing load in Figure 19, confirming that the rear bearing load is attributed to the sudden angular deformation at Joint C. As the operational speed increases, the angular deformation at Joint C continues to vary within $[0,4] \times 10^{-4}$ rad, with multiple phase changes occurring (conversely, the angular deformation of Joint D remains relatively small and can be practically ignored). This indicates that the angular deformation of the bolt joint exhibits non-deterministic behavior due to the discontinuity in the mechanical behavior at the joint interface. Moreover, it is noteworthy that when the angular deformation of Joint C reaches larger magnitudes, the amplitude and phase change rates of the rear bearing load become significantly higher. This is because the larger angular deformation of Joint C results in a larger additional PAI slant of the turbine element, leading to higher rates of variation in the rotor's dynamic responses with respect to the operational speed.

Obtain $\widetilde{\alpha}_C(t)$ by experiment data as shown in Figure 20, and substitute $\widetilde{\alpha}_C(t)$ into Equation (13), and the excitation characteristics of the tested rotor can be corrected. A simulation of the rotor's dynamic response is then performed based on this distributed excitation. The experimental results, when compared with the theoretical results shown in Figure 21, are in excellent agreement. This provides strong confirmation of the accuracy of the underlying mechanism by which the distributed rotational inertia load excitation model and the sudden angular deformation of the joints influence the rotor's dynamic response.

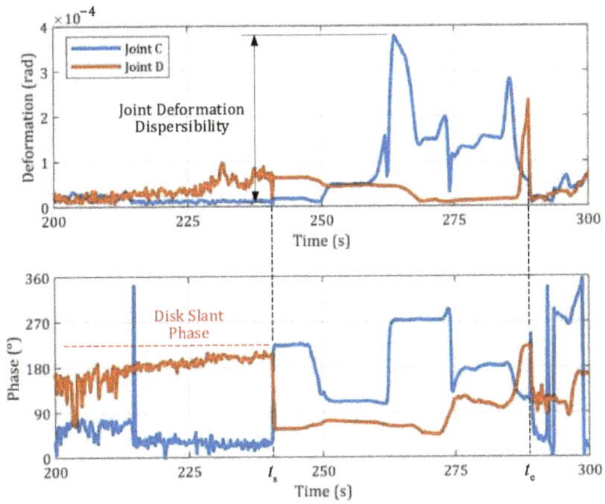

Figure 20. Rotor joints' angular deformation: the top graph shows the amplitude of the joints' angular deformation, and the bottom graph represents the phase of the joints' angular deformation.

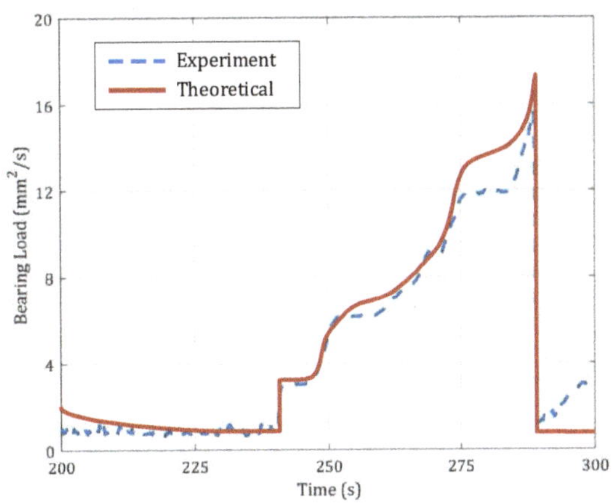

Figure 21. Comparison between experimental results and theoretical results.

6. Conclusions

This paper provides a comprehensive investigation of the high-speed jointed rotor system, with a primary focus on the influence of rotor motion state and joint mechanical characteristics on rotor dynamics. To achieve this, a distributed rotational inertia load excitation model has been developed, which considers both CM offset and PAI slant as forms of mass asymmetry, while also accounting for their axial distribution characteristics. Through simulations and subsequent experimental tests, the dynamic response of the rotor and the factors that significantly influence it have been thoroughly analyzed. The key conclusions derived from this study are as follows:

(1) At high operational speeds, bending deformation compromises the integrity of the rotor, leading to relative angular deformations between mass elements. This results in local mass asymmetry and distributed rotational inertia load excitation, significantly affecting the rotor's dynamic behavior in the high-speed range.

- In the low-speed range, rotor bending deformation can be neglected, and the rotor can be approximated as a rigid body. The rotor's CM offset and PAI slant dominate the whole rotor's dynamic response. However, as the operational speed exceeds the first/second critical speeds, the influence of rotational inertia forces/moments caused by the whole rotor diminishes with increasing speed.

- In the high-speed range, local mass asymmetry, particularly in the thin-disk turbine element, becomes more evident. The rotational inertia moment generated by the turbine element rapidly increases with the operational speed, even in the supercritical state, becoming the primary excitation source for rotor dynamic response. Excessive PAI slant of the turbine element can lead to a continuous increase and potential over-scale of the rotor's dynamic response.

(2) Angular deformation of the joints plays a critical role in influencing the rotor mass asymmetry state, leading to changes in rotational inertia load excitation distribution within the rotor and subsequently altering its dynamic response. Considering the discontinuous and uncertain nature of the mechanical characteristics of the joint interfaces under varying loads, the rotor's dynamic response exhibits significant sudden changes and dispersion features.

- Bending stiffness loss in the joints between mass elements weakens the rotor's integrity, intensifying the effect of the locally generated rotational inertia load and exacerbating the phenomenon of vibration following and dynamic response amplitude at high operational speeds.
- Under complex load environments, sudden angular deformation of the joints due to interfacial slip induces a sudden increase in angular displacement of surrounding mass elements, especially for the turbine element. The additional PAI slant of the turbine element causes a sudden increase in rotor dynamic response and vibration following. When the rotational inertia moment is significant enough to cause joint interface slip again, the angular deformation of the joints abruptly decreases, and the PAI of the turbine element returns to its original angular position, leading to a sudden drop in rotor dynamic response.

Nevertheless, the oversimplified representation of inertial and elastic characteristics within the rotor dynamic model may compromise the reliability of the derived rotor dynamic outcomes. Consequently, future research should incorporate more sophisticated mechanical models that duly consider the bending flexibility of shafts and long rotors.

Author Contributions: Conceptualization, F.W.; Software, X.C.; Validation, F.W.; Investigation, F.W.; Resources, F.W.; Data curation, X.C.; Writing—original draft, F.W.; Writing—review & editing, J.H.; Supervision, F.W.; Project administration, X.C. All authors have read and agreed to the published version of the manuscript.

Funding: The authors gratefully acknowledge the financial support provided by the National Science and Technology Major Project (Y2019-VIII-0011-0172) and the Science Center for Gas Turbine Project (P2021-A-I-002-002).

Data Availability Statement: Data is unavailable due to privacy.

Conflicts of Interest: The authors declare that they have no known competing financial interest or personal relationships that could have appeared to influence the work reported in this paper.

References

1. Jeffcott, H.H. The lateral vibration of loaded shafts in the neighbourhood of a whirling speed—The effect of want of balance. *Lond. Edinb. Dublin Philos. Mag. J. Sci.* **1919**, *37*, 304–314. [CrossRef]
2. Rao, J.S. *History of Rotating Machinery Dynamics*; Springer: New York, NY, USA, 2011.
3. Smith, D.M. The motion of a rotor carried by a flexible shaft in flexible bearings. *Proc. Roy. Soc.* **1933**, *142*, 92–118.
4. Den Hartog, J.P. *Mechanical Vibrations*; McGraw-Hill Book Company: New York, NY, USA, 1940.
5. Timoshenko, S.P. *Vibration Problems in Engineering*; D. Van Nostrand Company Inc.: New York, NY, USA, 1955.
6. Carnegie, W. Rotary inertia and gyroscopic effects in overhung shaft systems. *Bull. Mech. Eng. Educ.* **1964**, *3*, 191.
7. Kikuchi, K. Analysis of unbalance vibration of rotating shaft system with many bearings and disks. *Bull. JSME* **1970**, *13*, 864–872. [CrossRef]
8. Cruz, W.; Arzola, N.; Araque, O. Modeling and experimental validation of the vibration in an unbalance multi-stage rotor. *J. Mech. Eng. Sci.* **2019**, *13*, 5703–5716. [CrossRef]

9. Datz, J.; Karimi, M.; Marburg, S. Effect of Uncertainty in the Balancing Weights on the Vibration Response of a High-Speed Rotor. *J. Vib. Acoust.* **2021**, *143*, 061002. [CrossRef]
10. Lu, Z.Y.; Chen, Y.S.; Li, H.L.; Hou, L. Reversible model-simplifying method for aero-engine rotor systems. *J. Aerosp. Power* **2016**, *31*, 57–64.
11. Nicholas, J.C.; Gunter, E.J.; Allaire, P.E. Effect of residual shaft bow on unbalance response and balancing of a single mass flexible rotor—Part I: Unbalance response. *J. Eng. Power.* **1976**, *98*, 171–181. [CrossRef]
12. Nicholas, J.C.; Gunter, E.J.; Allaire, P.E. Effect of Residual Shaft Bow on Unbalance Response and Balancing of a Single Mass Flexible Rotor—Part II: Balancing. *J. Eng. Power.* **1976**, *98*, 182–187. [CrossRef]
13. Yanhong, M.A.; Haizhou, L.I.U.; Wangqun, D.E.N.G.; Hai, Y.A.N.G.; Jie, H.O.N.G. Vibration response analysis of rotor system with initial thermal deformation. *J. Beijing Univ. Aeronaut. Astronaut.* **2019**, *45*, 227–233.
14. Hong, J.; Yan, Q.; Sun, B.; Ma, Y. Vibration Failure Analysis of Multi-Disk High-Speed Rotor Based on Rotary Inertia Load Model. In Proceedings of the Turbo Expo: Power for Land, Sea, and Air, Rotterdam, The Netherlands, 13–17 June 2022; Volume 86076.
15. Ishida, Y.; Toshio, Y. *Linear and Nonlinear Rotordynamics: A Modern Treatment with Applications*; John Wiley & Sons: Hoboken, NJ, USA, 2013.
16. Liu, Y.; Liu, H.; Wang, N. Effects of typical machining errors on the nonlinear dynamic characteristics of rod-fastened rotor bearing system. *J. Vib. Acoust.* **2017**, *139*, 011004. [CrossRef]
17. Han, Z.; Wang, D.; Wang, Y.; Hong, J.; Cheng, R. Dynamic effects of the initial skewness of the inertial principal axis on the rotor with bolted joint. *Mech. Syst. Signal Process.* **2023**, *200*, 110564. [CrossRef]
18. Qin, Z.Y.; Han, Q.K.; Chu, F.L. Analytical model of bolted disk-drum joints and its application to dynamic analysis of jointed rotor. *Proc. Inst. Mech. Eng. Part C J. Mech. Eng. Sci.* **2013**, *228*, 55808444. [CrossRef]
19. Luan, Y.; Guan, Z.Q.; Cheng, G.D.; Liu, S. A simplified nonlinear dynamic model for the analysis of pipe structures with bolted flange joints. *J. Sound Vib.* **2012**, *331*, 325–344. [CrossRef]
20. Zhuo, M.; Yang, L.H.; Yu, L. Contact Stiffness Calculation and Effects on Rotordynamic of Rod Fastened Rotor. In Proceedings of the ASME 2016 International Mechanical Engineering Congress and Exposition, American Society of Mechanical Engineers, Phoenix, AR, USA, 11–17 November 2016.
21. Hong, J.; Chen, X.; Wang, Y.; Ma, Y. Optimization of dynamics of non-continuous rotor based on model of rotor stiffness. *Mech. Syst. Signal Process.* **2019**, *131*, 166–182. [CrossRef]
22. Ma, Y.; Ni, Y.; Chen, X. Bending stiffness loss of rod-rabbet joints and its effect on vibration response of rotor systems. *Acta Aeronaut. Astronaut. Sin.* **2021**, *42*, 223861.
23. Chen, X.; Ma, Y.; Hong, J. Vibration suppression of additional unbalance caused by the non-continuous characteristics of a typical aero-engine rotor. In *International Conference on Rotor Dynamics*; Springer: New York, NY, USA, 2018; pp. 34–48.
24. Liu, S.; Ma, Y.; Zhang, D.; Hong, J. Studies on dynamic characteristics of the joint in the aero-engine rotor system. *Mech. Syst. Signal Process.* **2012**, *29*, 120–136. [CrossRef]
25. Liu, Y.; Liu, H.; Fan, B.W. Nonlinear dynamic properties of disk-bolt rotor with interfacial cutting faults on assembly surfaces. *J. Vib. Control* **2018**, *24*, 4369–4382. [CrossRef]
26. Liu, Y.; Liu, H.; Fan, B. Numerical analysis on the static–dynamic coupling influences of parallelism flaw in disc-rod rotor ball bearing system. *Proc. Inst. Mech. Eng. Part K J. Multi-Body Dyn.* **2018**, *232*, 103–112. [CrossRef]
27. Barragan, J.M. *Engine Vibration Monitoring and Diagnosis Based on On-Board Captured Data*; MTU Aero Engines Gmbh Munich: Munich, Germany, 2003.
28. Zhao, B.; Sun, Q.; Zhao, R.; Yang, Y.; Sun, K.; Mu, X. Mass-eccentricity nonlinear evolution mechanism of combined rotor in fretting slip process. *Tribol. Int.* **2023**, *179*, 108195. [CrossRef]
29. Zhao, B.; Sun, Q.; Yang, Y.; Sun, K.; Liu, Z. Study on interface non-uniform slip of combined rotor considering real preload distribution. *Tribol. Int.* **2022**, *169*, 107482. [CrossRef]
30. Wang, C.; Zhang, D.; Zhu, X.; Hong, J. Study on the stiffness loss and the dynamic influence on rotor system of the bolted flange joint. In Proceedings of the Turbo Expo: Power for Land, Sea, and Air, Dusseldorf, Germany, 16–20 June 2014; Volume 45769.
31. Nizametdinov, F.R.; Romashin, Y.S.; Berne, A.L.; Leontyev, M.K. Investigation of bending stiffness of gas turbine engine rotor flanged connection. *J. Mech.* **2020**, *36*, 729–736. [CrossRef]
32. Hong, J.; Xu, X.; Su, Z.; Ma, Y. Joint stiffness loss and vibration characteristics of high-speed rotor. *J. Beijing Univ. Aeronaut. Astronaut.* **2019**, *45*, 18–25.

symmetry

MDPI

Article

Effects of KoBo-Processing and Subsequent Annealing Treatment on Grain Boundary Network and Texture Development in Laser Powder Bed Fusion (LPBF) AlSi10Mg Alloy

Przemysław Snopiński

Department of Engineering Materials and Biomaterials, Silesian University of Technology, 18A Konarskiego Street, 44-100 Gliwice, Poland; przemyslaw.snopinski@polsl.pl

Abstract: It is well known that the properties of polycrystalline metals are related to grain boundaries (GBs), which are fundamental structural elements where crystallographic orientations change abruptly and often exhibit some degree of symmetry. Grain boundaries often exhibit unique structural, chemical, and electronic properties that differ from bulk crystalline domains. Their effects on material properties, including mechanical strength, corrosion resistance, and electrical conductivity, make grain boundaries a focus of intense scientific investigation. In this study, the microstructural transformation of an AlSi10Mg alloy subjected to KoBo extrusion and subsequent annealing is investigated. A notable discovery is the effectiveness of a strain-annealing method for grain boundary engineering (GBE) of the LPBF AlSi10Mg alloy. In particular, this study shows a significant increase in the population of coincidence site lattice boundaries (CSL), which embody the symmetry of the crystal lattice structure. These boundaries, which are characterised by a high degree of symmetry, contribute to their special properties compared to random grain boundaries. The experimental results emphasise the crucial role of strain-induced boundary migration (SIBM) in the development of a brass texture in the microstructure of the alloy after annealing. In addition, the presented results demonstrate the feasibility of applying GBE to materials with high stacking fault energy (SFE), which opens up new possibilities for optimizing their properties.

Keywords: strain-induced boundary migration; AlSi10Mg; CSL boundaries; crystallographic texture

1. Introduction

Grain boundary engineering (GBE) is a processing strategy aimed at enhancing the properties of polycrystalline materials through the manipulation of grain boundaries (GBs) in terms of distribution and character. Originating from Watanabe's concept of "GB design and control" in 1984 [1], GBE has become a powerful technique for improving inherent properties [2,3], particularly in face-cantered cubic metals with low to medium stacking fault energy (SFE) such as copper [4], nickel-based alloys [5], austenitic steels [6,7] and metal-matrix composites [8,9].

The core of GBE involves thermomechanical processing (TMP) to increase the prevalence of low-coincidence site lattice (CSL, $\Sigma \leq 29$) boundaries, focusing on $\Sigma3$ boundaries (twin boundaries, TB) to disrupt random boundary connections [4]. Recent research has extended GBE to high-SFE materials like aluminium alloys. Shamanian et al. (2016) observed an increase in $\Sigma3$ boundaries during friction stir processing (FSP) of A413 cast aluminium alloy [10], and Jia et al. (2018) investigated the formation of $\Sigma = 3\{110\}$ incoherent twin boundaries (ITBs) in an Al-8Zn alloy [11]. Similarly, Pradeep et al. (2021) studied AA5053 subjected to friction stir processing, reporting an increased percentage of $\Sigma = 3n$ (n = 1, 2, 3) CSL boundaries [12].

In laser powder bed fusion (LPBF)-produced alloys, grain boundary character distribution (GBCD) modification often involves post-processing heat treatments. In this context, Li et al. treated an Inconel 718 superalloy produced by selective laser melting and found that the internal residual stress provides a sufficient activation energy for driving recrystallisation-based GBE during annealing. They produced an alloy with a high fraction of Σ3 GBs (64.89%) and a stable grain size [13]. In another study, Dong et al. [14] investigated a similar strategy based on isothermal annealing to modify GBCD in SLM Co-Cr alloy. They obtained a large fraction (more than 80%) of special boundaries (Σ3, Σ9, and Σ27) after complete recrystallisation.

An alternative GBE strategy involves plastic deformation followed by annealing. Such a method has been proposed as effective for SLM 304 austenitic stainless steel, for example, [15]. The authors of the cited manuscript found that laser peening followed by annealing at 1260 K for 48 h can lead to an optimal grain boundary character distribution with over 80 CSL%. An alternative approach was recently proposed by the author of this manuscript for the LPBF AlSi10Mg alloy [16]. It was shown that the grain boundary character distribution can be modified by post-processing during KoBo extrusion.

The KoBo-extrusion method is a severe plastic deformation (SPD) technique that enables deformation at a high strain rate. In this SPD method, the die is oscillated cyclically during the extrusion process, causing the deformation path to change continuously [17]. This unique deformation mode ensures a grain-refined structure without dislocation accumulation and work hardening. Another feature that distinguishes this technique from other SPD methods is the high plasticity of KoBo-processed metals, as it offers higher degree of deformation without the need for annealing treatment. This is because KoBo extrusion generates a large amount of heat, which intensifies the rearrangement of dislocations and the diffusion-controlled structural processes. As a result, materials with unique structural and mechanical properties can be produced [18,19].

In this paper, a novel post-processing approach to improve the special grain boundary population in the LPBF AlSi10Mg alloy is investigated. The LPBF sample is subjected to a two-step process involving high strain rate KoBo extrusion followed by annealing, which leads to a remarkable increase in the population of coincidence site lattice boundaries (CSL). These CSL boundaries, characterised by their unique crystal lattice symmetry, play a crucial role in the microstructural transformation of the alloy, which is particularly evident in the development of a distinct brass structure after annealing. The novelty lies in the limited research on GBE strategies for LPBF materials and in the application of the KoBo-extrusion process. Demonstrating the effectiveness of GBE on high stacking fault energy (SFE) materials, as observed in this study, opens up innovative possibilities for optimizing the properties of LPBF materials. Despite the known importance of GBE for controlling grain boundary character distribution (GBCD) and improving material properties, there is currently a knowledge gap in understanding GBE strategies for high-SFE materials, such as aluminium alloys, produced by LPBF. The results presented here aim to close this gap, representing the first investigation of GBE specifically tailored to aluminium alloys produced by LPBF.

2. Methodology

The specimens used in this experimental work were produced from an AlSi10Mg alloy powder via selective laser melting (SLM) with the following process parameters: laser power = 175 W; layer thickness = 20 μm; scanning speed = 1400 mm/s; and scan rotation = 67°. The cylindrically shaped samples had a total length of 50 mm and a diameter of 60 mm. Further details on the chemical composition and microstructural features of this alloy can be found in Refs. [20,21].

The SLM samples were post-processed via a KoBo-extrusion method, depicted in Figure 1, of which the principal description can be found in Ref. [22]. The extrusion process was conducted on a custom-made horizontal hydraulic press. The SLM samples were reversibly twisted with a constant angle of ±8° just before entering the die, where the cross section was reduced from 60 mm to 4 mm. The following process parameters were applied:

- punch speed of 0.2 mm/s;
- die rotation frequency of 5 Hz;
- maximal measured temperature close to the extrusion die (~280 °C);
- sample cooling with room-temperature water.

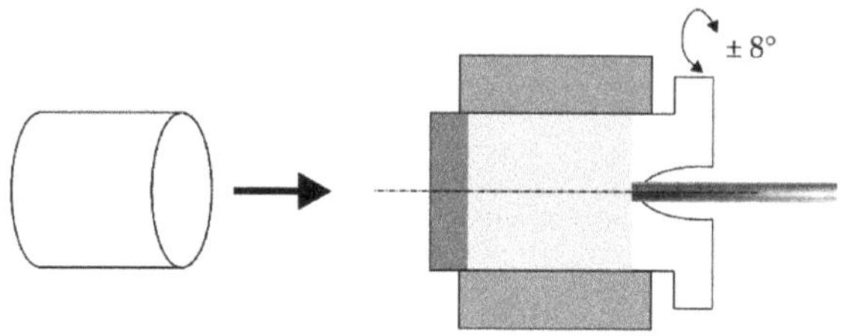

Figure 1. Schematic illustration of the KoBo-extrusion process.

Following the KoBo-extrusion process, the midsection of the extruded profile was cut into 300 mm long segments that were then annealed in a resistance furnace for 1 h at 250 °C, 300 °C, 350 °C, 400 °C, 450 °C, or 500 °C under Ar atmosphere.

All specimens underwent the same metallographic preparation process with an automatic polishing machine LaboPol-20 (Struers, Copenhagen, Denmark) to ensure high surface quality. The material was first ground with 800 and 1200 grit SiC water paper, then polished with 6, 3, and 1 μm diamond pastes and finally mirror-polished with 0.04 μm colloidal silica suspension (OP-U suspension, Struers, Copenhagen, Denmark) for 1 h. The microstructure was characterised in detail via electron backscatter diffraction EBSD in a scanning electron microscope Zeiss Supra 35 (Carl Zeiss NTS GmbH, Oberkochen, Germany) equipped with an EDAX NT EBSD detector. EBSD scans were carried out on the sections perpendicular to the extrusion direction to evaluate the microstructural changes. Scans were registered in the centre area of the samples with a step size of 100 nm. TSL OIM Analysis 7 software was used to analyse the scanned raw data.

TEM analysis was performed in the Titan 80-300 FEI S/TEM microscope (Fei, Hillsboro, OR, USA), which was operated at 300 kV. To achieve ultra-thin lamellae suitable for HRTEM examination, ion milling was employed. Argon ions were used to selectively mill away material from the sample surface, gradually thinning the lamellae to ~80 nm. The TEM lamellae were extracted from the KoBo-processed sample along the extrusion direction.

The TEM lamella was additionally analysed via transmission Kikuchi Diffraction (TKD) to obtain crystallographic information that allowed characterisation of grain boundaries with improved resolution. The TKD map was acquired with a step size of 20 nm.

3. Results and Discussion

The main microstructural characteristics of the LPBF AlSi10Mg samples in the X–Y and X–Z planes were investigated using an EBSD technique. Figure 2a shows the inverse pole figure (IPF-Z) map taken in the X–Y plane. In this figure, the low-angle grain boundaries (LAGBs) with a misorientation angle ($2° < \theta < 15°$) and the high-angle grain boundaries (HAGBs) with a misorientation angle ($15° < \theta < 65°$) are represented by a red and green colour, respectively. As can be seen, the equiaxed grains with a preferred <001> crystallographic orientation dominated the alloy microstructure. According to statistical data, LAGBs accounted for 22.1% of the total boundary fraction, while HAGBs accounted for the remaining 77.9%. A grain boundary character-distribution map plotted based on the Brandon criterion is shown in Figure 2b. In this image, four main types (Σ3, Σ9, Σ27a, and Σ27b) of Coincidence Site Lattice (CSL) boundaries are distinguished using green, red, yellow, and orange colour coding, respectively. According to the statistical results, the Σ3, Σ9, and Σ27 CSL boundaries occupied volume fractions of ~1.5%, ~2.0% and ~4.0%, respectively. It can be also noted that a large fraction of CSL boundaries showed a curved geometry as opposed to the typical straight twin boundaries in low SFE materials.

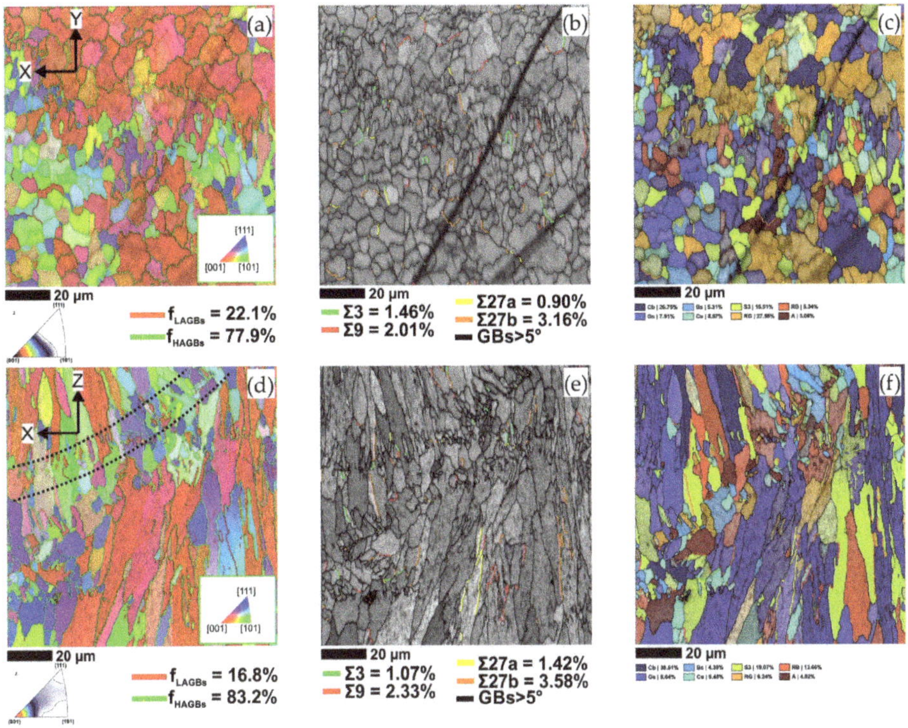

Figure 2. EBSD maps of the LPBF AlSi10Mg alloy. (**a**) IPF-Z map taken on the X–Y plane; (**b**) band contrast map taken on the X–Y plane; (**c**) texture component map with superimposed special grain boundaries taken on the X-Y plane; (**d**) IPF-Z map taken on the X–Z plane; (**e**) band contrast map taken on the X–Z plane; and (**f**) texture component map with superimposed special grain boundaries taken on the X–Z plane.

Figure 2d shows the IPF-Z map taken in the X–Z plane. EBSD investigation revealed typical columnar grain growth and confirmed the development of a strong <001>//BD texture corresponding to the <001>{001} preferred growth direction [23]. Because of the preferential growth of elongated grains towards the centre of the melt pool and the existence of small equiaxed grains, some of the melt pool boundary was also revealed, as indicated by the black dashed lines. According to statistical data, LAGBs accounted for 16.8% of the total boundary fraction, while HAGBs accounted for the remaining 83.2%. Figure 2e shows the grain boundary character-distribution map superimposed on the band contrast map. In the X–Z plane, the Σ3, Σ9, and Σ27 CSL boundaries occupied volume fractions of ~1.0%, ~2.3%, and ~5.0%, respectively, which were similar to the X–Y plane. The CSL boundaries were mainly distributed in grains that grew preferentially along the build direction, indicating that the crystal growth orientation could affect the twin formation ability [24]. According to the literature [25,26], the development of CSL boundaries in LPBF samples could stem from grain boundary migration and potential dynamic recrystallisation during sample fabrication, alongside alternative mechanisms involving icosahedral short-range order (ISRO) mediated nucleation.

Table 1 summarises the main microstructural characteristics of the LPBF AlSi10Mg sample. The calculation of the volume fraction of the different ideal texture components was based on the well-accepted definition system, that is, cube (Cb) {001}<100>; goss (Gs) {011}<100>; brass (Bs) {011}<211>; S {123}<634>, copper (Cu) {112}<111>, rotated goss (RG) {011}<011>, rotated brass (RB) {110}<556> and A {110}<111>. As can be seen in Figure 2c, the grain orientations are dominated by the cube, S, and rotated goss orientations. Furthermore, Figure 2f, representing the X–Z plane, provides clear evidence of selective grain growth, highlighting, in particular, the crystallographic orientations of cube, S, and rotated brass texture components.

Table 1. Texture characteristics of the LPBF AlSi10Mg sample analysed using EBSD.

Texture Component	Cube {001}<100> (%)	Goss {011}<100> (%)	Brass {110}<112> (%)	Copper {112}<111> (%)	S {123}<634> (%)	Rotated Goss (RG) {011}<011> (%)	Rotated Brass (RB) {110}<556> (%)	A {110}<111> (%)
X–Y plane	26.75	7.91	5.31	8.57	15.51	27.55	5.34	3.06
X–Z plane	38.51	8.64	4.38	5.48	19.07	6.24	13.66	4.02

In the next step, the microstructure of the KoBo-processed AlSi10Mg sample was analysed in detail. Figure 3a shows the IPF-Z map of the extruded profile taken in the ND-TD (extrusion) plane. It is evident that the coarse LPBF microstructure evolved into an ultra-fine-grained (UFG) with an average grain size of about 0.8 μm. In this UFG microstructure, high-angle boundaries (HAGBs) dominated, accounting for about 81% of the total grain boundary fraction.

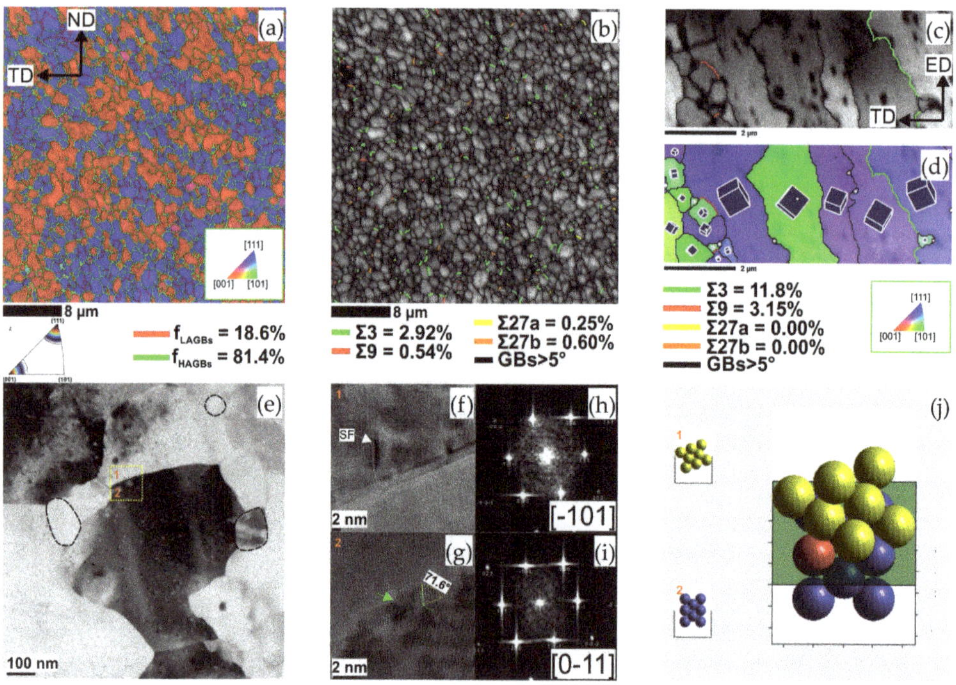

Figure 3. Microstructures of the KoBo-processed LPBF AlSi10Mg alloy sample. (**a**) IPF-Z image; (**b**) band contrast map; (**c**) EBDS TKD band contrast map; (**d**) EBSD TKD IPF-Y map; (**e**) bright field TEM image; (**f**) HRTEM image of the boundary area highlighted by a yellow square in (**e**)—note that

the TEM foil was oriented with the aim of revealing the atomic arrangement of grain labelled as (1); (**g**) HRTEM image of the boundary area highlighted by a yellow square in (**e**)—note that the TEM foil was oriented with the aim of revealing the atomic arrangement of grain labelled as (2); (**h**) FFT of the boundary labelled as (1) in (**e**); (**i**) FFT of the boundary labelled as (2) in (**e**); and (**j**) simulated atomic arrangement of the boundary area confirming the CSL Σ3 boundary.

Figure 3b shows the CSL boundary map overlaid on the EBSD band contrast map. It reveals that the Σ3-CSL boundaries accounted for approximately 3% by volume, while the Σ9, Σ27a, and Σ27b boundaries together accounted for no more than 1.4%.

Further crystallographic information was obtained from the TKD mapping. In the TKD pattern quality map (Kikuchi band contrast), four primary types of coincidence site lattice boundaries (CSL) were analysed, as shown in Figure 3c. The green boundary on the Kikuchi band contrast map reflected a misorientation of approximately 62° around the <111> direction, which corresponded to the Σ3 CSL boundary. In addition, a boundary with a misorientation of 41° around the <110> direction (coded red) was identified, which corresponds to the Σ9 CSL boundary. Quantitative analysis of the grain boundary character distribution revealed that the Σ3 and Σ9 CSL boundaries occupied volume fractions of 11.8% and 3.2%, respectively.

For a more in-depth analysis of grain characteristics, an IPF-Y map was generated, as shown in Figure 3d. Together with the conventional EBSD data, it was found that the KoBo-processed material consisted of UFG grains with widths ranging from 200 to 1700 nm. In addition, the TKD map confirmed that most of the grains were oriented along the <111> direction, which is typical of extruded alloys.

Figure 3e shows the bright field (BF) TEM image of the KoBo-processed sample. This image confirms that both the grains and the particles were highly refined. The ultra-fine-grained (UFG) grains are visibly encircled by numerous Si particles, each exhibiting an average size of approximately 150 nm (as indicated by the black dashed outlines).

Further observations in HRTEM mode, shown in Figure 3f,g, revealed that the grain boundary, indicated by a yellow dashed square in Figure 3e, corresponded to {111} an Σ3-CSL boundary, commonly known as the coherent twin boundary [27]. These images also revealed multiple facets (indicated by green arrows) that lead to elastic distortions due to different displacements between the two grain boundaries (faceting of GB is thought to be a process that aims to minimise the total free energy) [28]. The grain boundary model shown in Figure 3j confirmed the relative crystal orientation representing the Σ3-CSL boundary.

The effect of annealing temperature on the population of low-angle grain boundaries (LAGBs) and high-angle grain boundaries (HAGBs) in the KoBo-processed sample was examined using electron backscatter diffraction (EBSD) and the results are summarised in Figure 4. As stated in the previous section, the population of LAGBs in KoBo-processed sample was about 19% LAGBs. EBSD analysis showed that the population of LAGBs increased after annealing at temperatures ranging from 250 °C to 400 °C. This effect is illustrated in Figure 4a, which confirms that the LAGB population rose to about 28% after annealing at 250 °C. Although the population of LAGBs increased, the grain size of ~0.8 μm remained unchanged. The population of low-angle grain boundaries continued to increase to about 31% as the annealing temperature rose to 300 °C (Figure 4b). Furthermore, the grains started to grow, resulting in an average grain size of ~0.9 μm. After annealing at 350 °C, the grains became more elongated and thicker, as seen in Figure 4c. This change in the grain morphology led to an increase in the average grain size to ~1.1 μm. At the same time, the population of LAGBs rose to about 32%. This implies that LAGBs were formed during the recovery process, possibly due to polygonisation [29]. Annealing at higher temperatures resulted in a decrease in the population of LAGBs. For the sample annealed at 400 °C, the population of LAGBs decreased to about 29% (Figure 4d). The average grain size in this condition was ~1.2 μm. The microstructure underwent the most significant evolution when the annealing temperature exceeded 450 °C. As shown in Figure 4e, the population of LAGBs decreased to around 26%, and grains with <101> preferred orientation appeared in the microstructure. The microstructure also showed an average grain size of ~1.0 µm, with several small equiaxed grains. These grains are likely to have formed due to recrystallisation, which could stem from the coalescence of subgrains (polygons) and the migration of local high-angle boundaries [30]. The largest grain size of ~1.3 µm was observed at 500 °C. In this case, the LAGBs turned into HAGBs, reducing the LAGBs fraction sharply to around 22% (Figure 4f). EBSD data indicated that grain growth was very restricted in the temperature range studied. This phenomenon could be explained by the dispersed Si particles, which were created during the KoBo-extrusion process, and which prevented the grain boundaries from moving, thus hindering grain growth.

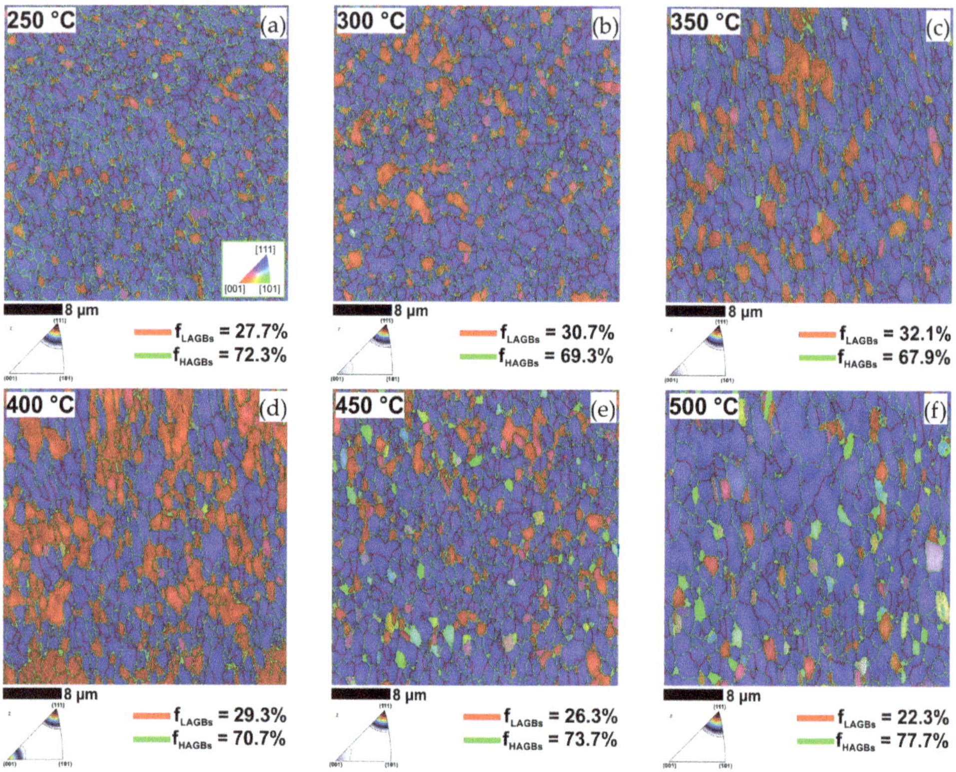

Figure 4. IPF-Z maps of the KoBo-processed samples (**a**) annealed at 250 °C for 1 h, (**b**) annealed at 300 °C for 1 h, (**c**) annealed at 350 °C for 1 h, (**d**) annealed at 400 °C for 1 h, (**e**) annealed at 450 °C for 1 h, or (**f**) annealed at 500 °C for 1 h.

Figure 5 shows the CSL boundary map superimposed on the band contrast maps for the annealed samples. EBSD analysis revealed a moderate increase in the fraction of CSL-Σ3 boundaries to about 4% after annealing at temperatures up to 400 °C. Despite this increase, the volume fraction of Σ9, Σ27a, and Σ27b boundaries, which indicate the optimisation of the grain boundary structure, remained relatively low at around 1% in this temperature range. The population of the low-Σ-CSL GBs increased after annealing at 450 °C. Under these conditions, the Σ3 and Σ9 boundaries occupied volume fractions of 5.76% and 1.16%, respectively. The Σ27a and Σ27b boundaries showed a consistent trend and together accounted for 1.34%, similar to the Σ9 boundaries. Following Randle's Σ3 regeneration model [31,32], it can be concluded that the Σ27 boundaries were generated by the Σ3 + Σ9 = Σ27 relationship. Consequently, the percentage of Σ27 boundaries appeared to be closely related to the population of Σ9 boundaries.

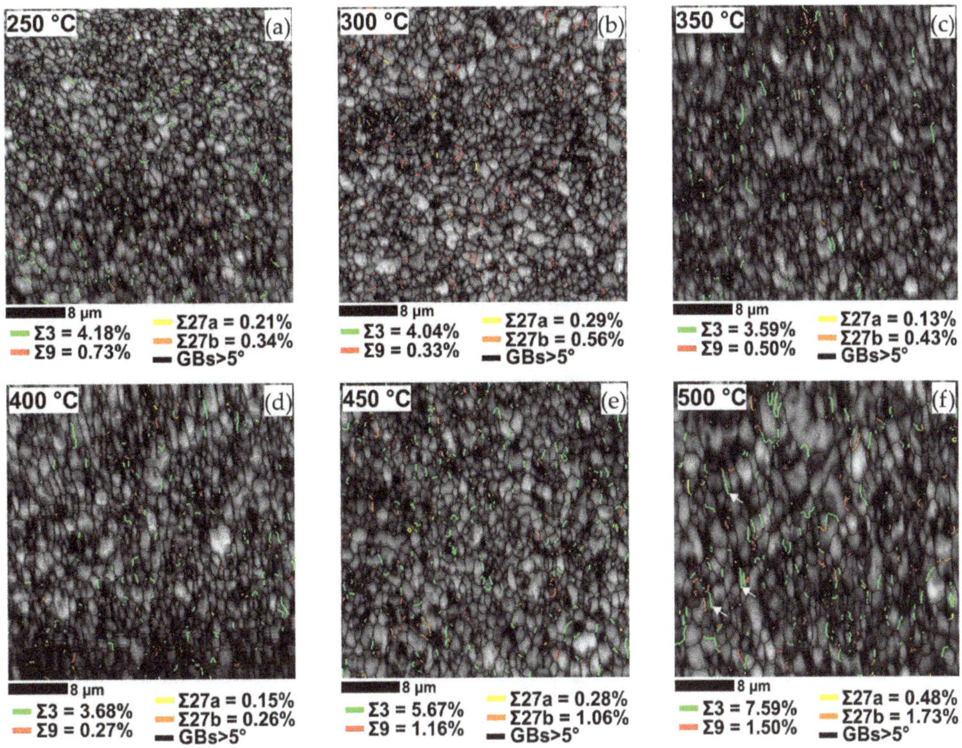

Figure 5. CSL boundary maps superimposed on band contrast maps for KoBo-processed samples (**a**) annealed at 250 °C for 1 h, (**b**) annealed at 300 °C for 1 h, (**c**) annealed at 350 °C for 1 h, (**d**) annealed at 400 °C for 1 h, (**e**) annealed at 450 °C for 1 h, or (**f**) annealed at 500 °C for 1 h.

When the annealing temperature was further increased to 500 °C, more thermal activation energy was provided for recrystallisation, resulting in a higher percentage of special low-Σ-CSL GBs. Under these conditions, Σ3 and Σ9 boundaries accounted for 7.59% and 1.50%, respectively, while Σ27a and Σ27b boundaries contributed 0.48% and 1.73%, respectively. A noteworthy observation in this analysis was the distinctive well-defined and parallel-sided appearance of certain CSL-Σ3 boundaries, which is commonly associated with coherent twin boundaries formed during recrystallisation [33]. These boundaries are highlighted by white arrows in Figure 5f.

Additionally, it is crucial to highlight that the annealing twins (CSL Σ3 boundaries), follow a similar trend to HAGBs, as depicted in Figures 4f and 5f. The notable increase in the population of HAGBs and annealing Σ3 twin boundaries within the alloy microstructure suggests that KoBo extrusion supplied ample activation energy for grain boundary migration. This, in turn, facilitated the GBE in LPBF AlSi10Mg alloy.

Figure 6 displays the special grain boundary maps superimposed on the texture component maps, and Table 2 summarises the results of the quantitative analysis.

The moderate temperature of ~280 °C during the extrusion process provided the conditions for the occurrence of dynamic recovery (DRV), dynamic recrystallisation (DRX), precipitation, and the formation of microtextures. For this reason, the main textural components identified in the KoBo-processed sample were cube, brass, and rotated brass orientations, as shown in Figure 6a. According to study [34], the cube texture is formed as a result of DRX, as this is a typical recrystallisation texture, whereas brass is a typical deformation texture that can occur due to the simultaneous presence of tensile and compressive stresses during the hot extrusion process [35]. It is worth mentioning that a similar texture was reported by Yang et al. [36], who found that in the extruded Al–Mg–Si alloy, the main textures were brass {110}<112> and recrystallisation cube {100}<001>.

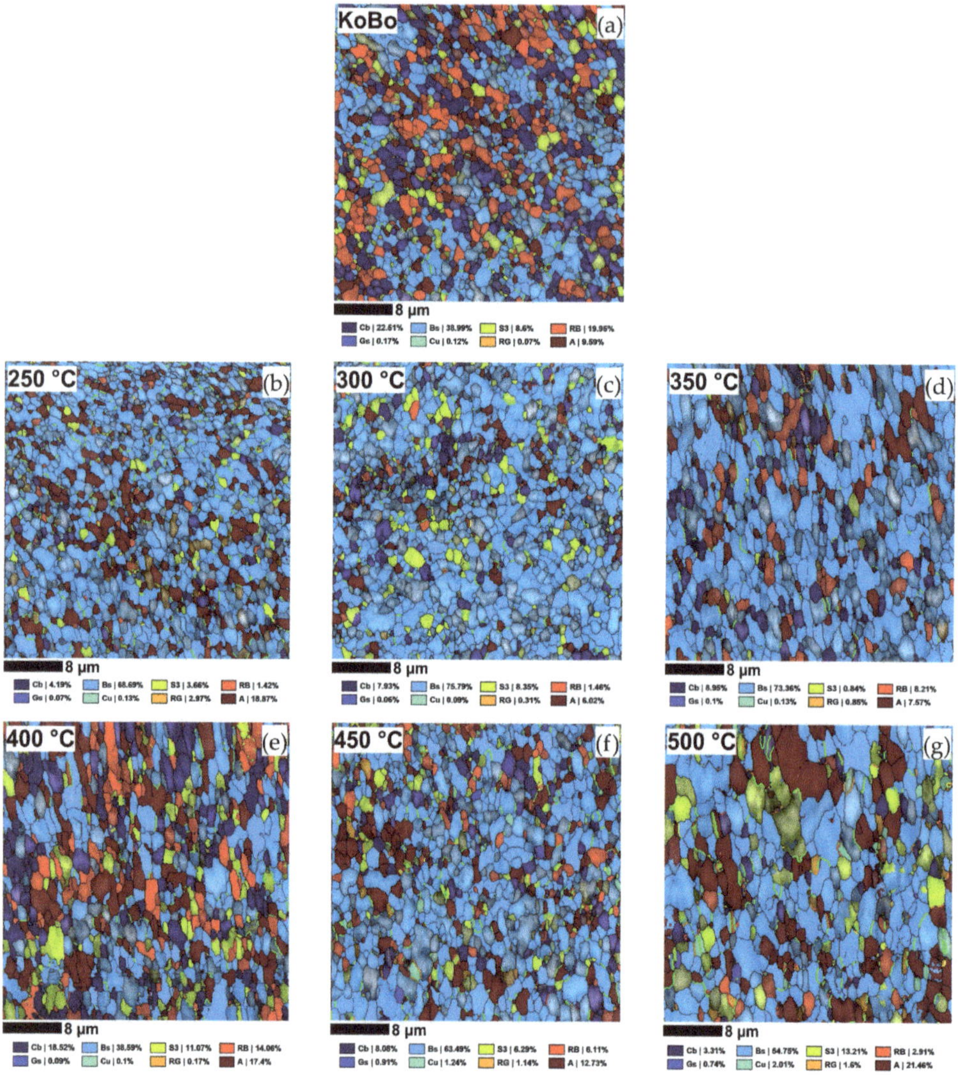

Figure 6. Texture components map with superimposed special grain boundaries for KoBo-processed and annealed samples. The colour coding is shown below the maps. (**a**) Kobo-processed, (**b**) annealed at 250 °C for 1 h, (**c**) annealed at 300 °C for 1 h, (**d**) annealed at 350 °C for 1 h, (**e**) annealed at 400 °C for 1 h, (**f**) annealed at 450 °C for 1 h, or (**g**) annealed at 500 °C for 1 h.

With reference to the literature [37–39] and the microstructural observations in Figure 3, it can be assumed that the refined second phase precipitates acted as "strong pinning points" that caused a retarding force for recrystallisation, resulting in a larger population of the brass deformation texture.

Table 2. Texture characteristics of the KoBo-processed and annealed AlSi10Mg samples. The values in the table correspond to the texture map shown in Figure 5.

Texture Component	Cube {001}<100> (%)	Goss {011}<100> (%)	Brass {110}<112> (%)	Copper {112}<111> (%)	S {123}<634> (%)	R. Goss (RG) {011}<011> (%)	R. Brass (RB) {110}<556> (%)	A {112}<110> (%)
KoBo	22.51	0.17	38.99	0.12	8.6	0.07	19.95	9.59
Annealed at 250 °C	4.19	0.07	68.69	0.13	3.66	2.97	1.42	18.87
Annealed at 300 °C	7.93	0.06	75.79	0.09	8.35	0.31	1.46	6.02
Annealed at 350 °C	8.95	0.1	73.36	0.13	0.84	0.85	8.21	7.57
Annealed at 400 °C	18.52	0.09	38.59	0.1	11.07	0.17	14.06	17.4
Annealed at 450 °C	8.08	0.91	63.49	1.24	6.29	1.14	6.11	12.73
Annealed at 500 °C	3.31	0.74	54.75	2.01	13.21	1.6	2.91	21.46

After annealing at 250 °C, the fractions of individual texture components changed noticeably, as shown in Figure 6b. In this condition, brass-oriented grains dominated the microstructure and accounted for about 69% of the total texture. When the annealing temperature rose to 300 °C, the fraction of brass-oriented grains grew to 76%, as seen in Figure 6c. Additionally, more S-oriented grains emerged as part of the recrystallisation texture [40]. Further increasing the annealing temperature to 350 °C had no significant effect on the texture. According to data from Figure 6d, the brass-oriented grains in this condition constituted about 74% of the total grains.

After annealing at 400 °C, the dominant texture component was brass texture. It was also noticeable that some amount of brass texture transformed into the recrystallisation cube texture. Furthermore, S texture and rotated brass texture were clearly present in some grains, as displayed in Figure 6e. Annealing at 450 °C resulted in an enhancement of brass texture, which constituted about 63% of the total textures, as shown in Figure 6f. According to the literature [41,42], the enhancement of brass texture could be attributed to strain-induced boundary migration (SIBM). This was consistent with the texture component map showing the development of numerous CSL-Σ3 boundaries near the brass-oriented grains. This effect was more pronounced in the sample annealed at 500 °C, which had the highest fraction of CSL-Σ3 boundaries. As seen in Figure 6g, numerous CSL-Σ3 boundaries ranging from about 0.5 μm to 2.5 μm in length were located in close proximity to the brass-oriented grains.

In order to analyse the textural components in detail, the (111) pole figures were plotted. It can be seen, in Figure 7, that all pole figures have similar characteristic textures that are typical of the extruded alloys. From Figure 7a, it can be noted that strong cube, brass, and rotated brass textures with a maximum intensity of 10.59 random levels (rl) developed in the KoBo-processed sample. According to the literature [43], the deformation at elevated temperature is favourable for the formation of the Bs texture, which might have formed due to the dynamic recovery or activation of non-octahedral slip systems during the extrusion process. Furthermore, the highly heterogeneous flow of the metal [44] and the relatively high plastic strain during KoBo extrusion could promote strain-induced boundary migration (SIBM), which is beneficial for the formation of a brass texture [45]. Furthermore, the precipitate drag effect can enhance SIBM. It has been revealed that incoherent precipitates with a high degree of misfit can increase the grain boundary energy and decrease the drag force [46], thus accelerating SIBM.

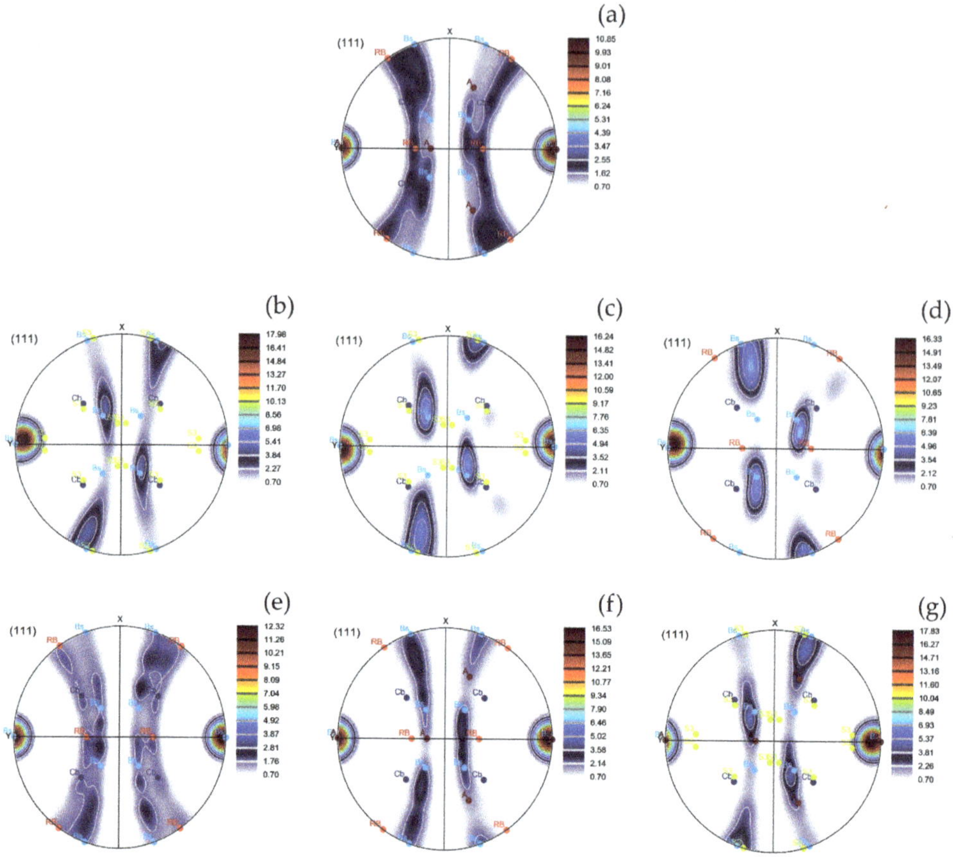

Figure 7. The (111) pole figures of the KoBo-processed and annealed samples. (**a**) Kobo-processed, (**b**) annealed at 250 °C for 1 h, (**c**) annealed at 300 °C for 1 h, (**d**) annealed at 350 °C for 1 h, (**e**) annealed at 400 °C for 1 h, (**f**) annealed at 450 °C for 1 h, or (**g**) annealed at 500 °C for 1 h.

After annealing at 250 °C, the brass texture was strengthened. In this condition, recovery dominated; therefore, the intensity of the brass texture was significantly increased to 17.98 Rl, as seen in Figure 7b. Referring to the work of Huang et al. [47] the brass texture is very stable; therefore, the brass-oriented subgrains can easily survive during annealing processes, while the grains with other orientations tend to be more easily recovered.

When the alloy was annealed at 300 and 350 °C, a large number of small subgrains formed, as seen in Figure 7c,d, which promoted the occurrence of the polygonisation of the brass-oriented parent grains [48]; this resulted in an additional minor enhancement of brass texture, as seen in Figure 7c,d.

As the annealing temperature was further increased to 400 °C, as shown in Figure 7e, the brass texture partially evolved into cube {100}<001> texture [47,49], which indicated that the recrystallisation process had occurred to some extent. This agreed with the microstructural observation that revealed an increase in the average grain size. Furthermore, some amount of brass texture translated into the rotated brass texture.

The relative stability of the deformed brass grains to annealing is highlighted in an increase in the relative volume fraction of this texture component in the sample annealed at 450 °C, as shown in Figure 7f. This stability of the brass component is attributed to its ability to deform with very few slip systems so that it is less affected than other deformation components during recrystallisation [50]. Furthermore, as the annealing temperature increased, the growth advantage for recrystallised cube grains disappeared. Therefore, the percentage volume fraction of the cube decreased from 18.5 to 8.08 as the annealing temperature increased from 400 to 450 °C.

Annealing at 500 °C resulted in the overall texture intensity strengthening, with the maximum brass component intensity of 17.8 rl, as shown in Figure 7g. Furthermore, there was a gradual increase in {112}<110> α-fibre orientation; however, this texture component is reported to be the last to recrystallise and be replaced [51].

This article presents experimental results indicating that a distinctive distribution of grain boundary characteristics and microtextures can be achieved through the innovative strain-annealing approach, hinting at the material's potential to exhibit unique properties. For example, a substantial volume fraction of brass texture, typically formed during cold rolling or shear deformation, was proved to be advantageous in enhancing the strength and fatigue resistance of aluminium alloys [52]. However, given its relatively rigid orientation, it has a minimal impact on ductility [53]. On the contrary, the cube texture enhances the ductility and formability of the alloy but at the expense of reduced strength [54].

By carefully controlling the distribution of grain boundary properties, it is also possible to develop a complex microstructure that aids in the recombination or annihilation of point defects caused by irradiation [55]. This control also helps to prevent the harmful segregation of elements crucial for the chemical and mechanical stability of the alloy [56]. Research into the structure of grain boundaries has shown that low-energy grain boundaries, particularly low ΣCSL grain boundaries (special grain boundaries), exhibit strong resistance to intergranular corrosion [57]. In addition, the presence of these grain boundaries in the microstructure contributes to increased resistance to creep and stress corrosion cracking [10].

4. Conclusions

Based on the microstructural observations presented in this study, the main conclusions can be formulated as follows:

- KoBo extrusion caused a significant grain refinement of LPBF AlSi10Mg alloy, resulting in a fine-grained microstructure with an average grain size of about 0.8 μm.
- The ultra-fine-grained microstructure produced by KoBo extrusion was preserved after annealing up to 500 °C.
- In the as-printed AlSi10Mg sample, the in situ-formed CSL boundaries accounted for about of 7% of the total boundary fraction.
- The grain boundary character distribution in LPBF AlSI10Mg alloy was optimised in the KoBo extrusion followed by annealing at 500 °C. After this thermomechanical treatment, the population of Σ3, Σ9 and Σ27 was increased to about 11%.
- Strain-induced boundary migration (SIBM) as well as the nucleation of twin boundaries promoted the formation of a strong brass texture in the KoBo-processed samples subjected to recrystallisation annealing.

In summary, the KoBo-extrusion process imparts a significant amount of strain to the LPBF AlSi10Mg alloy, making the strain-annealing approach effective in improving the CSL boundary fraction. Our research demonstrates that elevated annealing temperatures accelerate recrystallisation, leading to a higher density of CSL boundaries. The LPBF AlSi10Mg alloy, with its unique microstructure achieved through the presented thermomechanical treatment, holds great promise for aerospace and nuclear applications. The increased presence of low-CSL boundaries mitigates radiation-induced defects by promoting more stable grain boundaries, reducing the susceptibility of the material to radiation damage, and ultimately improving its overall resistance to the radiation environment.

Funding: The research was funded by the National Science Centre, Poland, based on decision number 2021/43/D/ST8/01946.

Data Availability Statement: Author agree to make data and materials supporting the results or analyses presented in their paper available upon reasonable request.

Conflicts of Interest: The author declares no conflict of interest.

References

1. Watanabe, T. An approach to grain boundary design for strong and ductile polycrystals. *Res. Mech.* **1984**, *11*, 47–84.
2. Cao, W.; Xia, S.; Bai, Q.; Zhang, W.; Zhou, B.; Li, Z.; Jiang, L. Effects of initial microstructure on the grain boundary network during grain boundary engineering in Hastelloy N alloy. *J. Alloys Compd.* **2017**, *704*, 724–733. [CrossRef]
3. Guan, X.J.; Shi, F.; Ji, H.M.; Li, X.W. A possibility to synchronously improve the high-temperature strength and ductility in face-centered cubic metals through grain boundary engineering. *Scr. Mater.* **2020**, *187*, 216–220. [CrossRef]
4. Yang, X.; Wang, P.; Huang, M. Grain boundary evolution during low-strain grain boundary engineering achieved by strain-induced boundary migration in pure copper. *Mater. Sci. Eng. A* **2022**, *833*, 142532. [CrossRef]
5. McCarley, J.; Helmink, R.; Goetz, R.; Tin, S. Grain Boundary Engineering of a Low Stacking Fault Energy Ni-based Superalloy. *Metall. Mater. Trans. A* **2017**, *48*, 1666–1677. [CrossRef]
6. Palaparti, D.P.R.; Vijayanand, V.D.; Mariappan, K.; Reddy, G.V.P. Correction: Correlating CSL Evolution and Strain Energy in Single Step Grain Boundary Engineered Austenitic Stainless Steel. *Metall. Mater. Trans. A* **2023**, *54*, 2506. [CrossRef]
7. Ogawa, Y. Temperature-sensitive ductilization in hydrogen-alloyed Fe-Cr-Ni austenitic steel by enhanced deformation twinning. *Scr. Mater.* **2024**, *238*, 115760. [CrossRef]
8. Dammu, S.; Singh, A.P.; Lala, S.R.F.; Srivastava, C. Grain Boundary Engineering in Electrodeposited Tin-Carbon Nanotube Composite Coatings for Enhanced Corrosion Resistance Performance. *Metall. Mater. Trans. A* **2023**, *54*, 3928–3939. [CrossRef]
9. Chen, X.-H.; Wang, F.; Zhang, F. Grain boundary engineering process for nano reinforced aluminum matrix composites. *J. Alloys Compd.* **2023**, *939*, 168834. [CrossRef]
10. Shamanian, M.; Mostaan, H.; Safari, M.; Szpunar, J.A. EBSD Study on Grain Boundary and Microtexture Evolutions During Friction Stir Processing of A413 Cast Aluminum Alloy. *J. Mater. Eng. Perform.* **2016**, *25*, 2824–2835. [CrossRef]
11. Jia, H.; Jin, S.; Li, Y. Formation of $\Sigma3\{110\}$ incoherent twin boundaries through geometrically necessary boundaries in an Al-8Zn alloy subjected to one pass of equal channel angular pressing. *J. Alloys Compd.* **2018**, *762*, 190–195. [CrossRef]
12. Pradeep, S.; Jain, V.K.S.; Muthukumaran, S.; Kumar, R. Microstructure and texture evolution during multi-pass friction stir processed AA5083. *Mater. Lett.* **2021**, *288*, 129382. [CrossRef]
13. Liu, B.; Ding, Y.; Xu, J.; Gao, Y.; Chu, C.; Hu, Y.; Chen, D. Grain boundary engineering activated by residual stress during the laser powder bed fusion of Inconel 718 and the electrochemical corrosion performance. *Mater. Charact.* **2023**, *204*, 113160. [CrossRef]
14. Dong, X.; Zhou, Y.; Qu, Y.; Wu, M.; Sun, Q.; Shi, H.; Peng, H.; Zhang, Y.; Xu, S.; Li, N.; et al. Recrystallization behavior and grain boundary character evolution in Co-Cr alloy from selective laser melting to heat treatment. *Mater. Charact.* **2022**, *185*, 111716. [CrossRef]
15. Tokita, S.; Kokawa, H.; Kodama, S.; Sato, Y.S.; Sano, Y.; Li, Z.; Feng, K.; Wu, Y. Suppression of intergranular corrosion by surface grain boundary engineering of 304 austenitic stainless steel using laser peening plus annealing. *Mater. Today Commun.* **2020**, *25*, 101572. [CrossRef]
16. Snopiński, P.; Matus, K.; Łagoda, M.; Appiah, A.N.S.; Hajnyš, J. Engineering an ultra-fine grained microstructure, twins and stacking faults in PBF-LB/M Al-Si alloy via KoBo extrusion method. *J. Alloys Compd.* **2024**, *970*, 172576. [CrossRef]
17. Korbel, A.; Błaż, L.; Bochniak, W.; Pawlyta, M.; Ostachowski, P.; Łagoda, M. Nano-Dimensional Elements in the Structure of Zinc Subjected to KOBO Extrusion. *Metallogr. Microstruct. Anal.* **2023**, *12*, 427–432. [CrossRef]
18. Dobkowska, A.; Zielińska, A.; Paulin, I.; Donik, Č.; Łojkowski, M.; Koralnik, M.; Adamczyk-Cieslak, B.; Paradowski, K.; Tkocz, M.; Kuc, D.; et al. Microstructure and properties of an AZ61 alloy after extrusion with a forward-backward oscillating die without preheating of the initial billet. *J. Alloys Compd.* **2023**, *952*, 169843. [CrossRef]
19. Liu, S.; Wang, Y.; Yarigarravesh, M.; Tayyebi, M.; Tayebi, M. Evaluation of whisker alignment and anisotropic mechanical properties of ZK60 alloy reinforced with SiCw during KOBO extrusion method. *J. Manuf. Process.* **2022**, *84*, 344–356. [CrossRef]
20. Snopiński, P.; Matus, K. Characterisation of Microstructure and Special Grain Boundaries in LPBF AlSi10Mg Alloy Subjected to the KoBo Extrusion Process. *Symmetry* **2023**, *15*, 1634. [CrossRef]
21. Snopiński, P.; Woźniak, A.; Łukowiec, D.; Matus, K.; Tański, T.; Rusz, S.; Hilšer, O. Evolution of Microstructure, Texture and Corrosion Properties of Additively Manufactured AlSi10Mg Alloy Subjected to Equal Channel Angular Pressing (ECAP). *Symmetry* **2022**, *14*, 674. [CrossRef]
22. Korbel, A.; Bochniak, W.; Ostachowski, P.; Błaż, L. Visco-Plastic Flow of Metal in Dynamic Conditions of Complex Strain Scheme. *Metall. Mater. Trans. A* **2011**, *42*, 2881–2897. [CrossRef]
23. Jodi, D.E.; Kitashima, T.; Singh, A.; Watanabe, M. High-temperature microstructural stability of pure Ni fabricated by laser powder bed fusion using Gaussian and flat-top beam profiles. *Mater. Charact.* **2023**, *200*, 112897. [CrossRef]
24. Xue, S.; Kuo, W.; Li, Q.; Fan, Z.; Ding, J.; Su, R.; Wang, H.; Zhang, X. Texture-directed twin formation propensity in Al with high stacking fault energy. *Acta Mater.* **2018**, *144*, 226–234. [CrossRef]

25. Zhu, Z.; Li, W.; Nguyen, Q.B.; An, X.; Lu, W.; Li, Z.; Ng, F.L.; Nai, S.M.L.; Wei, J. Enhanced strength–ductility synergy and transformation-induced plasticity of the selective laser melting fabricated 304L stainless steel. *Addit. Manuf.* **2020**, *35*, 101300. [CrossRef]
26. Monier, L.; Buttard, M.; Veron, M.; Blandin, J.-J.; Martin, G.; Villaret, F.; Shen, Y.; Yrieix, B.; Ernould, C.; Guyon, J.; et al. On the origin of grain refinement and twin boundaries in as-fabricated austenitic stainless steels produced by laser powder bed fusion. *Addit. Manuf.* **2023**, *61*, 103351. [CrossRef]
27. Brandon, D.G. The structure of high-angle grain boundaries. *Acta Metall.* **1966**, *14*, 1479–1484. [CrossRef]
28. Fang, Z.; Xiao, J.; Tan, S.; Deng, C.; Wang, G.; Mao, S.X. Atomic-scale observation of dynamic grain boundary structural transformation during shear-mediated migration. *Sci. Adv.* **2023**, *8*, eabn3785. [CrossRef]
29. Illgen, C.; Bohne, B.; Wagner, M.F.-X.; Frint, P. Thermal stability of SPD-processed aluminum alloys—Internal friction as an indication for recovery, recrystallization and abnormal grain growth. *J. Mater. Res. Technol.* **2022**, *17*, 1752–1759. [CrossRef]
30. Doherty, R.D. The Deformed State and Nucleation of Recrystallization. *Met. Sci.* **1974**, *8*, 132–142. [CrossRef]
31. Randle, V.; Hu, Y. The role of vicinal Σ3 boundaries and Σ9 boundaries in grain boundary engineering. *J. Mater. Sci.* **2005**, *40*, 3243–3246. [CrossRef]
32. Randle, V. Twinning-related grain boundary engineering. *Acta Mater.* **2004**, *52*, 4067–4081. [CrossRef]
33. Laleh, M.; Hughes, A.E.; Tan, M.Y.; Rohrer, G.S.; Primig, S.; Haghdadi, N. Grain boundary character distribution in an additively manufactured austenitic stainless steel. *Scr. Mater.* **2021**, *192*, 115–119. [CrossRef]
34. Goli, F.; Jamaati, R. Significant Texture Weakening of Al-Cu-Mg Alloy by Low Strain Asymmetric Cross Rolling Process. *J. Mater. Eng. Perform.* **2023**, *32*, 5576–5582. [CrossRef]
35. Li, Z.; Chen, L.; Zhang, X.; Zhao, G.; Zhang, C. Strengthening mechanism and anisotropy of mechanical properties of Si3N4p/Al-Mg-Si composites fabricated by sintering and extrusion. *Mater. Des.* **2021**, *210*, 110111. [CrossRef]
36. Yang, W.; Ji, S.; Li, Z.; Wang, M. Grain boundary precipitation induced by grain crystallographic misorientations in an extruded Al–Mg–Si–Cu alloy. *J. Alloys Compd.* **2015**, *624*, 27–30. [CrossRef]
37. Xiao, Z.; Yang, X.; Wang, J.; Fang, Z.; Guo, C.; Zhang, D.; Yang, Y.; Zhang, X. Influence of Fe addition on annealing behaviors of a phosphorus containing brass. *J. Alloys Compd.* **2017**, *712*, 268–276. [CrossRef]
38. Bo, G.; Jiang, F.; Dong, Z.; Wang, G.; Zhang, H. Revealing the influence of pre-precipitation microstructure on hot workability in an Al-Cu-Mg-Zr alloy. *Mater. Sci. Eng. A* **2019**, *755*, 147–157. [CrossRef]
39. Li, X.; Cui, X.; Liu, H.; Zhu, Z.; Liu, J.; Zhang, X.; Cui, H.; Li, H.; Pan, Y.; Feng, R.; et al. Study on the improvement and mechanism of AA6101 electrical conductivity by trace TM (Zr, V, Ti) elements-assisted boron treatment. *J. Alloys Compd.* **2023**, *939*, 168728. [CrossRef]
40. Hjelen, J.; Ørsund, R.; Nes, E. On the origin of recrystallization textures in aluminium. *Acta Metall. Mater.* **1991**, *39*, 1377–1404. [CrossRef]
41. Zhao, Q.; Liu, Z. Texture Evolution of Hot Rolled Al–Cu–Mg–Zr Alloy During Annealing. *Met. Mater. Int.* **2022**, *28*, 2947–2961. [CrossRef]
42. Wang, X.; Xiong, W.; Zheng, Y.; Zhang, J. The Correlation between Texture Evolution and Recrystallization Behavior during Rheologic Forming of 2195 Al–Li Alloy Cylindric Shell. *Metals* **2023**, *13*, 853. [CrossRef]
43. Li, S.; Zhao, Q.; Liu, Z.; Li, F. A Review of Texture Evolution Mechanisms During Deformation by Rolling in Aluminum Alloys. *J. Mater. Eng. Perform.* **2018**, *27*, 3350–3373. [CrossRef]
44. Gusak, A.; Danielewski, M.; Korbel, A.; Bochniak, M.; Storozhuk, N. Elementary model of severe plastic deformation by KoBo process. *J. Appl. Phys.* **2014**, *115*, 34905. [CrossRef]
45. Bate, P.S.; Huang, Y.; Humphreys, F.J. Development of the "brass" texture component during the hot deformation of Al–6Cu–0.4Zr. *Acta Mater.* **2004**, *52*, 4281–4289. [CrossRef]
46. Ihara, K.; Miura, Y. Dynamic recrystallization in Al-Mg-Sc alloys. *Mater. Sci. Eng. A* **2004**, *387–389*, 647–650. [CrossRef]
47. Huang, T.; Liu, F.; Liu, Z.; He, G. Evolution of Microstructure, Texture, and Hardness in an Al-Cu-Mg Alloy during Annealing. *J. Mater. Eng. Perform.* **2022**, *31*, 1419–1431. [CrossRef]
48. Humphreys, F.J.; Hatherly, M. Chapter 12—Recrystallization Textures. In *Hatherly MBT-R and RAP*, 2nd ed.; Humphreys FJ, Ed.; Elsevier: Oxford, UK, 2004; pp. 379–413. [CrossRef]
49. Zhao, Q.; Liu, Z.; Li, S.; Huang, T.; Xia, P.; Lu, L. Evolution of the Brass texture in an Al-Cu-Mg alloy during hot rolling. *J. Alloys Compd.* **2017**, *691*, 786–799. [CrossRef]
50. Alvi, M.H.; Cheong, S.W.; Suni, J.P.; Weiland, H.; Rollett, A.D. Cube texture in hot-rolled aluminum alloy 1050 (AA1050)—Nucleation and growth behavior. *Acta Mater.* **2008**, *56*, 3098–3108. [CrossRef]
51. Shen, F.; Zhou, Z.; Li, W.; Sun, Z.; Tian, J.; Xie, C.; Guo, J.; Liao, Z.; Yi, D.; Zhang, J.; et al. Micro-mechanism of texture evolution during isochronal annealing of as-annealed hot rolled Al-Cu-Mg sheet. *Mater. Des.* **2019**, *165*, 107575. [CrossRef]
52. Shen, F.; Yi, D.; Wang, B.; Liu, H.; Jiang, Y.; Tang, C.; Jiang, B. Semi-quantitative evaluation of texture components and anisotropy of the yield strength in 2524 T3 alloy sheets. *Mater. Sci. Eng. A* **2016**, *675*, 386–395. [CrossRef]
53. Zhou, W.; Huo, Q.; Wang, C.; Zhang, Y.; Zhang, Z.; Hu, S.; Zhao, S.; Nagaumi, H.; Yang, X. Effects of texture component, dislocation evolution and precipitation on the creep resistance of T5-treated Al-Mg-Si alloy with different Cu contents. *Mater. Charact.* **2022**, *189*, 111981. [CrossRef]

54. Hirsch, J.; Al-Samman, T. Superior light metals by texture engineering: Optimized aluminum and magnesium alloys for automotive applications. *Acta Mater.* **2013**, *61*, 818–843. [CrossRef]
55. Sakaguchi, N.; Endo, M.; Watanabe, S.; Kinoshita, H.; Yamashita, S.; Kokawa, H. Radiation-induced segregation and corrosion behavior on Σ3 coincidence site lattice and random grain boundaries in proton-irradiated type-316L austenitic stainless steel. *J. Nucl. Mater.* **2013**, *434*, 65–71. [CrossRef]
56. Barnard, L.; Barr, C.M.; Hattar, K.; Morgan, D.; Nathaniel, J.E.; Szlurfarska, I.; Morgan, D.; Taheri, M.L. Grain boundary character dependence of radiation-induced segregation in a model Ni–Cr alloy. *J. Mater. Res.* **2015**, *30*, 1290–1299. [CrossRef]
57. Hu, Y.; Bai, Q.; Xia, S.; Liu, K.; He, Q.; Xu, G. Applying Grain Boundary Engineering and Stabilizing Heat Treatment to 321 Stainless Steel for Enhancing Intergranular Corrosion Resistance. *J. Mater. Eng. Perform.* **2023**. [CrossRef]

symmetry

MDPI

Article

A Microstructural Study of Cu-10Al-7Ag Shape Memory Alloy in As-Cast and Quenched Conditions

Lovro Liverić [1], Wojciech Sitek [2], Przemysław Snopiński [3], Wojciech Maziarz [4] and Tamara Holjevac Grgurić [5,*]

1 Faculty of Engineering & Centre for Micro- and Nanosciences and Technologies, University of Rijeka, Vukovarska 58, 51 000 Rijeka, Croatia; lovro.liveric@riteh.uniri.hr
2 Scientific and Didactic Laboratory of Nanotechnology and Material Technologies, Faculty of Mechanical Engineering, Silesian University of Technology, 44-100 Gliwice, Poland; wojciech.sitek@polsl.pl
3 Department of Engineering Materials and Biomaterials, Faculty of Mechanical Engineering, Silesian University of Technology, 44-100 Gliwice, Poland; przemyslaw.snopinski@polsl.pl
4 Institute of Metallurgy and Materials Science, Polish Academy of Sciences, 25 Reymont Str., 30-059 Krakow, Poland; maziarz.w@imim.pl
5 School of Medicine, Catholic University of Croatia, Ilica 242, 10 000 Zagreb, Croatia
* Correspondence: tamara.grguric@unicath.hr

Abstract: Shape memory alloys (SMAs) represent an exceptional class of smart materials as they are able to recover their shape after mechanical deformation, making them suitable for use in actuators, sensors and smart devices. These unique properties are due to the thermoelastic martensitic transformation that can occur during both thermal and mechanical deformation. Cu-based SMAs, especially those incorporating Al and Ag, are attracting much attention due to their facile production and cost-effectiveness. Among them, Cu-Al-Ag SMAs stand out due to their notably high temperature range for martensitic transformation. In this study, a Cu-based SMA with a new ternary composition of Cu-10Al-7Ag wt.% was prepared by arc melting and the samples cut from this casting alloy were quenched in water. Subsequently, the phase composition and the development of the microstructure were investigated. In addition, the morphology of the martensite was studied using advanced techniques such as electron backscatter diffraction (EBSD) and transmission electron microscopy (TEM). The analyzes confirmed the presence of martensitic structures in both samples; mainly 18R ($\beta_{1'}$) martensite was present but a small volume fraction of ($\gamma_{1'}$) martensite also was noticed in the as-quenched sample. The observation of fine, twinned martensite plates in the SMA alloy with symmetrically occurring basal plane traces between the twin variants underlines the inherent correlation between microstructural symmetry and the properties of the material and provides valuable insights into its behavior. The hardness of the quenched sample was found to be lower than the as-cast counterpart, which can be linked to the solutioning of Ag particles during the heat treatment.

Keywords: shape memory alloys; Cu-Al-Ag alloys; heat treatment; microstructure; phase transformations; martensitic transformations

1. Introduction

Cu-based shape memory alloys (SMAs) exhibit remarkable properties, such as super elasticity and shape memory effect, attributed to a diffusionless reversible martensitic transformation [1–3]. Due to their affordability and ease of processing, Cu SMAs are utilized in various fields, including medicine, electronics and aerospace [4–7]. Among these alloys, copper-aluminum-based alloys (Cu-Al) are characterized by their exceptional mechanical properties, excellent electrical and thermal conductivity, and relatively low cost compared to other SMAs, such as Ni-Ti alloys [4,7].

Recently, ternary Cu-Al-Mn SMAs have attracted considerable attention. This increased interest stems from the addition of manganese in binary Cu-Al alloys, which serves to stabilize the β-phase region and extend it towards lower aluminum concentrations. As a result, the β-phase exhibits increased stability against diffusive decomposition [8–12]. As the alloy cools, the β-phase undergoes transformations with disordered order and progresses from β(A2) to β_2(B2) to β_1($L2_1$) [12–16]. These transformations lead to cubic (b.c.c.) superlattice structures such as the Fe_3Al type (DO_3), the CsCl type (B2) and the Cu_2MnAl type ($L2_1$). At lower temperatures, spinodal decomposition can occur between the DO_3 (Cu_3Al) and $L2_1$ (Cu_2AlMn) phases [16]. The transition from A2 to $L2_1$ cannot be suppressed at high aluminum concentrations, resulting in the $L2_1$ phase, known as Heusler alloy (Cu_2AlMn), undergoing a metastable transformation to 6M martensite. Conversely, at lower aluminum concentrations, typically below 16%at., A2 transforms either into A1 (disordered face-centered cubic structure f.c.c.) or into a 2M structure [10–12]. Depending on the chemical composition, various martensite structures can be observed in Cu-based SMA, including hexagonal 2H and rhombohedral structures 9R, 18R, 6R and 3R [11,12,17–19].

To date, there have been only a few studies on the addition of Ag to binary Cu-Al SMAs and ternary Cu-SMA alloys. The addition of Ag to SMA alloys of the Cu-Al system improves their hardness, corrosion resistance, aging properties and magnetic properties, as stated by Silva et al. [13]. Santos et al. [14] attributes the increased microhardness of ternary alloys with 3%at. Ag to the formation of bainite. Cu-Al-Ag SMAs are characterized by the highest temperature of martensitic transformation, primarily between 200 °C and 600 °C, depending on the Al/Ag ratio, so that they are used for high-temperature applications in certain areas [20,21]. Furthermore, the addition of 4%wt. Ag in Cu-Al alloys affects the phase transformations, can promote the formation of pearlitic phases and can change the stability of this pearlitic phase [22]. The precipitation of Ag in Cu-Al-Ag alloys exhibits a zero-order reaction, indicating constant Ag diffusion rates and an interface-controlled formation of Ag-rich phases [2]. Furthermore, the addition of Ag to Cu-based SMAs significantly improves the martensitic transformation, ductility and operating temperature range compared to conventional Cu-Ni-Ti and Cu-Al-Ni SMAs [13,23]. In the study [24], the influence of silver content on the microstructure, martensitic transformation and shape memory effect of Cu-Al-Ag alloys is investigated. The potential for optimizing stability and improving shape memory performance through thermal cycle training and specific adjustments of the silver content for use at high temperatures is demonstrated. The presence of Ag in the Cu-11%wt. Al alloy has a significant influence on the cooling rate, phase formation and martensitic transformation temperatures, providing insights into the optimization of these alloys for specific applications [25]. In addition, Krishna [26], who investigated Cu-XAl-5Mn-2Ag alloys, found that the concentration of aluminum influences martensite formation and has a direct impact on the shape memory effect. Manasijević [27] proved that Cu-Al-Mn-Ag alloys with a higher manganese content yield complete martensite, while those with a lower manganese content exhibit α-phase precipitation.

This investigation will further contribute to the development of Cu-Al-Ag alloys with improved properties and expand their range of applications in various fields. In the paper, an alloy with the composition Cu-10%wt. Al-7%wt. Ag is produced in an electric arc furnace by melting high-purity metals. Morphological and microstructural changes in the alloys were investigated in both the as-cast and quenched states by optical (OM), scanning electron microscopy (SEM) and transmission electron microscopy (TEM). The influence of heat treatment on the microstructure of the investigated alloy was examined using the SEM and TEM-EDS and SEM-EBSD methods. Phase identification was carried out using X-ray diffraction (XRD) and high-resolution X-ray diffraction (XRD), and the hardness of the alloy was determined using the micro-Vickers hardness test.

2. Materials and Methods

The Cu-10Al-7Ag SMA alloy was prepared by melting raw metals (Mateck Material-Technologie & Kristalle, Jülich, Germany): copper (99.9%), aluminum (99.5%), and silver (99.99%). The metals were melted in an electric-arc furnace that had been previously vacuumed and subjected to multiple cycles of argon pumping. The samples underwent more than four remelting processes for better homogenization and were cast in cylindrical molds with dimensions of 8 × 12 mm. After casting, the samples were quenched in chamber furnaces (OVER, Zagreb, Croatia) at 900 °C for 30 min and rapidly cooled in water.

For metallographic analysis, the samples were prepared by cutting, cold mounting, and grinding with 320#, 600#, 800#, 1200#, and 2400# SiC abrasives. Subsequently, they were polished with 1 μm and ≤0.2 μm diamond paste using Citopress-20 and Tegramin-30 (Struers, Willich, Germany). The prepared samples were etched with a solution of 2.5 g $FeCl_3$, 48 mL CH_3OH, and 10 mL H_2O.

The microstructures were analyzed using an Axio Vert A1 optical microscope (Carl Zeiss NTS GmbH, Oberkochen, Germany), a FEG QUANTA 250 scanning electron microscope (Thermo Fisher, formerly FEI, Waltham, MA, USA) equipped with an energy dispersive X-ray spectroscopy detector (EDS) (Oxford Instruments, Abingdon, UK), and two transmission electron microscopes, Tecnai G2 F20 (200 kV) (TEM) with an energy dispersive X-ray microanalyzer (EDX), a HAADF detector for the scanning transmission technique (STEM), a field emission gun (FEG) and high-resolution Gatan UltraScan and wide-angle SIS Megaview cameras (Thermo Fisher, formerly FEI, USA) and Titan 80–300, FEI S/TEM microscope equipped with spherical aberration correctors in both the probe and imaging optics. The microscope was operated at 300 kV (Thermo Fisher, formerly FEI, USA). The TEM instruments provided SAED, bright and dark field TEM, and HAADF images of a lamellar sample (approximately 10 × 8 μm) prepared by the focused ion beam (FIB) method using the Helios 650HP SEM/FIB dual beam microscope (Thermo Fisher, formerly FEI, USA) with Octane Pro EDS detector (EDAX Inc., Mahwah, NJ, USA).

Electron backscatter diffraction (EBSD) analyses were conducted using a Supra 35 scanning electron microscope (SEM) from Carl Zeiss NTS GmbH, Oberkochen, Germany. The SEM operated at an acceleration voltage of 15 kV, a working distance of 17 mm, a tilt angle of 70°, and utilized step sizes ranging from 0.4 to 0.06 μm. Samples for EBSD analysis were prepared following standard metallographic techniques, involving mechanical grinding with SiC papers, polishing with diamond pastes, and a final polishing step for 1 h using 0.04 μm colloidal silica.

X-ray diffraction (XRD) analysis was performed using a Bruker D8 Advance diffractometer (Bruker Corporation, Billerica, MA, USA) with CuKα radiation. The scans were collected in the 2θ range of 20–90°, with a 2θ step size of 0.02° and a counting time of 0.6 s, under the accelerating voltage of 40 kV and current of 25 mA.

The crystal structure and overall texture of samples in both cast and quenched states were analyzed through high-energy X-ray diffraction measurements at DESY in Hamburg, Germany, utilizing the P07B beamline (87.1 keV, λ = 0.0142342 nm). For phase analysis, diffraction patterns were recorded in continuous mode using a 2D Mar345 Image Plate detector. To ensure textureless measurements, all samples underwent a 180° rotation about the ω-axis during X-ray exposure. To maintain the Bragg condition for satellite reflections, samples were continuously rotated around the ω sample axis by $\omega < \pm 10°$, with a beam size of 1 × 1 mm². Subsequently, the 2D patterns were integrated using the Fit2D software (https://www.esrf.fr/computing/scientific/FIT2D/index.html) and presented in a graph of relative intensity vs. 2Theta angle.

The microhardness of the alloys was determined using the FM-ARS-9000 fully automatic microhardness measurement system (Future-Tech, Kawasaki-City, Japan). A load of HV 100 g and a dwell time of 15 s were applied during the measurements. Prior to microhardness testing, each sample underwent meticulous preparation, involving thorough cleaning, grinding, and polishing to ensure result accuracy. The Vickers microhardness values were calculated as the average of five individual measurements taken from each sample.

3. Results

3.1. Microstructure of the Cu-10Al-7Ag SMA Alloy

Figure 1 shows the optical micrographs (OM) of the Cu-10Al-7Ag SMA alloy in the as-cast state. From the observation of the microstructure, it is clear that the alloy has the potential to exhibit shape memory properties. As can be seen in Figure 1a, the as-cast sample microstructure consists of coarse grains with a measured average initial size of ~460 μm. The higher magnification OM in Figure 1b documents the presence of martensite needles and indicates the existence of secondary phases (bright precipitates) randomly distributed in the alloy matrix. The precipitation of these secondary phases at room temperature may have occurred due to their limited solubility in the parent matrix. It is worth noting that these precipitate phases could potentially have a detrimental effect on the shape memory properties of the alloy.

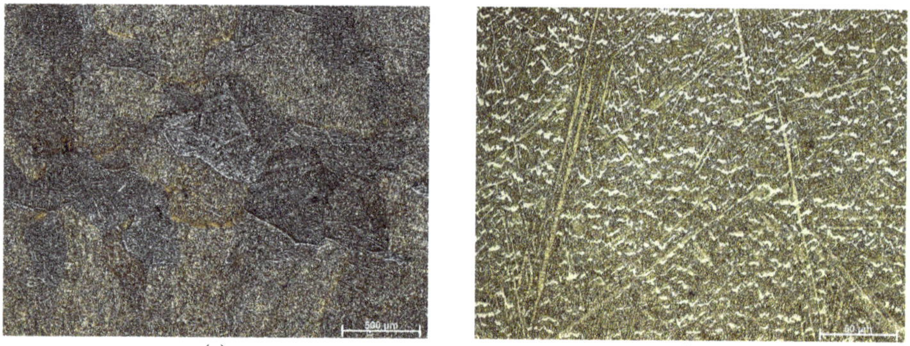

(**a**) (**b**)

Figure 1. Microstructure of the Cu-10Al-7Ag SMA alloy in the as-cast condition: (**a**) low-magnification polarized light OM image, (**b**) high-magnification bright field OM image.

Figure 2 shows the effect of the heat treatment on the microstructure of the Cu-10Al-7Ag SMA alloy. The high cooling rate led to an increased dislocation density [28], which provided numerous heterogeneous nucleation sites for martensite formation with a zigzag and arrow morphology (Figure 2a). A similar martensite variant was also observed by other researchers in the following studies [29]. The higher magnification image also shows the more uniform distribution of silver precipitates within the martensitic matrix, Figure 2b. In addition, it can be seen that the size of the grains decreased to ~400 μm after quenching. The finer grains of SMA can help to reduce the brittleness and improve the shape memory properties of the samples [30].

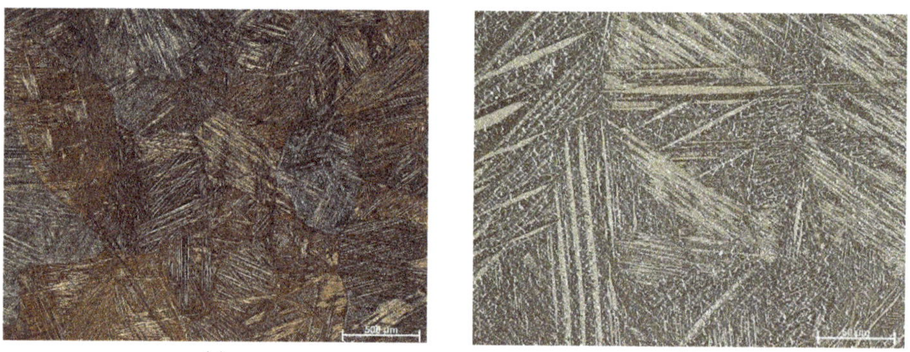

(**a**) (**b**)

Figure 2. Microstructure of the Cu-10Al-7Ag SMA alloy in the quenched condition: (**a**) low-magnification polarized light OM image, (**b**) high-magnification bright field OM image.

The microstructures and phase compositions were then examined using SEM/EDX. The SEM images of as-cast and quenched Cu-10Al-7Ag are shown in Figure 3, while the corresponding EDX results are listed in Tables 1 and 2, respectively. It is worth noting that both images were taken at the same magnification to emphasize the microstructural differences between the two states.

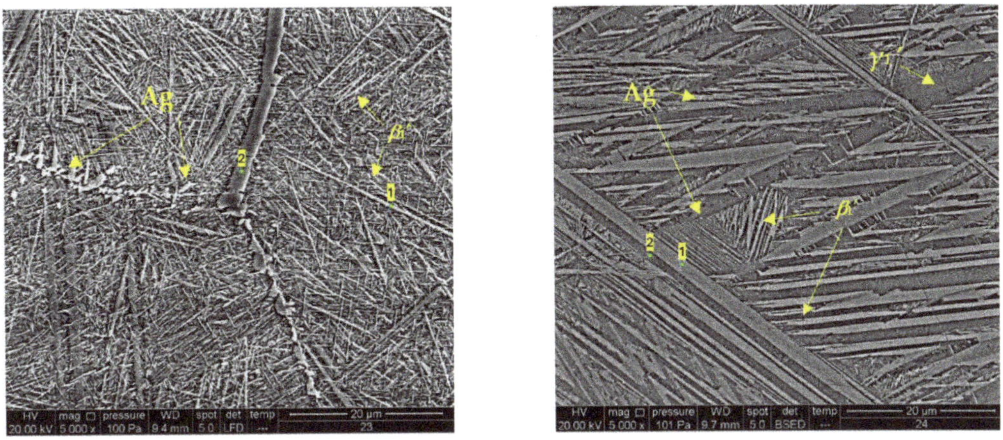

(**a**) (**b**)

Figure 3. SEM microstructures of the Cu-10Al-7Ag SMA alloy (**a**) as-cast condition, (**b**) quenched.

Table 1. Results of EDS chemical composition microanalysis of as-cast sample.

	Al	Cu	Ag
Point 1	9.21 ± 0.20	80.87 ± 0.55	6.78 ± 0.24
Point 2	10.28 ± 0.42	81.51 ± 0.63	8.21 ± 0.52

Table 2. Results of EDS chemical composition microanalysis of quenched sample.

	Al	Cu	Ag
Point 1	9.70 ± 1.11	80.41 ± 1.70	9.89 ± 1.42
Point 2	10.30 ± 0.90	81.2 ± 1.10	8.50 ± 1.05

As shown in Figure 3a, the microstructure of the as-cast sample consists of V-type $\beta_{1'}$-martensite and several fine precipitates. The EDX results in Table 1 show that the precipitates in the grain interior are enriched in Cu and Al, while the precipitates in the form of an almost continuous layer along a grain boundary region are enriched with silver.

After quenching, the precipitates, which form a continuous layer along the grain boundary, have disappeared. In this state, the Ag precipitates are uniformly distributed in the matrix of Cu-10Al-7Ag alloy, as shown in Figure 3b. It is worth noting that quenching resulted in the formation of coarse martensite plates, which may indicate the presence of another type of martensite, possibly the $\gamma_1{}'$, 2H type. In addition, the Cu-10Al-7Ag alloy was found to exhibit predominantly the self-accommodating zig-zag morphology typical of 18R martensite in the quenched state. The martensitic transformation observed in the Cu-10Al-7Ag SMA alloy illustrates the fundamental role of symmetry in determining the shape memory properties of the alloy. The transition from the highly symmetric β-phase to the martensitic phases with lower symmetry (e.g., 18R, 2H) underlines how closely symmetry changes are linked to the functional behavior of the alloy.

Both conventional and high-energy X-ray diffraction techniques were used to further analyze the phase composition. The conventional X-ray diffractogram is shown in Figure 4a. Phase identification revealed the presence of $\beta_1{}'$-martensite and silver phases in both samples.

It is noteworthy that in the quenched sample, the peaks of the $\beta_1{}'$-martensite phase almost disappear at higher 2theta angles. This phenomenon could be due either to a reduction in the content of the $\beta_1{}'$-martensite phase after quenching or to changes in the macrotexture. However, it is also evident that the peaks corresponding to the $\beta_1{}'$-martensite phase are more pronounced in the quenched state around the 2theta angle of 45 degrees. Moreover, the peaks associated with the Ag phase in the as-cast state are clearly recognizable as precipitates that are unevenly distributed at the grain border, but disappear after quenching, which is consistent with the observations of the microstructure in SEM and as a secondary phase on Figure 1b in OM.

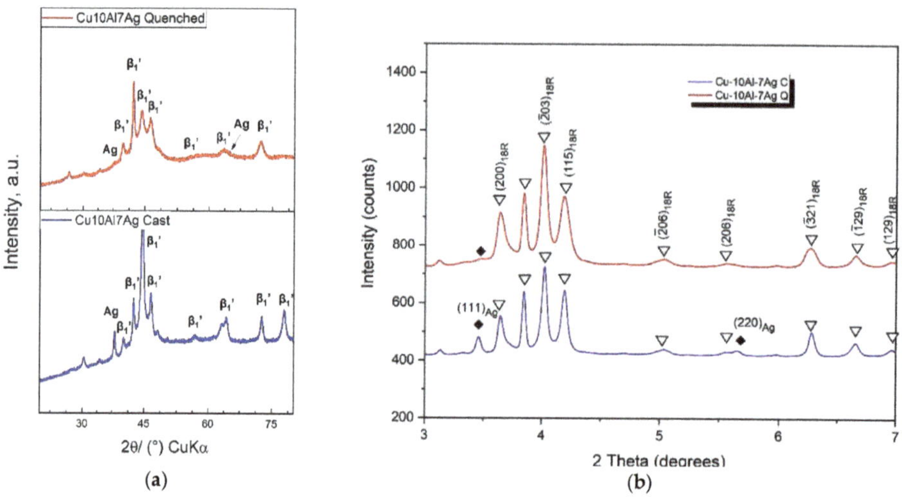

Figure 4. XRD diffractogram of Cu-10Al-7Ag SMA alloy: (**a**) regular XRD diffractogram, (**b**) high-energy X-ray diffraction measurements from synchrotron.

To gain insight into the grain structure, an EBSD analysis was performed. Figure 5a shows the IPF map, which includes the low-angle grain boundaries (red), the high-angle grain boundaries (green) and the twin boundaries (black). The colors in the IPF Z-map correspond to the Z-direction within the reference frame of the sample, as shown in Figure 5b. General observation EBSD analysis revealed differences from the microstructural observations in the SEM, particularly in the identification of twin boundaries and the presence of small round particles. These differences may be due to the different capabilities of the two techniques, with SEM Figure 3 revealing variations in composition and EBSD Figure 5 providing data on crystallographic orientation.

Figure 5a confirms that the martensite phase has a distinct zigzag morphology, which is consistent with the optical microscopy results. The color represents different martensite orientations. The grain boundary distribution histogram shows that high-angle grain boundaries predominate and account for about 80% of the grain boundary fraction (HAGBs + twin boundaries). It is also worth noting that a large proportion of the HAGBs are twin boundaries, which is reflected in an increased proportion of boundaries with a misorientation angle of 60 degrees. These boundaries are clearly visible in the IPF-Z EBSD map as they mainly separate the parallel $\beta_1{}'$-martensite plates. Significantly, twin boundaries are present in both straight, parallel shapes and curved configurations, indicating the coexistence of both coherent and incoherent twin boundaries within the microstructure. The high proportion of these boundaries indicates a well-defined microstructural mechanism that underpins the alloy's excellent shape memory and pseudoelastic properties. According to the grain size distribution histograms in Figure 6, the microstructure in Figure 6b consists of martensite plates with a dominant size of about 1 μm.

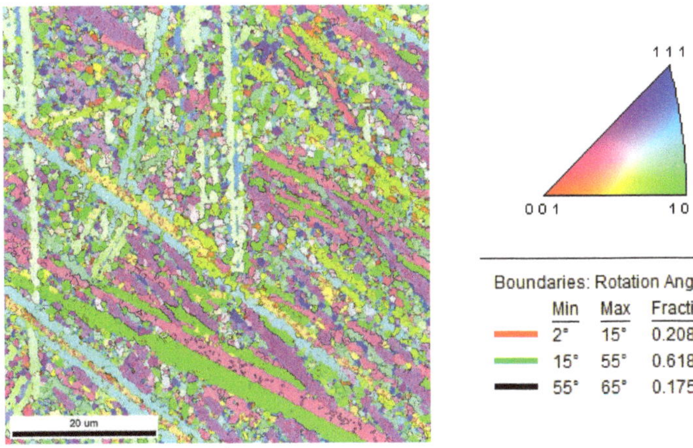

Boundaries: Rotation Angle

	Min	Max	Fraction	Number	Length
(orange)	2°	15°	0.208	21566	2.49 mm
(green)	15°	55°	0.618	64168	7.41 mm
(black)	55°	65°	0.175	18131	2.09 mm

(a) (b)

Figure 5. Results of the EBSD investigation of cast sample: (**a**) IPF-Z image, (**b**) IPF map reference frame and the statistical information about boundary fraction.

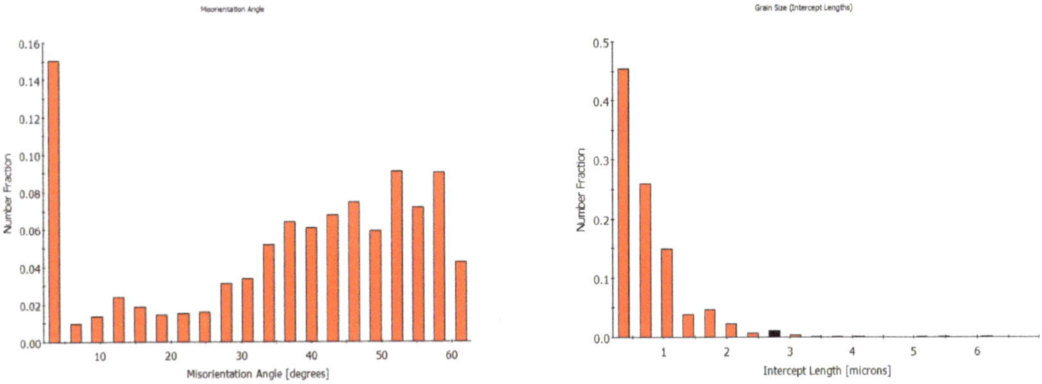

(a) (b)

Figure 6. Results of the EBSD investigation of cast sample: (**a**) histogram of grain boundary misorientation distribution, (**b**) histogram of grain size distribution.

Further insight into the dislocation properties was gained by plotting the Kernel Average Misorientation Map. It is known that the KAM value directly reflects the geometrically necessary dislocation density, i.e., the higher the KAM value, the higher the GND density. Figure 7 shows that there is no large accumulation of dislocations in the as-cast state. The highest KAM values are seen in the large martensite plates, indicating a higher GND accumulation in this area, which is probably due to the presence of dense networks of misfit dislocations inside plates. The average KAM value measured is 0.60°.

Figure 8 shows the IPF-Z map of the quenched sample. It is noticeable that the microstructure appears much finer in the quenched state than in the as-cast state. Figure 9 shows the histograms of the EBSD investigation of the grain boundary misorientation and the grain size distribution of the quenched sample. Most of the grains have a size of less than 1 μm, Figure 9b. Analysis of the grain boundary distribution histogram shows a notable shift where low angle grain boundaries (LAGBs) now dominate and account for approximately 51% of the total, Figure 9a. This increase in the proportion of LAGBs can be attributed to dislocation multiplication that occurs during heat treatment [31]. Of particular interest is the observation that the fraction of twin boundaries increased to about 21% after quenching, as indicated by the histogram of the boundary distribution, which shows a peak around a misorientation angle of 60 degrees, Figure 9. This suggests that the fraction of the twinned 18R phase is increased after quenching. Such mechanisms favor the increase in the proportion of LAGBs because they reduce the total energy of the system by organizing the dislocations in more stable configurations. The presence of more LAGBs usually indicates a material that can better resist plastic deformation, which is particularly important for applications that require high elasticity under mechanical stress.

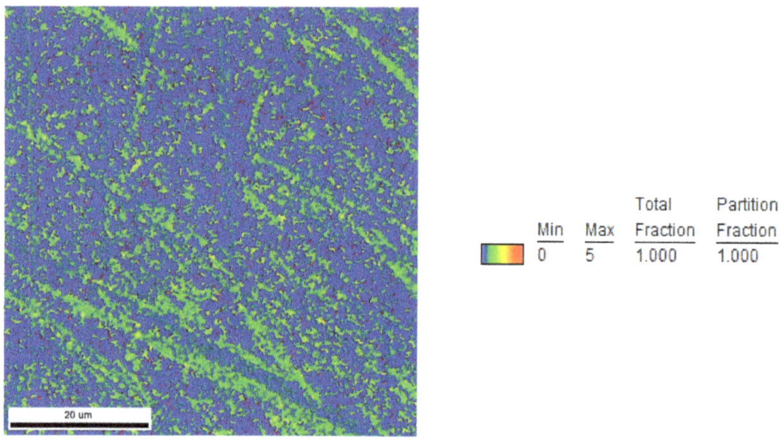

(a) (b)

Figure 7. Kernel Average Misorientation map of Cu-10Al-7Ag SMA cast sample (**a**) and corresponding legend showing color coding in KAM map (**b**).

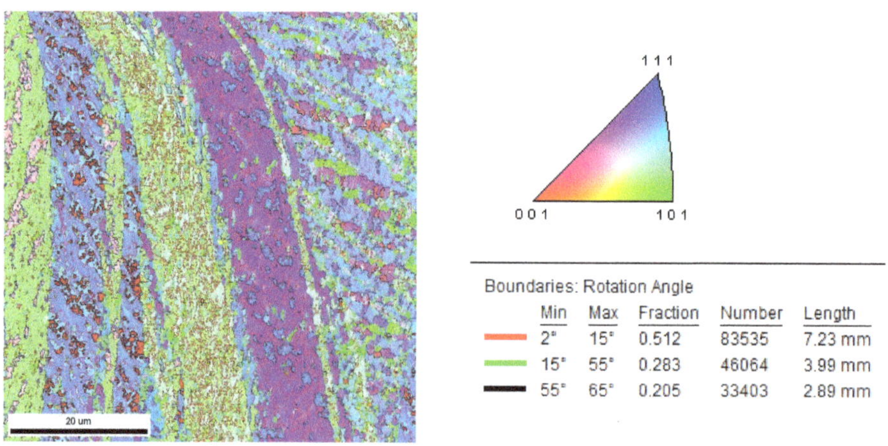

(a) (b)

Figure 8. Results of the EBSD investigation of Cu-10Al-7Ag SMA quenched sample: (**a**) IPF-Z image, (**b**) IPF map reference frame and the statistical information about boundary fraction.

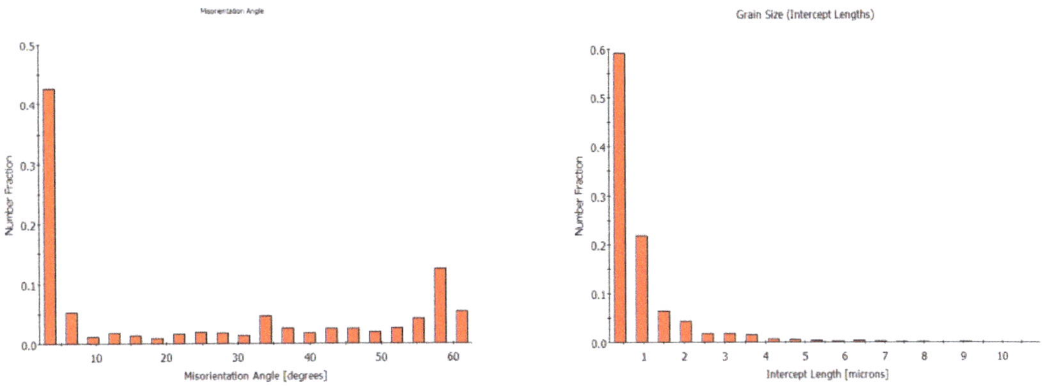

(**a**) (**b**)

Figure 9. Results of the EBSD investigation of Cu-10Al-7Ag SMA quenched sample: (**a**) histogram of grain boundary misorientation distribution, (**b**) histogram of grain size distribution.

Figure 10 shows the KAM map of the Cu-10Al-7Ag SMA alloy sample in its quenched state. It can be seen that the increased KAM values are mainly concentrated around the Low Angle Grain Boundaries (LAGBs) and reach higher average values for the quenched sample (0.91°) than for the as-cast sample (0.60°). This increase in the KAM value is obviously due to the increased presence of geometrically necessary dislocations. According to the work of Yi et al. [32], the thermal cycle involving the martensitic transformation must coincide with the occurrence of dislocations at the interface. In addition, the phase transformation leads to the formation of subgrain boundaries, which contributes to the propagation of dislocations. Consequently, a significantly higher number of generated dislocations can be assumed in the quenched sample compared to its as-cast counterpart, which is consistent with the electron backscatter diffraction (EBSD) data.

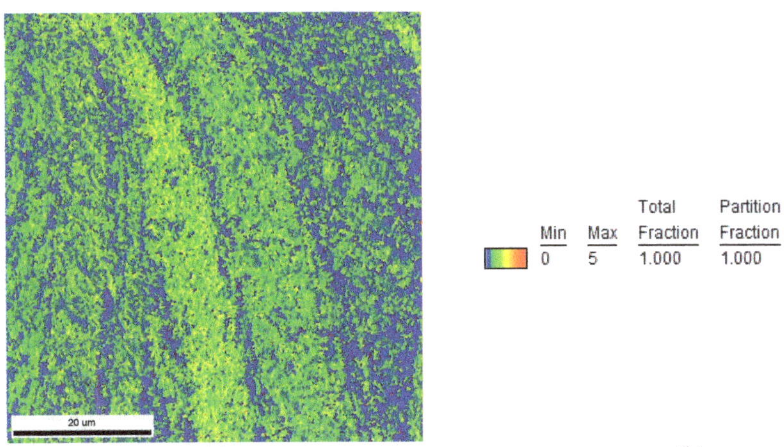

(**a**) (**b**)

Figure 10. Kernel Average Misorientation map of Cu-10Al-7Ag SMA quenched sample (**a**) and corresponding legend showing a color coding in KAM map (**b**).

STEM observations were carried out to show the distribution of the most important alloying elements. Figure 11 shows the STEM-HAADF image together with the corresponding elemental distribution maps. As can be observed, the elemental maps show a random (homogeneous) distribution of Ag precipitates in the as-cast matrix of the sample. These precipitates have a length of about 0.5 to several microns. This image also shows that the alloy matrix is rich in Cu and Al.

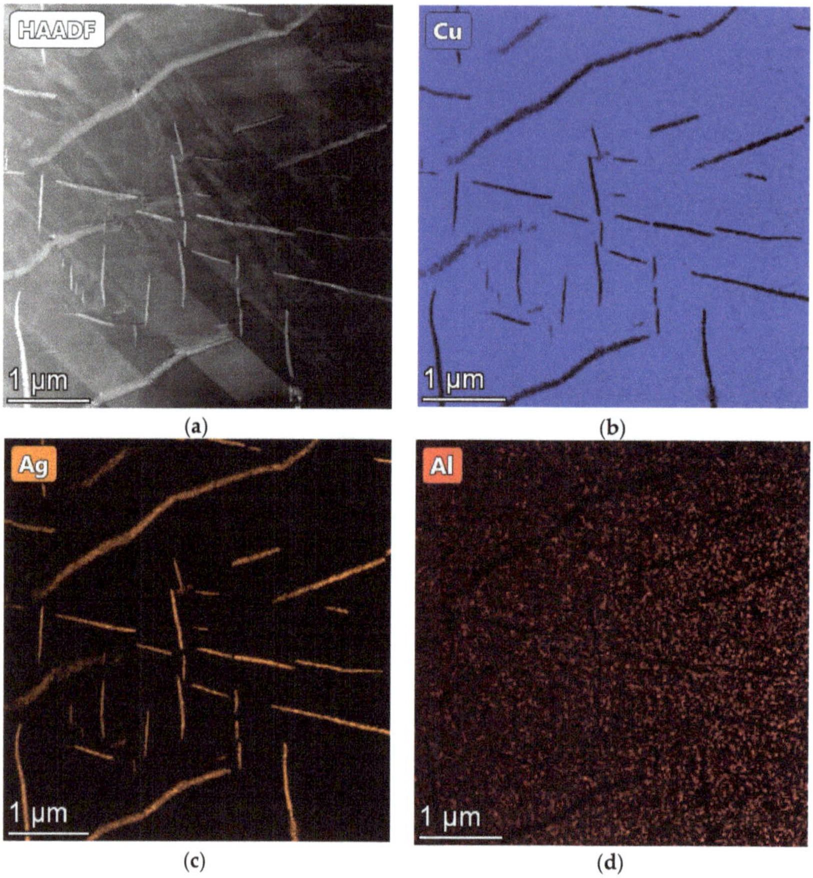

Figure 11. STEM HAADF and EDS elemental maps of the Cu-10Al-7Ag SMA alloy in as-cast condition (**a**) STEM-HAADF image, (**b**) Cu element distribution map, (**c**) Ag element distribution map, (**d**) Al element distribution map.

Examination of the microstructural changes using SEM and EBSD techniques sheds light on the symmetrical characteristics inherent in martensitic transformations. The patterns of twin boundaries and the transformation-induced changes in the microstructure highlight the fundamental symmetry principles that govern these phase transitions in the alloys.

Figure 12 shows the results of the quantitative chemical analysis performed for the indicated areas, showing the composition of the three microstructural constituents observed in the as-cast sample. It can be seen that the general composition of the as-cast sample (matrix plus Ag precipitates) differs slightly from the nominal composition. The measured composition is depleted in both Al and Ag, which is most likely due to high heterogeneity of the as-cast sample. This heterogeneity can be mainly related to the different density of Ag precipitates in the microstructure, as they also contain a considerable amount of Cu and a small amount of Al, as can be seen in Figure 12.

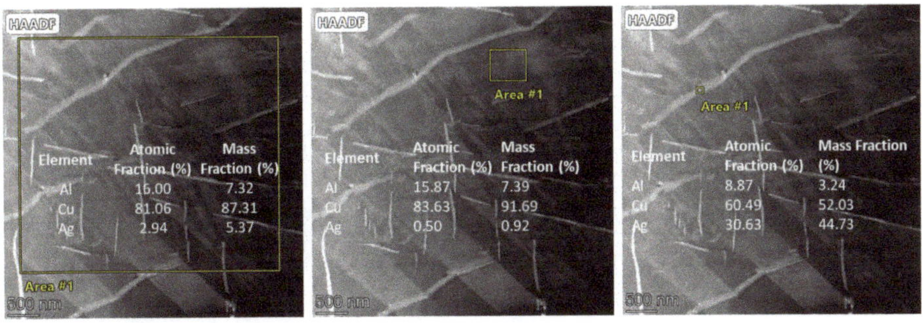

Figure 12. STEM HAADF and results of chemical analyses performed in the indicated areas of the Cu-10Al-7Ag SMA alloy in as-cast condition.

Figure 13 shows the bright field (BF), dark field (DF) and corresponding selected area diffraction pattern (SADP) of specimen in the as-cast condition. It reveals the clearly contrasting martensite plates having a width of 150–800 nm which were identified as 18R phase. SADP vell fits to the [010] zone axis of the R18 martensite with very low monoclinic distortion of β = 88.7°. This pattern exhibits a contrast with a 0.619 nm periodicity typical of the 18R stacking sequence that is visible in the DF image taken from the (0012) spot. The streaks between stacking spots could be attributed to the presence of many structural faults inside the martensite variants, probably produced by the internal stresses.

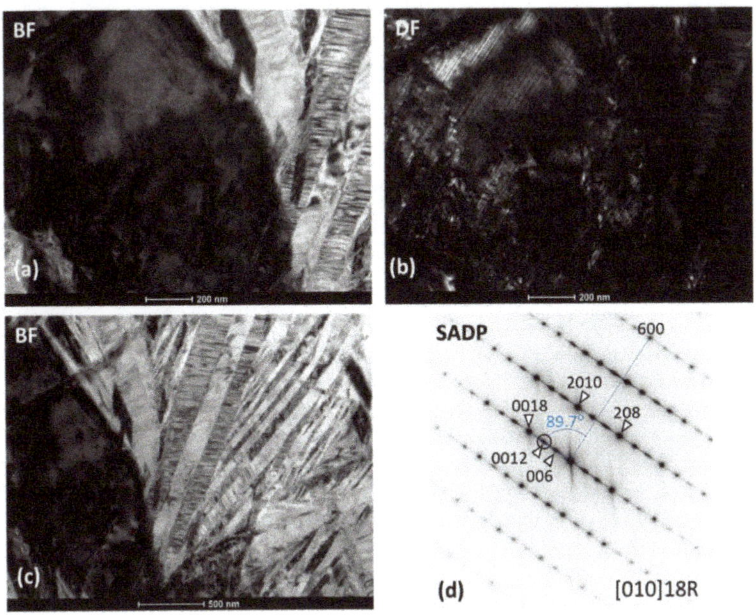

Figure 13. TEM images of the Cu-10Al-7Ag SMA alloy in cast condition: (**a**) bright-field TEM image taken at higher magnification and (**b**) dark field TEM image revealing 18R stacking sequence, (**c**) bright field TEM image taken at low magnification and corresponding (**d**) dark-selected area diffraction pattern (SADP).

To gain greater insight into the complex microstructure of the SMA alloy, observations in the scanning transmission mode (STEM) were performed. The BF-STEM image, Figure 14a, confirms that the alloy has a martensitic structure, with fine, twinned martensite plates present throughout. In Figure 14a, the angles between the traces of the basal planes in the twin variants and the twin boundary were measured to be 57°, which indicates that the basal plane traces appear symmetrically in the twin variants. In addition, the corresponding STEM-HAADF (Z-contrast) image in Figure 14b reveals minor chemical composition fluctuations within the martensite plates. However, it should be mentioned here that this change in contrast may be related to the different orientation of the martensite plates.

(a) (b)

Figure 14. STEM (**a**) and HAADF (**b**) images showing the martensite matrix morphology in Cu-10Al-7Ag SMA alloy in cast condition.

Further observations were made on the quenched sample. Figure 15 shows the STEM-HAADF image with the corresponding element maps. Figure 15a reveals significant evolution of the microstructure after quenching. It can be clearly seen that the precipitates of the Ag phase have a different size and morphology. Figure 15b shows Ag nanoprecipitates with the mean size of about 10 nm, and more evenly distributed due to the rapid cooling rate restricting their growth and resulting in a finer dispersion. A similar microstructure was recently described by Silva et al. [33] for a Cu-11%Al-10%Mn-3%Ag alloy. In addition, there are some undissolved Ag phase clusters in the microstructure, indicating that complete dissolution was not achieved after quenching.

These morphological differences between the as-cast and quenched states have a significant impact on the mechanical and functional properties of the alloy, as smaller, more uniformly distributed precipitates affect the mechanical properties and alter the thermomechanical properties by more effectively retaining dislocations and distributing stress more evenly across the matrix.

Figure 16a–d is a set of TEM images of the as-quenched sample showing BF image and corresponding DF images and SAED pattern. The TEM observations confirm the presence of the $\beta_1{}'$-martensite phase. Two types of martensite plates can distinguish with thickness range from 100 to 500 nm and few nanometers, respectively (fine plates marked by ovals in the DF images). Such configuration of the martensite plates is due to the self-accommodation character of the martensitic microstructure formed during thermoelastic martensitic transformation. The electron diffraction pattern registered from two variants of martensite plates being in twin orientation was indexed as 18R structure with the [010] zone axis, Figure 16c. Dark field images (DF-1 and DF-2) are taken from (0018) spots of two variants being in a twin relationship. However, apart from the main spots taken for the DF image, additional small spots also must be considered (indicated by the arrows) for the formation of the DF image responsible for the fine martensite plates visible in DF images.

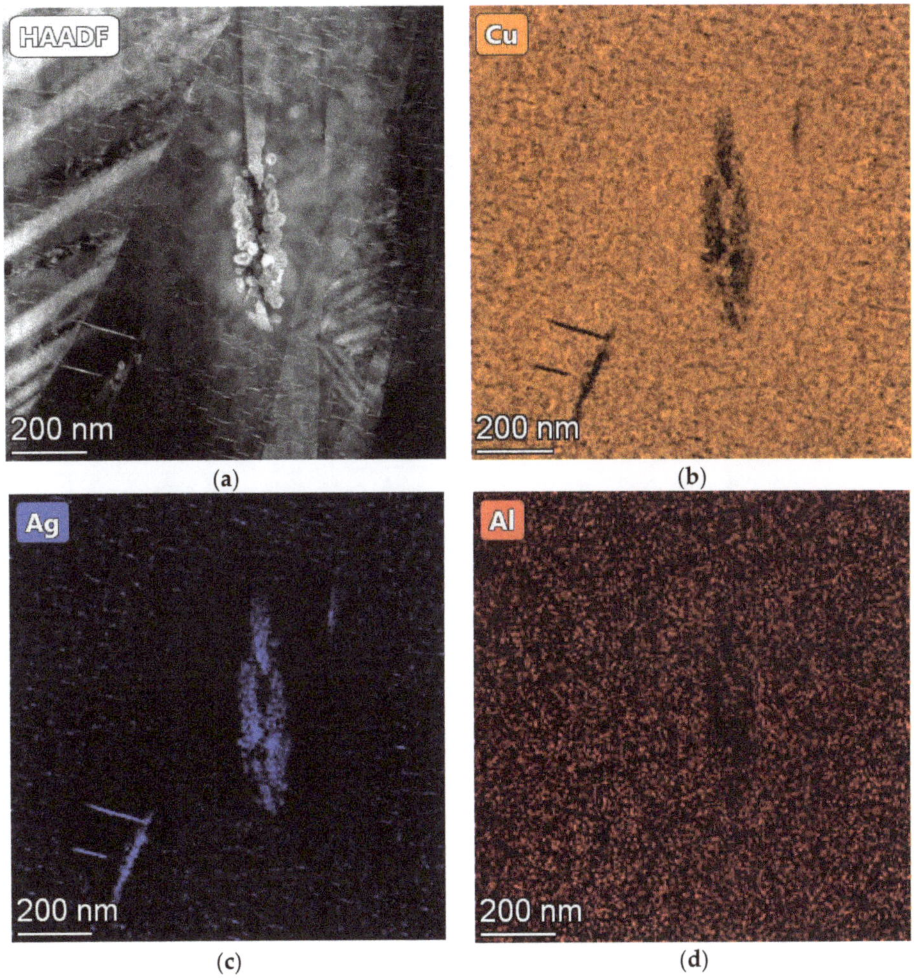

Figure 15. STEM HAADF and EDS maps of the Cu-10Al-7Ag SMA alloy in quenched condition. (**a**) STEM-HAADF image, (**b**) Cu element distribution map, (**c**) Ag element distribution map, (**d**) Al element distribution map.

The TEM images provided above reveal a predominance of the β' phase. However, upon careful analysis of the TEM lamellae, a small volume fraction amount of thermally induced 2H-martensitic Al_4Cu_9 ($\gamma_{1'}$) phase, with a plate-like morphology was observed (see Figure 17) in the quenched sample. It is worth noting that this observation is consistent with the results of the SEM analysis but was not confirmed by X-ray diffraction.

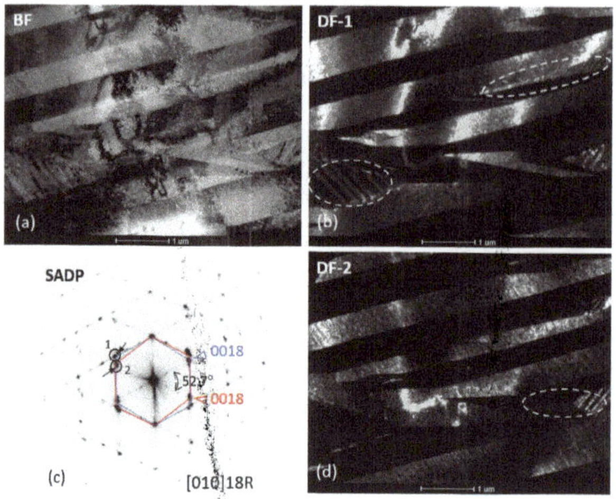

Figure 16. TEM images of the Cu-10Al-7Ag SMA alloy in quenched condition: (**a**) bright-field TEM image, (**b**,**d**) dark-field TEM images and (**c**) selected area diffraction pattern (SADP).

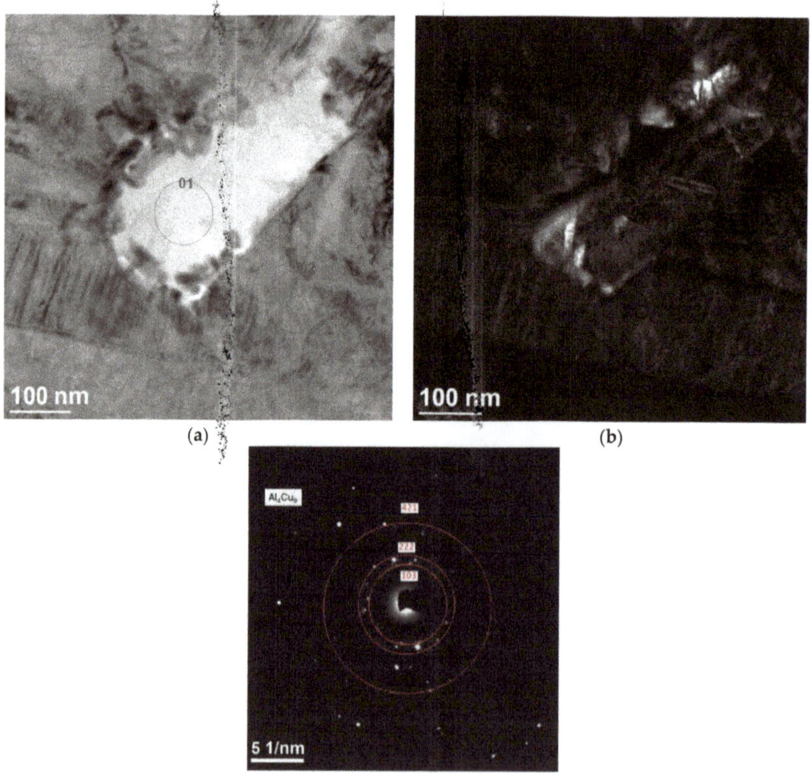

Figure 17. TEM images of the Cu-10Al-7Ag SMA alloy in quenched condition: (**a**) bright-field TEM image, (**b**) dark-field TEM image, and (**c**) selected area diffraction pattern (SADP) obtained from the 01-circle area shown in (**a**) indicating the presence of Al_4Cu_9 ($\gamma_{1'}$) phase. Note that around the Al_4Cu_9 ($\gamma_{1'}$), the $\beta_{1'}$ martensite plates are visible.

According to the literature, the martensitic transformation in Cu-Al alloys occurs during rapid quenching. By cooling at high cooling rates, β phase changes into β_1 parent austenite phase of $L2_1$ (a type of DO_3 superlattice) [34]. The $D0_3$ ordered parent β austenitic phase transforms then into modulated 2H or 18R martensite, or a combination of both, depending on various factors such as alloy composition, temperature, applied stress magnitude, and crystal orientation [35].

According to the study of Sedlák et al., within a certain range of aluminum and manganese content, both $\beta_{1'}$ and $\gamma_1{}'$ martensite phases coexist, which is in line with the findings presented in our article. The $\gamma_1{}'$ phase remains stable even upon heating back up to room temperature due to thermal hysteresis [36].

3.2. *Effect of Quenching on the Alloy Hardness*

The Cu-10Al-7Ag alloy in the as-cast state exhibits higher microhardness values compared to the quenched state (Table 3). This is typical behavior for Cu-SMA materials and is consistent with previous studies of SMAs. [11–14,37–41]

Table 3. Microhardness of Cu-Al-Ag SMA alloy.

Hardness	**HV**					
Cu-10Al-7Ag Alloy	**1**	**2**	**3**	**4**	**5**	**Average**
As-cast state	225.68	287.79	323.21	278.72	323.34	278.85
Quenched state	186.57	250.05	184.20	222.53	181.80	210.84

The slight variance in the hardness between the tested values, in both as-cast and quenched conditions can be ascribed to the formation of mechanically induced martensite in stressed areas, occurring under high indentation loads [42].

Studies have shown that the introduction of Ag particles into binary Cu-Al alloys increases their microhardness by influencing the nucleation rate and the activation energy of the eutectoid decomposition reaction [23,43].

The Ag additions improve the microhardness of Cu-Al alloys by hindering dislocation movement, refining the grain structure and influencing the phase behavior.

Nevertheless, the Ag particles are located within the Cu/Al metal matrix. Therefore, in this study, the presence of Ag particles, although it contributes to the strengthening of the matrix, is considered marginal compared to other sources of hardening, such as grain size, or does not change the microhardness values [13,23].

4. Conclusions

In this article, the microstructure of the Cu-10Al-7Ag SMA alloy in as-cast and quenched conditions was investigated throughout. The following conclusions can be drawn from this study:

In the Cu-10Al-7Ag SMA alloy, stable martensite formed even in the as-cast state without additional heat treatment, which indicates its potential for exhibiting shape memory properties.

The hardness of the quenched sample was found to be lower than the as-cast counterpart, which can be linked to the solutioning of Ag particles during the heat treatment.

Using the SEM, SEM-EBSD, XRD and TEM techniques, the presence of type 18R martensite was confirmed in both the as-cast and quenched samples, while the small volume fraction presence of type 2H martensite was noticed using the TEM technique in quenched sample.

Understanding the role of alloy composition, especially with the addition of Ag, in martensitic transformation is crucial for designing SMAs with tailored properties. While the presence of the martensite phase is necessary for the shape memory effect, it alone does not guarantee this behavior. Our findings reveal that not all SMAs in the martensite phase exhibit the shape memory effect, highlighting the critical role of additional factors such as composition, microstructure, and thermal history in influencing shape memory behavior. Further investigations could explore the influence of other alloying elements and optimize heat treatment processes to enhance the performance of Cu-Al-Ag alloys.

This research contributes to the broader field of SMAs, shedding light on the interplay between the material state, microstructure, and phase composition. The findings provide valuable insights for engineering applications where shape memory behavior is desired.

Author Contributions: Conceptualization, L.L. and T.H.G.; methodology, L.L., P.S. and T.H.G.; validation, L.L., W.S., P.S., W.M. and T.H.G.; formal analysis, L.L., P.S., W.M. and T.H.G.; investigation, L.L., W.S., P.S., W.M. and T.H.G.; resources, L.L., W.S., P.S. and W.M.; writing—original draft preparation, L.L. and P.S.; writing—review and editing, L.L., W.S., P.S., W.M. and T.H.G.; visualization, L.L., P.S. and T.H.G.; supervision, W.S., W.M. and T.H.G. All authors have read and agreed to the published version of the manuscript.

Funding: This work has been supported in part by University of Rijeka under the project: uniri-iskusni-tehnic-23-293.

Data Availability Statement: Data will be made available upon request.

Conflicts of Interest: The authors declare no conflicts of interest.

References

1. Mazzer, E.M.; da Silva, M.R.; Gargarella, P. Revisiting Cu-Based Shape Memory Alloys: Recent Developments and New Perspectives. *J. Mater. Res.* **2022**, *37*, 162–182. [CrossRef]
2. Silva, R.A.G.; Adorno Ii, A.T.; Carvalho Ii, T.M.; Ii, A.G.M.; Ii, C.M.A.S. Precipitation Reaction in Alpha-Cu-Al-Ag Alloys. *Rev. Matéria V* **2011**, *16*, 747–753. [CrossRef]
3. Grgurić, T.H.; Manasijević, D.; Kožuh, S.; Ivanić, I.; Anžel, I.; Kosec, B.; Bizjak, M.; Bajsić, E.G.; Balanović, L.; Gojić, M. The Effect of the Processing Parameters on the Martensitic Transformation of Cu-Al-Mn Shape Memory Alloy. *J. Alloys Compd.* **2018**, *765*, 664–676. [CrossRef]
4. Alaneme, K.K.; Okotete, E.A. Reconciling Viability and Cost-Effective Shape Memory Alloy Options—A Review of Copper and Iron Based Shape Memory Metallic Systems. *Eng. Sci. Technol. Int. J.* **2016**, *19*, 1582–1592. [CrossRef]
5. Ren, C.X.; Wang, Q.; Hou, J.P.; Zhang, Z.J.; Yang, H.J.; Zhang, Z.F. Exploring the Strength and Ductility Improvement of Cu–Al Alloys. *Mater. Sci. Eng. A* **2020**, *786*, 139441. [CrossRef]
6. Najah, S.; Al-Humairi, S. Cu-Based Shape Memory Alloys: Modified Structures and Their Related Properties. In *Recent Advancements in the Metallurgical Engineering and Electrodeposition*; Springer: Berlin/Heidelberg, Germany, 2020.
7. Dobrzański, L.A.; Dobrzański, L.B.; Dobrzańska-Danikiewicz, A.D.; Dobrzańska, J. Nitinol Type Alloys General Characteristics and Applications in Endodontics. *Processes* **2022**, *10*, 101. [CrossRef]
8. Silva, R.A.G.; Paganotti, A.; Adorno, A.T.; Santos, C.M.A.; Carvalho, T.M. Characteristics of the Cu-18.84 at.%Al-10.28 at.%Mn-1.57 at.%Ag Alloy after Slow Cooling from High Temperatures. *J. Therm. Anal. Calorim.* **2015**, *121*, 1233–1238. [CrossRef]
9. Mallik, U.S.; Sampath, V. Effect of Alloying on Microstructure and Shape Memory Characteristics of Cu–Al–Mn Shape Memory Alloys. *Mater. Sci. Eng. A* **2008**, *481–482*, 680–683. [CrossRef]
10. Santos, C.M.A.; Adorno, A.T.; Oda, N.Y.; Sales, B.O.; Silva, L.S.; Silva, R.A.G. Phase Transformations and Aging of the Cu72.9Al15.0Mn10.5Ag1.6 Alloy. *J. Alloys Compd.* **2016**, *685*, 587–592. [CrossRef]
11. Liverić, L.; Holjevac Grgurić, T.; Mandić, V.; Chulist, R. Influence of Manganese Content on Martensitic Transformation of Cu-Al-Mn-Ag Alloy. *Materials* **2023**, *16*, 5782. [CrossRef]
12. Liverić, L.; Grgurić, T.H.; Bajsić, E.G.; Kršulja, M. Effect of Silver Addition on Cu-Based Shape Memory Alloys. *Chem. Biochem. Eng. Q.* **2023**, *37*, 153–161. [CrossRef]
13. Silva, R.A.G.; Paganotti, A.; Gama, S.; Adorno, A.T.; Carvalho, T.M.; Santos, C.M.A. Investigation of Thermal, Mechanical and Magnetic Behaviors of the Cu-11%Al Alloy with Ag and Mn Additions. *Mater. Charact.* **2013**, *75*, 194–199. [CrossRef]
14. Santos, C.M.A.; Adorno, A.T.; Paganotti, A.; Silva, C.C.S.; Oliveira, A.B.; Silva, R.A.G. Phase Stability in the Cu-9 Wt%Al-10 Wt%Mn-3 Wt%Ag Alloy. *J. Phys. Chem. Solids* **2017**, *104*, 145–151. [CrossRef]
15. Pilz, C.B.; Matsumura, E.L.; Paganotti, A.; Cornejo, D.R.; Silva, R.A.G. Microstructure and Phase Stability of CuAlMnAgZr Multicomponent Alloys. *Mater. Chem. Phys.* **2020**, *241*, 122343. [CrossRef]
16. Velazquez, D.; Romero, R. Spinodal Decomposition and Martensitic Transformation in Cu–Al–Mn Shape Memory Alloy. *J. Therm. Anal. Calorim.* **2017**, *130*, 2007–2013. [CrossRef]
17. Zárubová, N.; Novák, V. Phase Stability of CuAlMn Shape Memory Alloys. *Mater. Sci. Eng. A* **2004**, *378*, 216–221. [CrossRef]
18. Qian, S.; Geng, Y.; Wang, Y.; Pillsbury, T.E.; Hada, Y.; Yamaguchi, Y.; Fujimoto, K.; Hwang, Y.; Radermacher, R.; Cui, J.; et al. Elastocaloric Effect in CuAlZn and CuAlMn Shape Memory Alloys under Compression. *Philos. Trans. R. Soc. A Math. Phys. Eng. Sci.* **2016**, *374*, 20150309. [CrossRef] [PubMed]

19. Suru, M.G.; Lohan, N.M.; Pricop, B.; Mihalache, E.; Mocanu, M.; Bujoreanu, L.G. Precipitation Effects on the Martensitic Transformation in a Cu-Al-Ni Shape Memory Alloy. *J. Mater. Eng. Perform.* **2016**, *25*, 1562–1569. [CrossRef]
20. Fernández, J.; Isalgue, A.; Franch, R. Effect of Thermal Cycling on CuAlAg Shape Memory Alloys. *Mater. Today Proc.* **2015**, *2*, S805–S808. [CrossRef]
21. Ferreira, R.O.; Silva, L.S.; Silva, R.A.G. Thermal Behavior of As-Annealed CuAlMnAgZr Alloys. *J. Therm. Anal. Calorim.* **2021**, *146*, 595–600. [CrossRef]
22. Adorno, A.T.; Benedetti, A.V.; Da Silva, R.A.G.; Blanco, M. Influence of the Al Content on the Phase Transformations in Cu-Al-Ag Alloys. *Eclet. Quim.* **2003**, *28*, 33–38. [CrossRef]
23. Seifollahzadeh, P.; Alizadeh, M.; Szabó, Á.; Gubicza, J.; El-Tahawy, M. Microstructure and Mechanical Behavior of Cu–Al–Ag Shape Memory Alloys Processed by Accumulative Roll Bonding and Subsequent Annealing. *Crystals* **2022**, *12*, 1167. [CrossRef]
24. Guilemany, J.M.; Fernandez, J.; Zhang, X.M. TEM Study on the Microstructure of Cu-Al-Ag Shape Memory Alloys. *Mater. Sci. Eng. A* **2006**, *438–440*, 726–729. [CrossRef]
25. Silva, R.A.G.; MacHado, E.S.; Adorno, A.T.; Magdalena, A.G.; Carvalho, T.M. Completeness of B-Phase Decomposition Reaction in Cu-Al-Ag Alloys. *J. Therm. Anal. Calorim.* **2012**, *109*, 927–931. [CrossRef]
26. Krishna, T.S.V.; Rao, D.S. Effect of Aluminium on Microstructure and Shape Memory Effect in Cu-Al-Ag-Mn Shape Memory Alloys. *Aust. J. Mech. Eng.* **2022**, 1–17. [CrossRef]
27. Manasijevićmanasijević, D.; Holjevac, T.; Grgurić, G.; Balanovićbalanović, L.; Stamenković, U.S.; Gorgievski, M.; Gojićgojić, M. Effect of Mn Content on the Microstructure and Phase Transformation Temperatures of the Cu-Al-Mn-Ag Shape Memory Alloys. *Chem. Biochem. Eng. Q.* **2020**, *58*, 293–299. [CrossRef]
28. Izadinia, M.; Dehghani, K. Structure and Properties of Nanostructured Cu-13.2Al-5.1Ni Shape Memory Alloy Produced by Melt Spinning. *Trans. Nonferrous Met. Soc. China* **2011**, *21*, 2037–2043. [CrossRef]
29. Gurau, G.; Gurau, C.; Fernandes, F.M.B.; Silva, R.J.C.; Marin, F. Using High Speed High Pressure Torsion for Cu–13Al–4Ni Shape Memory Alloy Processing. *Trans. Indian Inst. Met.* **2021**, *74*, 2459–2469. [CrossRef]
30. Shafeeq, M.M.; Gupta, G.K.; Malik, M.M.; Sampath, V.; Modi, O.P. Influence of Quenching Methods on Martensitic Transformation and Mechanical Properties of P/M Processed Cu–Al–Ni–Ti Shape Memory Alloys. *Powder Metall.* **2016**, *59*, 271–280. [CrossRef]
31. Yi, J.; Wan, L.; Shu, T.; Zhang, X.; Liu, F.; Cheng, G.J. Unleashing Multi-Scale Mechanical Enhancement in NiTi Shape Memory Alloy via Annular Intra-Laser Deposition with Homogenized Ti2Ni Nanoprecipitates. *Acta Mater* **2024**, *262*, 119418. [CrossRef]
32. Suresh, K.S.; Bhaumik, S.K.; Suwas, S. Effect of Thermal and Thermo-Mechanical Cycling on the Microstructure of Ni-Rich NiTi Shape Memory Alloys. *Mater. Lett.* **2013**, *99*, 150–153. [CrossRef]
33. Silva, R.A.G.; Paganotti, A.; Jabase, L.; Adorno, A.T.; Carvalho, T.M.; Santos, C.M.A. Ag-Rich Precipitates Formation in the Cu–11%Al–10%Mn–3%Ag Alloy. *J. Alloys Compd.* **2014**, *615*, S160–S162. [CrossRef]
34. Canbay, C.A.; Karaduman, O.; Ünlü, N.; Baiz, S.A.; Özkul, İ. Heat Treatment and Quenching Media Effects on the Thermodynamical, Thermoelastical and Structural Characteristics of a New Cu-Based Quaternary Shape Memory Alloy. *Compos. B Eng.* **2019**, *174*, 106940. [CrossRef]
35. Ge, Y.; Vronka, M.; Veřtát, P.; Karlik, M.; Hannula, S.P.; Heczko, O. Deformation Twinning with Different Twin-Boundary Mobility in 2H Martensite in Cu–Ni–Al Shape Memory Alloy. *Acta Mater.* **2022**, *226*, 117598. [CrossRef]
36. Sedlák, P.; Seiner, H.; Landa, M.; Novák, V.; Šittner, P.; Mañosa, L. Elastic Constants of Bcc Austenite and 2H Orthorhombic Martensite in CuAlNi Shape Memory Alloy. *Acta Mater.* **2005**, *53*, 3643–3661. [CrossRef]
37. Kožuh, S.; Gojić, M.; Ivanić, I.; Holjevac Grgurić, T.; Kosec, B.; Anžel, I. The Effect of Heat Treatment on the Microstructure and Mechanical Properties of Cu-Al-Mn Shape Memory Alloy. *Kem. U Ind.* **2018**, *67*, 11–17. [CrossRef]
38. Jain, A.K.; Hussain, S.; Kumar, P.; Pandey, A.; Dasgupta, R. Effect of Varying Al/Mn Ratio on Phase Transformation in Cu–Al–Mn Shape Memory Alloys. *Trans. Indian Inst. Met.* **2016**, *69*, 1289–1295. [CrossRef]
39. Silva, R.A.G.; Gama, S.; Paganotti, A.; Adorno, A.T.; Carvalho, T.M.; Santos, C.M.A. Effect of Ag Addition on Phase Transitions of the Cu-22.26 at.%Al-9.93 at.%Mn Alloy. *Thermochim. Acta* **2013**, *554*, 71–75. [CrossRef]
40. Gholami-Kermanshahi, M.; Wu, Y.Y.; Lange, G.; Chang, S.H. Effect of Alloying Elements (Nb, Ag) on the Damping Performance of Cu–Al–Mn Shape Memory Alloys. *J. Alloys Compd.* **2023**, *930*, 167438. [CrossRef]
41. Taha, A.S.; Hammad, F.H. Application of the Hall-Petch Relation to Microhardness Measurements on Al, Cu, Al-MD 105, and Al-Cu Alloys. *Phys. Status Solidi A* **1990**, *119*, 455–462. [CrossRef]
42. Ćorić -Mladen Franz, D. Properties of Thermally Treated CuZn27Al3 Shape Memory Alloy. *Stroj. Vestn. J. Mech. Eng.* **2009**, *55*, 623–632.
43. Adorno, A.T.; Silva, R.A.G. Isothermal Decomposition Kinetics in the Cu–9%Al–4%Ag Alloy. *J. Alloys Compd.* **2004**, *375*, 128–133. [CrossRef]

symmetry

MDPI

Article

Direct and Inverse Kinematics of a 3RRR Symmetric Planar Robot: An Alternative of Active Joints

Jordy Josue Martinez Cardona [1], Manuel Cardona [2], Jorge I. Canales-Verdial [3] and Jose Luis Ordoñez-Avila [1,*]

[1] Faculty of Engineering, Universidad Tecnológica Centroamericana (UNITEC), San Pedro Sula 21112, Honduras; josuecar108@unitec.edu
[2] Research Department, Universidad Don Bosco, Soyapango 1774, El Salvador; manuel.cardona@udb.edu.sv
[3] Mechatronics Department, Central New Mexico Community College, Albuquerque, NM 87106, USA; jcanalesverdia@cnm.edu
* Correspondence: jlordonez@unitec.edu

Abstract: Existing direct and inverse kinematic models of planar parallel robots assume that the robot's active joints are all at the bases. However, this approach becomes excessively complex when modeling a planar parallel robot in which the active joints are within one single kinematic chain. To address this problem, our article unveils an alternative for a 3RRR symmetric planar robot modeling technique for the derivation of the robot workspace and the analysis of its direct and inverse kinematics. The workspace was defined using a system of inequalities, and the direct and inverse kinematics models were generated using vectorial analysis and an optimized geometrical approach, respectively. The resulting models are systematically presented and validated. Two final model renditions are delivered supplying a thorough equation analysis and an applicability discussion based on the importance of the robot's mobile platform orientation. The advantages of this model are discussed in comparison to the traditional modeling approach: whereas conventional techniques require the solution of complex eighth-degree polynomials for the analysis of the active joint configuration of these robots, these models provide an efficient back-of-the-envelope analysis approach that requires the solution of a simple second-degree polynomial.

Keywords: parallel robot model; planar robot; active joint; workspace; model optimization

1. Introduction

Kinematic models supply a practical mathematical analysis toolbox to predict robotic operation performance. Robot kinematics can be defined as an analytical method that describes a robot's spatiotemporal motion by modeling the relationship between a robot's joint positions and its end-effector's position and orientation coordinates [1]. Direct kinematic models begin with the joint position variables and yield the resulting end-effector's position and orientation coordinates. Moreover, direct kinematics harness geometric methods to derive models by the implementation of homogeneous transformation matrices, the Denavit–Hartenberg algorithm, and the use of quaternions [2]. In contrast, inverse kinematics reverse the approach starting first with a desired end-effector position and orientation and then ascertains the required robot joint position values. Nonetheless, despite being a counterpart to direct kinetics, inverse kinematic models can also be derived via geometric methods; in this case, however, it is through means of homogeneous transformation matrices, kinematic decoupling, and Screws theory [2,3].

Robot kinematic models exist to comprehend a robot's motion and conduct an operation performance analysis. These models must facilitate the engineer's decision making and design the action-planning process. However, traditional modeling approaches for 3RRR parallel planar robots [3–12] do not provide simple models for robot configurations with active joints that are all placed within one single kinematic chain. The traditional models with the active joints located at the bases actually render the modeling of single kinematic chain active joint robots as a complex task that requires solving computing-intensive calculations. Therefore, the originality of this paper is the kinematic modeling of 3RRR parallel planar robots with single kinematic chain active joints. These kinematics are demystified by providing a simplified approach that reduces the computational task from a burdensome eighth-degree polynomial model to a simple back-of-the-envelope solvable mathematical model.

1.1. Planar Parallel Robots

An industrial robot is an automatically controlled, reprogrammable, and multipurpose manipulator with at least three independent movements (degrees of freedom); for use in manufacturing processes, this manipulator has the ability to move a variety of functional objects through previously configured movements for the purpose of performing different tasks [13]. Robots can be classified into open and closed kinematic chain robots. Open-chain robots, which have links that are connected in series, have become ubiquitous in industrial robotic applications. In the case of the 3RRR, which has three independent closed kinematic chains, the characteristics of symmetry allow for the mathematical modeling of one of the chains and generalization to the other two. The subset of planar robots in which the motion is restricted to a single 2D plane is called planar parallel robots. Considering the R label for revolute joints and P label for prismatic joints, the possible configurations for parallel planar robots are depicted in Figure 1. The PPP configuration is not feasible due to the lack of independence. The end effectors are situated at the centroid of the internal triangle of the robot, enabling it to perform tasks in soft robotics [14,15], traditional pick-and-place operations [16], and even medical applications [17].

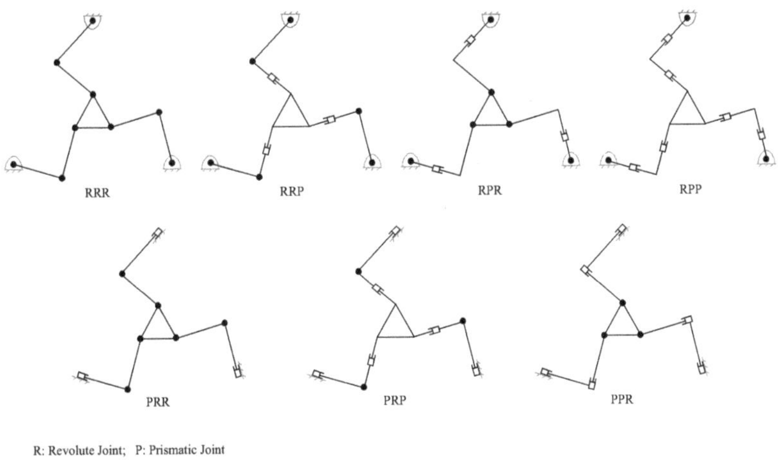

Figure 1. Possible parallel robots configurations.

The particular 3RRR symmetric planar robot studied in this work is shown in Figure 2. This robot contains six chain links that may be of different or equal sizes and a triangular chain that joins the three interdependent kinematic chains and acts as the tool center point. Moreover, this planar robot's movement is restricted to 2D-planar motion. Generally, parallel robots that do not exhibit singularities are isotonic. Therefore, using Grubler's formula, it is determined that this robot contains only three degrees of freedom: x, y, and ϕ, where x and y describe the robot's position with respect to the Cartesian plane and ϕ corresponds to the orientation angle that the end-effector forms with the horizontal. Given this robot configuration, kinematic models are required to describe the robot's operation motion.

The planar robots studied in this research apply the similarity law (six chains of the same size); this means the robot has geometic symmetry [18–21]. Also, symmetry is present in the Jacobian matrix condition index, which can be evidenced by graphically observing the distribution of the condition index throughout the workspace [3]. Additionally, the symmetry of parallel robots is present in the differential mathematical model, which is represented through Jacobian matrices [18].

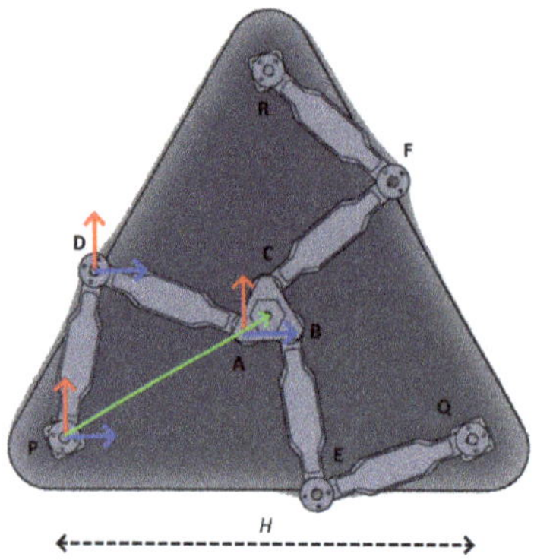

Figure 2. 3RRR symmetric planar robot geometry.

1.2. Existing 3RRR Symmetric Planar Robot Kinematic Models

Mathematical models exist in the literature for the analysis of articulated parallel robots. Table 1 shows an approach comparison among the modeling methods used by researchers, including the active joints for the mathematical modeling techniques, which were implemented to derive direct and inverse kinematics. Geometric models have the advantage of being more conventional, allowing the planar robot to move with speed and precision. The main disadvantage of geometric models is the computational expense due to the complexity of their kinematic equations. The advantage of screw theory is that it enables the development of a simple dynamic model with which controllers can be developed.

Table 1. Comparison of articulated 3RRR symmetric planar robot study (N/A: is not desribed in the paper).

References	Active Joints	Direct Kinematics	Inverse Kinematics
[3]	P, Q, R	N/A	Geometry Method
[4]	P, Q, R	N/A	Screws theory
[5]	P, Q, R	Geometry Method	Geometry Method
[6]	P, Q, R	N/A	State Variables
[7]	P, Q, R	N/A	Geometry Method
[8]	P, Q, R	Transformation matrix	Geometry Method
[9]	P, Q, R	Geometry Method	Geometry Method
[10]	P, Q, R	N/A	Geometry Method
[11]	P, Q, R	Numeric Method	Geometry Method
[12]	P, Q, R	N/A	Geometry Method
[22]	All Joints	Geometry Method	N/A

Among the referenced literature, it is observed that most articles base their analyses under the assumption that the active joints are all placed at the kinematic chain bases in P, Q, and R in Figure 2. The only exception is [22], which studies two additional configurations but limits its analysis to the singularities. It is also observed that the geometric method is the most common approach to derive both direct and inverse kinematic models. However, this approach yields eighth-degree polynomial-based models, which must be solved via the implementation of large-scale computational resources. Moreover, solving these conventional mathematical models generates eight possible solutions to configure the position angles of the robot's active joints, which also increments workspace robot singularities.

Non-articulated planar parallel robot modeling has also been actively studied in P, D, and A joints [23–28] and cable-driven planar robots [29–32]. Furthermore, the applications of these planar parallel robots can range from window-washing applications [29] to orthopedics [23,25] and even neurorehabilitation [33]. Hence, the models derived in this paper may be applied to all the referenced literature in this section and broaden the body of knowledge of the planar parallel robot.

1.3. 3RRR Symmetric Planar Robot Active Joints

Robotic joints are the mechanical connections between chain links. A robotic joint can be categorized according to its kinematic design into revolute (R), prismatic (P), screw (H), cylindrical (C) universal (U), spherical (S), or parallelogram (Pa) joints. Each of these joints can be further classified into active or passive joints. Active joints, notated by an underscored joint abbreviation, exert a controlled force to generate a robotic position shift. Active joints utilize motors to control robotic motion and orientation. In contrast, passive joints move through externally actuated forces transmitted by kinematic chains.

This study focuses on the 3RRR symmetric planar robot shown in Figure 2. Existing literature models [3–12] position active joints at the base joints, i.e., P, Q, and R, whereas passive joints are at points A, B, C, D, E, and F. This study proposes a mathematical model of a planar parallel robot where the active joints are rather located at joints P, D, and A, all of which belong to a single kinematic chain. The kinematic models derived in this work are then compared and contrasted to the conventional models in Table 1, and a discussion of the advantages of our derived models is provided.

2. Methods

This work aims to develop the forward kinematics and inverse kinematics of a 3RRR planar robot with its active joints in a kinematic chain. To achieve this, several concepts are developed, starting with the definition of the workspace and followed by orientation points, inverse kinematics, and finally forward kinematics, which validate all previous models as shown in Figure 3.

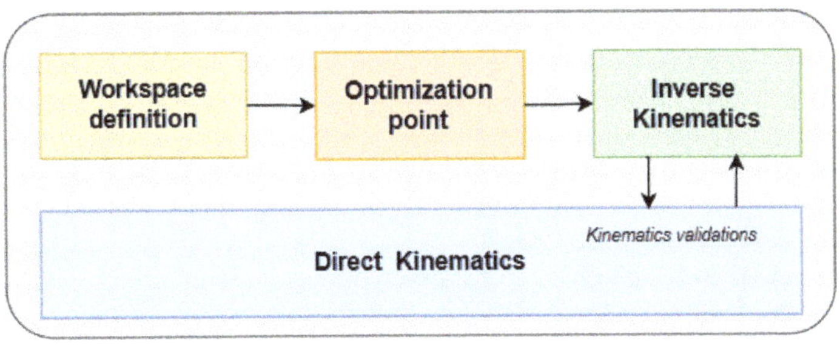

Figure 3. Phases of the kinematic model.

Also, some assumptions and methods are taken for the kinematic model derivation, including the following:

1. The location of the robot's active joints was taken at joints P, D, and A, as shown in Figure 2.
2. Joints Q, E, B, C, F, and R in Figure 2 are all passive joints.
3. The mathematical methods implemented for model derivation are the law of cosines, trigonometric properties, geometric relations, the Pythagoras theorem, derivatives, vectors, and matrices.

4. The mathematical models obtained are the workspace, the direct kinematics models, and the inverse kinematics models.
5. The workspace is defined using a system of inequalities provided by the constraints from the lengths of the fully extended kinematic chains.
6. The derivation of the direct kinematics model is approached via vectorial analysis.
7. The inverse kinematic model is derived through the geometric method and optimization techniques.
8. Model validation is achieved using MATLAB-based software corroboration.

The following parameter definitions were also conducted for model derivation:

Take H as the distance between the base joints in the planar robot, i.e., the length of segments $\bar{PQ}$, $\bar{QR}$, and $\bar{PR}$. Moreover, to reduce the robot singularities, all chains are assumed to be of equal length l and the end-effector is assumed to have the fixed dimensions of an equilateral triangle platform [3]. Hence, the dimensions of the chains and the end-effector sides can be calculated by (1) and (2):

$$h = \frac{1}{10}H \tag{1}$$

$$l = \frac{2}{5}H \tag{2}$$

where l is the length of each of the robot's six chains and h is the length of the three sides of the mobile platform's end-effector [3]. Furthermore, the distance from vertex point A to the centroid point G of the end-effector (as shown in Figure 4) must also be considered.

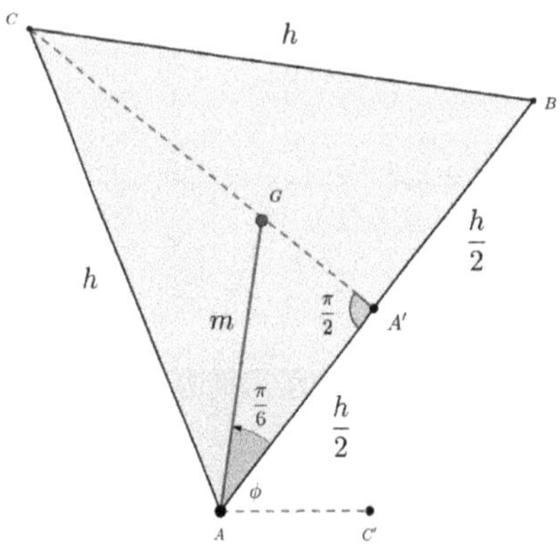

Figure 4. Planar parallel robot's mobile platform end-effector [32].

Given that point G is the centroid of the equilateral-triangle-shaped end-effector, ΔABC, let segment $\bar{AG}$ be equal to m, as shown in Figure 4. Furthermore, when the segment $\bar{CG}$ is extended, it bisects segment $\bar{AB}$ at point A'. Moreover, given that the orthocenter, incenter, and centroid of an equilateral triangle are all the same point, then $\bar{AA'} = \bar{A'B} = \frac{h}{2}$, $\angle CA'A = \frac{\pi}{2}$, and $\angle GAA' = \frac{\pi}{6}$. Furthermore, since $\Delta AGA'$ is a right triangle, the definition of the cosine trigonometric function (3) is used to find m in (4):

$$\cos\left(\frac{\pi}{6}\right) = \frac{\bar{AA'}}{\bar{AG}} \tag{3}$$

Given that $\cos\left(\frac{\pi}{6}\right) = \frac{\sqrt{3}}{2}$, $\bar{AA'} = \frac{h}{2}$, and $\bar{AG} = m$, isolating m leads to

$$m = \frac{h}{\sqrt{3}} \tag{4}$$

Then, using Equation (1) and rationalizing, m becomes a function of H:

$$m = \frac{1}{10}H * \frac{1}{\sqrt{3}} \tag{5}$$

$$m = \frac{1}{10\sqrt{3}}H \tag{6}$$

$$m = \frac{\sqrt{3}}{30}H \tag{7}$$

This method is applicable to symmetric 3RRR planar robots, which use parameters H, h, and l as design parameters, where l and h depend on H at the moment of designing the robot. Additionally, the active joints are located in one of the robot's kinematic chains. To perform the calculations, MATLAB was used, in which functions were developed for both the direct kinematics and the inverse kinematics of the robot.

3. Results

3.1. Workspace Definition

Taking each of the robot's kinematic chains with a specific position in a Cartesian reference plane [34], let point G with coordinates (P_x, P_y) be the final position of the end-effector's centroid. Then, to derive the robot's workspace, consider the delimitations given by the kinematic chain constraints.

3.1.1. First Kinematic Chain Workspace Constraints

The first kinematic chain, composed of links $\bar{PD}$, $\bar{DA}$, and $\bar{AG}$, has active joints at all three of its articulations P, D, and A. Moreover, its base articulation P is set at position $(0,0)$. Then, letting G, with coordinates (P_x, P_y), be all the possible positions that the end-effector's centroid can reach, the inequality shown in expression (8) is derived. This expression acts as the first workspace constraint, where the end-effector's position is limited by the length of the fully extended first kinematic chain, and it is shown in Figure 5:

$$(P_x)^2 + (P_y)^2 < (2l + m)^2 \tag{8}$$

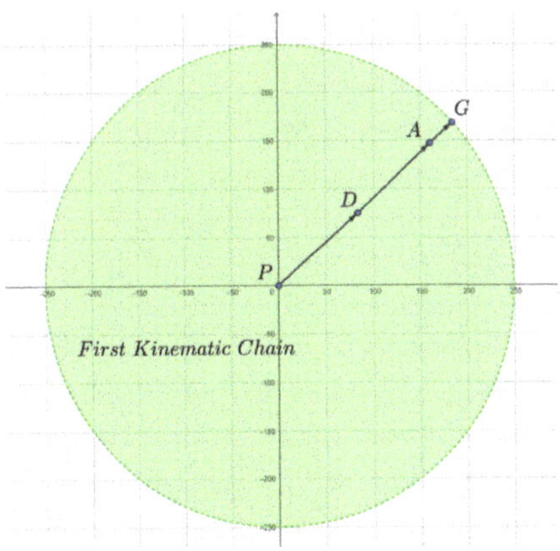

Figure 5. First kinematic chain workspace constraint.

Note that the workspace has been defined using less-than inequalities to avoid working mode misconfigurations. Whereas the conventional planar parallel robot kinematic models provide eight solutions to this robot configuration, only two of them have no singularities in their workspace [3]. In this research, by making the workspace definitions less-than inequatities, we prevent the robot from reaching potential working modes that, due to inertia, could result in a misconfiguration of the position of the kinematic chains. Hence, the models presented in this work yield two viable non-singular positions within the workspace: (1) a position when all kinematic chains are oriented elbow-down and (2) when they are oriented elbow-up.

3.1.2. Second Kinematic Chain Workspace Constraints

The second kinematic chain, composed of links $\bar{RF}$, $\bar{FC}$, and $\bar{CG}$, has a passive joint in all three of its articulations R, F, and C. Moreover, this chain's base joint, R, has a fixed position given by coordinates$\left(\frac{1}{2}H, \frac{\sqrt{3}}{2}H\right)$, as shown in Figure 6.

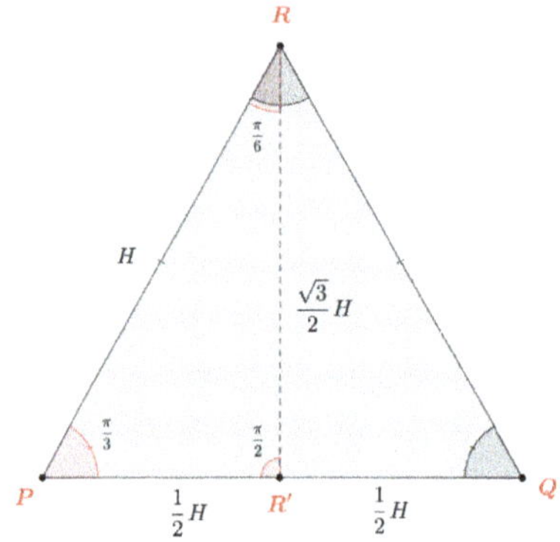

Figure 6. Planar parallel robot base joint positions.

Hence, the inequality shown in expression (9) further delimits the robot's workspace due to the second kinematic chain's length. As with the first kinematic chain, the distance from $\bar{CG}$ must be taken into consideration. However, given that ΔABC is an equilateral triangle, $\bar{AG} = \bar{BG} = \bar{CG} = m$; hence, (9) provides the second workspace constraint:

$$\left(P_x - \frac{H}{2}\right)^2 + \left(P_y - \frac{\sqrt{3}}{2}H\right)^2 < (2l + m)^2 \tag{9}$$

This expression further reduces the robot's workspace as shown in Figure 7 because the allowed final end-effector position, G, must satisfy both (8) and (9).

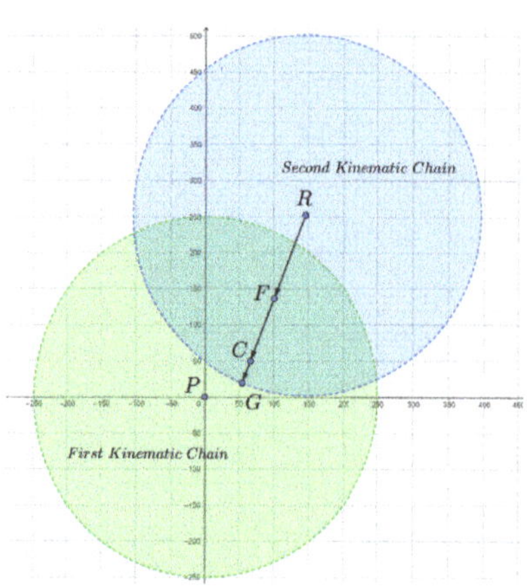

Figure 7. Robot's workspace constraints due to both the first and second kinematic chain lengths.

3.1.3. Third Kinematic Chain Workspace Constraints

The third kinematic chain, composed of links $\vec{QE}$, $\vec{EB}$, and $\vec{BG}$, has passive joints in all three of its articulations Q, E, and B. Moreover, its base joint Q has a fixed position given by coordinates $(H, 0)$, as shown in Figure 6. Hence, analogously to the workspace constraint expressions derived for the first two kinematic chains, the end-effector's final position G is delimited by expression (10):

$$(P_x - H)^2 + P_y^2 < (2l + m)^2 \tag{10}$$

Therefore, the end-effector's final position G, with coordinates (P_x, P_y), must satisfy all three inequality expressions (8)–(10), as shown in Figure 8.

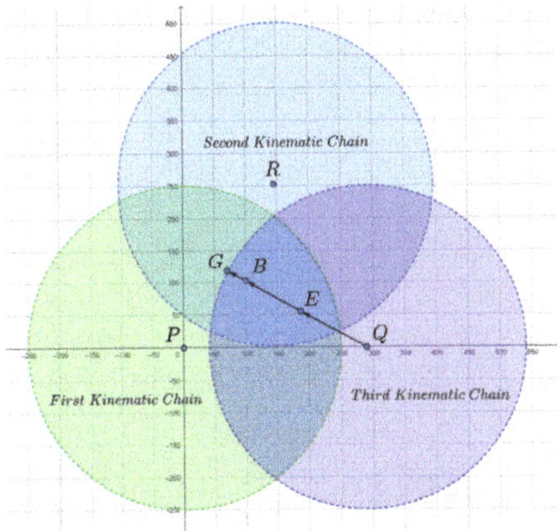

Figure 8. Robot's workspace constraints due to all three kinematic chain constraints.

Thus, this workspace description method offers a novel and straightforward approach. This system of inequalities can be visualized as the common area of three symmetric circles. These inequality systems will be utilized for both the direct and inverse kinematics of the study in the next two sections.

3.2. *Direct Kinematics Model*

The reference system is located at the active joint A according to Figure 9. Hence, P_x and P_y will be referenced there. To obtain the direct kinematics, a vector addition is used, where the links $\vec{AB}$, $\vec{BC}$, and $\vec{CG}$ will determine the vectors. The result of the vector $\vec{AG}$ is given in (11):

$$\vec{AB} + \vec{BC} + \vec{CG} = \vec{AG} \tag{11}$$

In Figure 9, each vector can be broken down into its components, as shown in (12)–(14):

$$\vec{AB} = (l\cos\theta_1)\hat{i} + (l\sin\theta_1)\hat{j} \tag{12}$$

$$\vec{BC} = [l\cos(\theta_1 + \theta_2)]\hat{i} + (l\sin(\theta_1 + \theta_2))\hat{j} \tag{13}$$

$$\vec{CG} = \left[mcos\left(\theta_1 + \theta_2 + \theta_3 + \frac{\pi}{6}\right)\right]\hat{i} + \left[m\sin\left(\theta_1 + \theta_2 + \theta_3 + \frac{\pi}{6}\right)\right]\hat{j} \tag{14}$$

If $\vec{AG} = P_x\hat{i} + P_y\hat{j}$ is the vector sum of (11), then the following Equations (15) and (16) are obtained for each component of the vector $\vec{AG}$:

$$P_x = l\cos\theta_1 + l\cos(\theta_1 + \theta_2) + mcos\left(\theta_1 + \theta_2 + \theta_3 + \frac{\pi}{6}\right) \tag{15}$$

$$P_y = l\sin\theta_1 + l\sin(\theta_1 + \theta_2) + m\sin\left(\theta_1 + \theta_2 + \theta_3 + \frac{\pi}{6}\right) \tag{16}$$

To obtain the orientation of the triangular chain ΔCDE, the angle $\angle ECJ$ is analyzed. From here, it is deduced that

$$\phi = \theta_1 + \theta_2 + \theta_3 \tag{17}$$

In conclusion, Equations (15)–(17) describe the 3-DOF that the 3RRR symmetric planar robot can have.

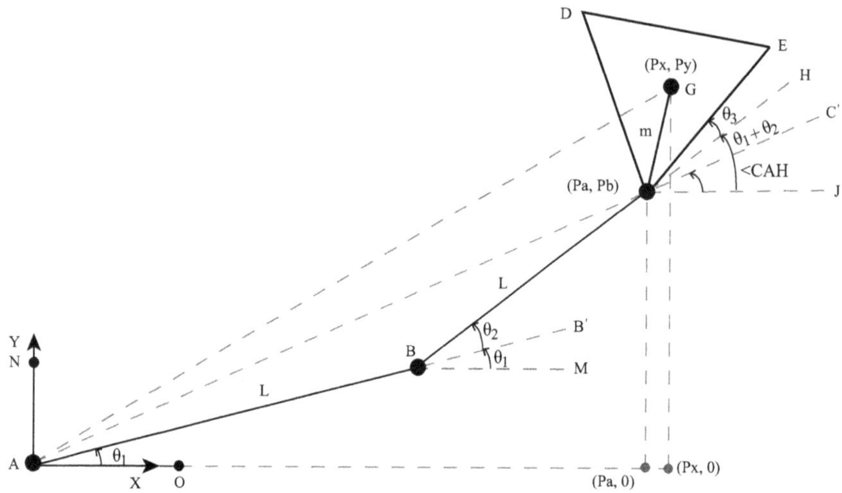

Figure 9. Kinematic chain geometry with active joints in A, B, and C.

3.3. Inverse Kinematics Model

The geometric method will be utilized to determine the inverse kinematics. To achieve this, we analyze Figure 9, depicting the motorized kinematic chain of the robot.

Point C with coordinates (P_a, P_b) is defined, which will help find the angles of the active joints θ_1, θ_2, and θ_3 in terms of these coordinates and the final position G with coordinates (P_x, P_y).

The first step is to analyze the chains AB and BC.

Solution for θ_1: Based on [35], if we analyze the angles forming $\angle NAO$, it can be expressed as (18):

$$\angle NAO = \angle NAC + \angle CAB + \theta_1 \tag{18}$$

Since $\angle NAO = \frac{\pi}{2}$ is an angle of the reference coordinate system, if θ_1 is isolated, then (19) occurs:

$$\theta_1 = \frac{\pi}{2} - \angle NAC - \angle CAB \tag{19}$$

To find angle $\angle NAC$, it is observed that the segments NA and CH are parallel since the sum of angles $\angle NAO$ and $\angle AHC$ equals π. Therefore, $\angle NAC = \angle ACH$. So, $\angle ACH$ is obtained by (20):

$$\cos(\angle ACH) = \frac{CH}{CA} \tag{20}$$

If $CH = P_b$, $CA = \sqrt{P_a^2 + P_b^2}$, and $\angle NAC = \angle ACH$, if $\angle ACH$ is isolated and substitutes the other variables, the result is (21):

$$\angle NAC = \arccos\left(\frac{P_b}{\sqrt{P_a^2 + P_b^2}}\right) \tag{21}$$

To obtain $\angle CAB$, the triangle ΔCAB is analyzed; using the cosine law on this triangle, the following is obtained (22):

$$CD^2 = CA^2 + AB^2 - 2(CA)(AB)\cos(\angle CAB) \tag{22}$$

If $CB = AB = l$ and $CA = \sqrt{P_a^2 + P_b^2}$, then (23):

$$l^2 = P_a^2 + P_b^2 + l^2 - 2l\sqrt{P_a^2 + P_b^2}\cos(\angle CAB) \tag{23}$$

If $\angle CAB$ is isolated, (24):

$$\angle CAB = \arccos\left(\frac{P_a^2 + P_b^2}{2l\sqrt{P_a^2 + P_b^2}}\right) \tag{24}$$

Therefore, replacing Equations (21) and (24) in (19), θ_1 is obtained in terms of known variables using (25):

$$\theta_1 = \frac{\pi}{2} - \arccos\left(\frac{P_b}{\sqrt{P_a^2 + P_b^2}}\right) - \arccos\left(\frac{P_a^2 + P_b^2}{2l\sqrt{P_a^2 + P_b^2}}\right) \tag{25}$$

Solution for θ_2 : The triangle ΔABC is analyzed [35], from which it is deduced that in (26),

$$\bar{AC}^2 = \bar{AB}^2 + \bar{BC}^2 - 2(\bar{AB})(\bar{BC})\cos(\angle ABC) \tag{26}$$

If $\bar{CA} = \sqrt{P_a^2 + P_b^2}$ and $\bar{CB} = \bar{AB} = l$, then (27):

$$P_a^2 + P_b^2 = l^2 + l^2 - 2(l)(l)\cos(\angle ABC) \tag{27}$$

Isolating for $\cos(\angle ABC)$ results in (28):

$$\cos(\angle ABC) = -\frac{P_a^2 + P_b^2 - 2l^2}{2l^2} \tag{28}$$

If $\angle ABC = 180 - \theta_2$, then (29):

$$\cos(180 - \theta_2) = -\frac{P_a^2 + P_b^2 - 2l^2}{2l^2} \tag{29}$$

By the trigonometric property $\cos(180 - \theta) = -\cos(\theta)$, $\cos(180 - \theta_2) = -\cos\theta_2$; therefore, (30).

$$-\cos(\theta_2) = -\frac{P_a^2 + P_b^2 - 2l^2}{2l^2} \tag{30}$$

Isolating for θ_2, it is deduced that (31):

$$\theta_2 = \arccos\left(\frac{P_a^2 + P_b^2 - 2l^2}{2l^2}\right) \tag{31}$$

Solution for θ_3: The triangle ΔACG is analyzed, and it is deduced that (32):

$$\bar{AG}^2 = \bar{AC}^2 + \bar{CG}^2 - 2(\bar{AC})(\bar{CG})\cos(\angle ACG) \tag{32}$$

If $\bar{AG} = \sqrt{P_x^2 + P_y^2}$, $\bar{CA} = \sqrt{P_a^2 + P_b^2}$ and $\bar{CG} = m$, then (33):

$$P_x^2 + P_y^2 = P_a^2 + P_b^2 + m^2 - 2m\sqrt{P_a^2 + P_b^2}\cos(\angle ACG) \tag{33}$$

Isolating for $\cos(\angle ACG)$ results in (34):

$$\cos(\angle ACG) = -\frac{P_x^2 + P_y^2 - P_a^2 - P_b^2 - m^2}{2m\sqrt{P_a^2 + P_b^2}} \tag{34}$$

If $\angle ACG = 180 - \angle GCC'$, then (35):

$$\cos(180 - \angle GCC') = -\frac{P_x^2 + P_y^2 - P_a^2 - P_b^2 - m^2}{2m\sqrt{P_a^2 + P_b^2}} \tag{35}$$

Since $\cos(180 - \angle GCC') = -\cos\angle GCC'$, then (36):

$$\cos(\angle GCC') = \frac{P_x^2 + P_y^2 - P_a^2 - P_b^2 - m^2}{2m\sqrt{P_a^2 + P_b^2}} \tag{36}$$

By trigonometry properties, it is known that $\tan^2(\angle GCC') + 1 = \sec^2(\angle GCC')$. Isolating for $\angle GCC'$, it is deduced that $\angle GCC' = \arctan(\pm\sqrt{\sec^2(\angle GCC') - 1})$. If $\sec(\angle GCC') = \frac{1}{\cos(\angle GCC')}$, using Equation (36), (37) is deduced:

$$\angle GCC' = \arctan\left(\pm\sqrt{\left(\frac{2m\sqrt{P_a^2 + P_b^2}}{P_x^2 + P_y^2 - P_a^2 - P_b^2 - m^2}\right)^2 - 1}\right) \tag{37}$$

This result generated two solutions for $\angle GCC'$. It would be necessary to validate which of the two solutions is the correct one. For this, the answers will be validated using the direct kinematics of the robot proposed in this research.

If the angles between the angle $\angle GCJ$ are analyzed, (38):

$$\theta_1 + \theta_2 + \theta_3 + \frac{\pi}{6} = \angle GCC' + \angle C'CJ \tag{38}$$

If $\angle C'CJ = \angle CAH$, and if θ_3 is isolated, then (39):

$$\theta_3 = \angle GCC' + \angle CAH - \theta_1 - \theta_2 - \frac{\pi}{6} \tag{39}$$

If $\angle CAH = \arctan\left(\frac{P_b}{P_a}\right)$, substituting in Equation (39), it is concluded that (40):

$$\theta_3 = \arctan\left(\frac{P_b}{P_a}\right) + \arctan\left(\pm\sqrt{\left(\frac{2m\sqrt{P_a^2 + P_b^2}}{P_x^2 + P_y^2 - P_a^2 - P_b^2 - m^2}\right)^2 - 1}\right) - \theta_1 - \theta_2 - \frac{\pi}{6} \tag{40}$$

Therefore, Equations (25), (31) and (40) describe the inverse kinematics of the robot. To reach position (P_x, P_y), the inverse kinematics will give two possible results, which must be validated using the direct kinematics to identify the correct angle configuration θ_1, θ_2, and θ_3.

Solution for P_a and P_b

Figure 10 represents the geometry of a mobile platform; all the parameters P_a and P_b are utilized. These parameters depend on the initial parameters P_x and P_y, selected from a mathematical equation. Additionally, P_a and P_b could be contingent upon the orientation angle ϕ, though it is essential to ascertain if the orientation of the robot truly impacts its application. An analysis will be conducted to determine the values of P_a and P_b in two scenarios: one where the orientation ϕ of the triangle ΔCDE is significant and another where it is not. In the latter case, the solutions will be based on a criterion involving the minimum distance from the active joints 1 and 3.

Solution for P_a and P_b where the orientation ϕ matters.

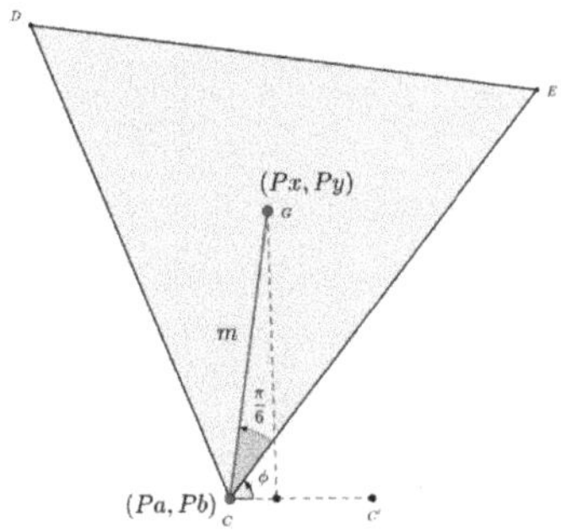

Figure 10. Mobile platform geometry considering ϕ.

It is assumed that P_x and P_y belong to the workspace. From Figure 10, if $\angle GCC' = \frac{\pi}{6} + \phi$, then the following Equations (41) and (42) can be deduced:

$$P_x - P_a = m\cos\left(\frac{\pi}{6} + \phi\right) \tag{41}$$

$$P_y - P_b = m\cos\left(\frac{\pi}{6} + \phi\right) \tag{42}$$

Isolating for P_a in (41) and P_b in Equation (42), the required values would be found to have the complete inverse kinematics considering the orientation ϕ (43) and (44):

$$P_a = P_x - m\cos\left(\frac{\pi}{6} + \phi\right) \tag{43}$$

$$P_b = P_y - m\sin\left(\frac{\pi}{6} + \phi\right) \tag{44}$$

Solution for P_a and P_b where the orientation ϕ does not matter:

Given that (P_x, P_y) is the desired position of the robot, in this case, ϕ is not considered in this analysis and the point (P_a, P_b) is taken using the following criterion: (P_a, P_b) is the minimum possible distance from active joint 1; in other words, from point A according to Figure 9. From this criterion, the problem needs a solution using optimization. Also, by geometry, (P_a, P_b) has a workspace given by the two chains $\bar{AB}$ and $\bar{BC}$ that can be defined by the following inequality (45):

$$P_a^2 + P_b^2 < 4l^2 \tag{45}$$

For this problem, a function to optimize and a restriction are needed. The optimization function is the distance between the active joints 1 and 3, which will be denoted by r. Putting it into an equation results in (46) as shown in Figure 11:

$$r^2 = P_a^2 + P_b^2 \tag{46}$$

The only restriction that this case has is the distance between point G, where the end-effector is located, and point C, where the active joint is located, so (46):

$$m^2 = (P_x - P_a)^2 + (P_y - P_b)^2 \tag{47}$$

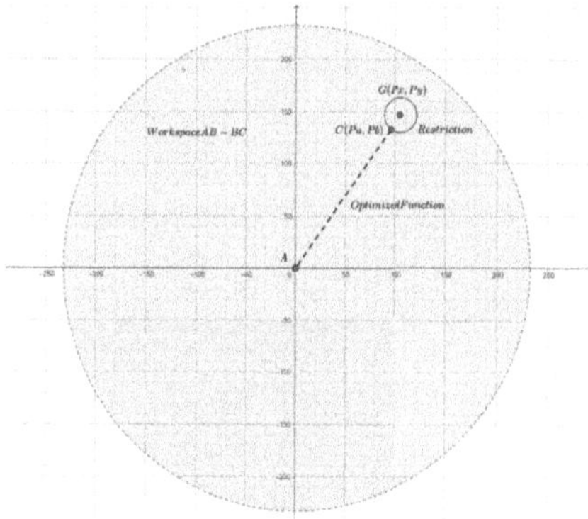

Figure 11. Optimization of P_a and P_b.

From here, P_b is isolated, taking the positive square root of the process, and replaced in (46). So, the function to optimize is (48):

$$r^2 = \left[P_y - \sqrt{m^2 - (P_x - P_a)^2}\right]^2 + P_a^2 \tag{48}$$

The derivative of r with respect to P_a is taken, and (49) is deduced:

$$\frac{dr}{dP_a} = \frac{P_y P_a + P_x\sqrt{m^2 - (P_x - P_a)^2} - P_y P_x}{\sqrt{m^2 - (P_x - P_a)^2}\sqrt{P_a^2 + \left(P_y - \sqrt{m^2 - (P_x - P_a)^2}\right)^2}} \tag{49}$$

Let $\frac{dr}{dP_a} = 0$, and the value of P_a that satisfies the equation is found by (50):

$$P_y P_a + P_x\sqrt{m^2 - (P_x - P_a)^2} - P_y P_x = 0 \tag{50}$$

By doing arithmetic operations, it follows that (51) occurs:

$$(P_x^2 + P_y^2)P_a^2 + (-2P_x P_y^2 - 2P_x^3)P_a + (P_x^2 P_y^2 - P_x^2 m^2 + P_x^4) = 0 \tag{51}$$

In order to work in a more organized manner, it is stated that $a = P_x^2 + P_y^2$, $b = -2P_x P_y^2 - 2P_x^3$, and $c = P_x^2 P_y^2 - P_x^2 m^2 + P_x^4$, so (52):

$$aP_a^2 + bP_a + c = 0 \tag{52}$$

By the quadratic equation, the roots of the equation are found by (53):

$$P_a = \frac{-b \pm \sqrt{b^2 - 4ac}}{2a} \tag{53}$$

Since there are two results of P_a, there must be two points: (P_{a1}, P_{b1}) and (P_{a2}, P_{b2}). The point P_b is determined with (54):

$$P_b = P_y - \sqrt{m^2 - (P_x - P_a)^2} \tag{54}$$

When isolating for P_b, the positive square root was taken. Now, the negative square will be taken and the same process will be performed, so P_b is now as shown in (55):

$$P_b = P_y + \sqrt{m^2 - (P_x - P_a)^2} \tag{55}$$

Consequently, the new function to optimize is the following (56):

$$r^2 = \left[P_y + \sqrt{m^2 - (P_x - P_a)^2}\right]^2 + P_a^2 \tag{56}$$

If the derivative of $\frac{dr}{dP_a}$ of Equation (56) is calculated, the derivative equals zero, and P_b is isolated, the solution for P_a is exactly the same as Equation (51). Since there are two solutions for P_a in that case, there are two more solutions: (P_{a3}, P_{b3}) and (P_{a4}, P_{b4}).

In brief, this proposed criterion will give four different answers obtained, as shown in Table 2.

Table 2. Solutions for optimization of P_a and P_b.

Solution	P_a	P_b
#1	$\frac{-b+\sqrt{b^2-4ac}}{2a}$	$P_y - \sqrt{m^2 - (P_x - P_a)^2}$
#2	$\frac{-b-\sqrt{b^2-4ac}}{2a}$	$P_y - \sqrt{m^2 - (P_x - P_a)^2}$
#3	$\frac{-b+\sqrt{b^2-4ac}}{2a}$	$P_y + \sqrt{m^2 - (P_x - P_a)^2}$
#4	$\frac{-b-\sqrt{b^2-4ac}}{2a}$	$P_y + \sqrt{m^2 - (P_x - P_a)^2}$

To find out which of the four possible points to use, (46) is used to determine which generates the smallest distance r. Realizing a rigorous analysis of finding the expression that generates the smallest result, it is found that solution 2 generates the smallest distance r. Then, to generate the smallest distance between joint 1 and joint 3, (57), and (58) must be used:

$$P_a = \frac{-b - \sqrt{b^2 - 4ac}}{2a} \tag{57}$$

$$P_b = P_y - \sqrt{m^2 - (P_x - P_a)^2} \tag{58}$$

The summary of the most relevant equations for the development of functions in Matlab is shown in Figure 12. In the inverse kinematics, an equation from the forward kinematics is used to determine the correct value of θ_3. And, $a = P_x^2 + P_y^2$, $b = -2P_xP_y^2 - 2P_x^3$, and $c = P_x^2P_y^2 - P_x^2m^2 + P_x^4$.

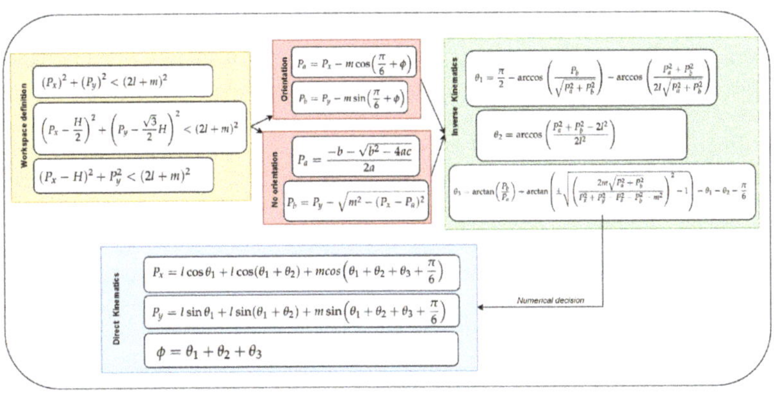

Figure 12. Results of the kinematic model when the orientation is needed.

4. Model Validation

4.1. Kinematics Model Implementations

Prior to employing direct and inverse kinematics, it is imperative that the desired point (P_x, P_y) of the 3RRR symmetric planar robot belongs to the workspace defined by the inequalities (8)–(10).

The inverse kinematics model, in the two approaches explained (considering the orientation ϕ of the mobile platform ΔCDE), will invariably produce two possible solutions for the angle configurations necessary to reach the desired point (P_x, P_y). The correct angle selection is determined by inputting these angle values into the direct kinematics to verify if the data obtained correspond to the primary values inputted in the inverse kinematics. An algorithm is employed to ascertain the correct angles.

For example, let $H = 136$, $P_x = 136$, $P_y = 16$, and $\phi = \frac{\pi}{4}$. In this scenario, the values of P_x and P_y satisfy the three inequalities of the workspace. Therefore, it is feasible for the robot to reach that location. Subsequently, if Equations (1), (2) and (4) are obtained using the aforementioned parameters, then $l = 92$, $m = 13.28$, and $h = 23$, respectively. With this data, the equations describing the inverse kinematics are employed. If Equations (25), (31) and (40) are utilized,

- If the orientation of the mobile platform ΔCDE is considered, then, employing Equations (43) and (44) to determine P_a and P_b, the two solutions in the form $(\theta_1, \theta_2, \theta_3)$ for this case would be $(-42.52°, 87.78°, -0.26°)$ and $(-42.52°, 87.78°, -147.52°)$, respectively.

There are two solutions, and to determine the correct one, the solutions are inputted into the direct kinematics to obtain the resulting values. For the first solution $(\theta_1, \theta_2, \theta_3) = (-42.52°, 87.78°, -0.26°)$, the direct kinematics yield $P_x = 136$, $P_y = 16$, and $\phi = 45°$. For the second solution $(-42.52°, 87.78°, -147.52°)$, the direct kinematics yield $P_x = 136.61$, $P_y = -9.47$, and $\phi = -102.26°$. Therefore, the correct configuration of angles is the first one.

- If the orientation of the mobile platform ΔCDE is not considered, then utilizing Equations (57) and (58) to determine P_a and P_b, respectively, yields the two identical solutions in the form $(\theta_1, \theta_2, \theta_3)$, which are both $(-41.52°, 95.55°, -77.77°)$.

If the solution is validated in the direct kinematics, then the result would be $P_x = 136$, $P_y = 16$, and $\phi = -23.29°$. Therefore, P_x and P_y are exactly as desired, but ϕ is different. However, this discrepancy is inconsequential as the angle ϕ is deemed unconsidered for the application.

To verify the proposed equations in the inverse kinematics and direct kinematics models, functions were implemented in MATLAB. These functions were used to ensure that the data from the inverse kinematics corresponded to those from the direct kinematics and vice versa. The conducted tests are presented in Tables 3 and 4, where the parameter H is varied across different values, thus causing the values of h, m, and l to adjust accordingly. This illustrates how the direct and inverse equations are utilized to showcase the values of certain desired points and their corresponding angles required to reach those positions. Table 3 presents the results using the model where the orientation ϕ was considered, and Table 4 demonstrates the outcomes using the model where ϕ was not considered and an optimization process was executed.

Table 3. θ_1, θ_2, and θ_3 obtained from certain parameters, and ϕ is considered.

H	P_x	P_y	ϕ	θ_1	θ_2	θ_1
250	125.5	101	21	−5.08°	85.34°	−58.76°
200	100	102	20	10.34°	69.66°	−60.00°
140	70	70	25	8.49°	71.23°	−54.73°
160	60	40	50	−31.03°	117.84°	−36.81°
250	40	30	10	−44.14°	159.49°	−105.34°

Table 3. *Cont.*

H	P_x	P_y	ϕ	θ_1	θ_2	θ_1
190	49	25	−10	−44.34°	146.23°	−111.89°
100	35	50	−20	14.64°	88.92°	−123.57°
110	50	60	25	14.42°	70.69°	−60.11°
95	40	50	45	10.18°	78.03°	−43.22°
81	35	52	37	28.89°	52.58°	−44.47°

Table 4. θ_1, θ_2, and θ_3 obtained from certain parameters, and ϕ is not considered.

H	P_x	P_y	θ_1	θ_2	θ_1
221	130	35	−31.35°	92.84°	−76.42°
181	120	35	−21.45°	75.42°	−67.71°
200	159	50	3.28°	28.34°	−44.17.81°
210	160	55	−1.81°	41.57°	−50.78°
150	61	90	22.36°	67.00°	−63.50°
135	61	82	24.29°	58.11°	−59.05°
100	40	49	6.70°	88.13°	−74.06°
107	40	51.7	6.01°	92.50°	−76.25°
93	65.9	35.6	7.53°	41.68°	−50.84°
351	100	100	−19.43°	128.87°	−94.43°

4.2. Analysis of the Optimization of (P_a, P_b)

Now for the optimization, it is crucial to understand the relationship between the two mathematical models found, both considering and not considering the orientation ϕ and how they can be compared. Basic statistical calculations will be employed to describe Table 5, which will contain the desired values P_x and P_y of the end-effector along with the two possible solutions. The standard deviation will be computed from each column of the table containing the angles. Additionally, it is noted that $H = 230$ and $\phi = 45°$ are constants. A total of 100 data points were collected to calculate the standard deviation. The table below presents a subset of 20 points.

Table 5. θ_1, θ_2, and θ_3 obtained from certain parameters, and ϕ is and is not considered (H = 230).

					ϕ Orientated			Not Orientated
P_x	P_y	ϕ	θ_1	θ_2	θ_3	θ_1	θ_2	θ_3
85	11	45°	−64.96°	127.35°	−17.39°	59.44°	133.63°	−96.81°
73	11	45°	−69.28°	135.55°	−21.27°	−62.21°	141.57°	−100.78°
60	23	45°	−61.6°	143.59°	−36.99o	−52.94°	147.83°	−103.91°
60	41	45°	−43.43°	139.82°	−51.39°	−36.82°	142.33°	−101.16°
60	100	45°	1.40°	111.22°	−67.63°	3.20°	111.66°	−85.83°
90	100	45°	−2.91°	96.225°	−48.31°	−0.76°	97.55°	−78.77°
131	100	45°	1.45°	65.78°	−22.239°	2.79°	69.12°	−64.56°
165	90	45°	12.21°	26.64°	6.14°	10.28°	36.64°	−48.32°
165	72	45°	−0.63°	41.51°	4.12°	−1.43°	50.02°	−55.01°
165	22	45°	−25.17°	56.84°	13.32°	−26.04°	67.28°	−63.64°
115	12	45°	−53.09°	105.34°	−7.25°	−50.24°	112.40°	−86.20°
115	51	45°	−31.25°	100.29°	−24.03°	−28.38°	104.59°	−82.29°
115	92	45°	−6.60°	83.94°	−32.33°	−4.60°	86.52°	−73.26°
117	90	45°	−7.53°	83.47°	−30.93°	−5.53°	86.21°	−73.10°
93	135	45°	19.17°	69.16°	−43.33°	20.39°	70.07°	−65.03°
130	130	45°	22.40°	40.77°	−18.17°	22.97°	44.05°	−52.02°
130	62	45°	−21.21°	84.88°	−18.67°	−19.21°	89.43°	−74.71°
91	11	45°	−62.77°	123.15°	−15.38°	−57.89°	129.57°	−94.78°
115	140	45°	25.58°	46.31°	−26.90°	26.45°	48.29°	−54.14°
70	71	45°	−20.13°	122.57°	−57.43°	−16.57°	123.96°	−91.98°

Then, if the standard deviation of each one of the columns that corresponds to an angle is included, the results are shown in Table 6.

Table 6. Standard deviation of the columns of angles

Angle Name	ϕ Orientation	No Orientation
θ_1	26.06	24.67
θ_2	32.46	31.62
θ_3	22.31	15.81

5. Discussion

In general, parallel robots offer several advantages over robotic arms, such as high speed and the ability to move masses, although they often have limitations in terms of their workspace and the complexity of kinematic equations [36]. This paper proposes a solution that preserves the advantages of parallel robots while reducing the complexity of kinematic equations by a vector analysis. The presented algorithm offers two alternatives: the traditional one, which requires knowledge of the orientation phi of the end-effector and is used in various applications as pick and place [37,38], or medical rehabilitation [39,40]. There is also an alternative that eliminates the need to know this orientation, using instead an optimization method, useful in monitoring applications like conventional or thermal cameras [41]. This means that the similarity law of planar 3RRR robots reduces the number of singularities obtained in kinematic models. Most models use P, Q, and R as active joints [3–12], which is the major difference from this kinematic model, which has active joints in P, D, and A. This complexity in kinematic models leads to the use of polynomial equations of degree eight for their solution. The workspace of this model is delimited by the areas common to the circles generated in P, Q, and R with the sum of the links as the radius. The main contribution of this work is the proposal of a new alternative for the implementation of planar 3RRR robots. This proposal has (15)–(17) as direct kinematics and (25), (31) and (40) as inverse kinematics. For applications where phi is not necessary, the application of inverse kinematic systems is proposed.

6. Conclusions

The development of a mathematical model for robots of this type can be challenging, with the aim of simplifying their behavior description. This research has led to the creation of a mathematical model that achieves this goal, offering a simpler alternative to the traditional approach. Despite being based on conventional methods, the mathematics of this model is notably more user-friendly. Basic mathematical tools and optimization techniques were employed in its development, aiming to streamline robot motion. An initial optimization analysis yielded promising standard deviations, yet further evaluation using more sophisticated statistical methods is warranted to fully assess the method's efficiency. Moreover, there is still much to explore, including the differential kinematics model and dynamic considerations, to draw comprehensive conclusions regarding the robot's positioning and configuration. As a future endeavor, the aim is to develop the dynamic model of this planar robot and conduct comparisons to assess its advantages and disadvantages relative to traditional models.

Author Contributions: Conceptualization, J.L.O.-A.; Methodology, J.L.O.-A.; Validation, M.C.; Formal analysis, J.J.M.C.; Investigation, J.J.M.C. and J.I.C.-V.; Writing—original draft, J.J.M.C.; Writing—review & editing, J.I.C.-V. and J.L.O.-A.; Visualization, M.C. and J.I.C.-V. All authors have read and agreed to the published version of the manuscript.

Funding: This work was supported by the Research Fund of Universidad Tecnológica Centroamericana (UNITEC), Honduras.

Data Availability Statement: Data are contained within the article.

Acknowledgments: This work was supported by the Research Fund of Universidad Tecnológica Centroamericana (UNITEC), Honduras.

Conflicts of Interest: The authors declare no conflict of interest.

References

1. Torres Vargas, L.Z. *Introducción a la Robótica*; Grupo Editorial Éxodo: Miguel Hidalgo, Mexico, 2012.
2. Barrientos, A. *Fundamentos de Robótica*, 2nd ed.; McGraw-Hill España: Madrid, Spain, 2012.
3. Cardona, M.N. Similarity law for the design and workspace optimization of 3RRR planar parallel robots. In Proceedings of the 2014 IEEE Central America and Panama Convention (CONCAPAN XXXIV), Panama City, Panama, 12–14 November 2014; pp. 1–6. [CrossRef]
4. Serrano, F.; Rodriguez, B.; Cardona, M. Obtención de un Modelo Dinámico Para un Robot 3RRR Basado en Teoría de Screws. *Rev. Iberoam. Autom. E Inform. Ind.* **2018**, *15*, 384–390. [CrossRef]
5. Rodelo, M.; Villa, J.; Duque, J.; Yime, E. Kinematic Analysis and Performance of a Planar 3RRR Parallel Robot with Kinematic Redundancy using Screw Theory. In Proceedings of the 2018 IEEE 2nd Colombian Conference on Robotics and Automation (CCRA), Barranquilla, Colombia, 1–3 November 2018; IEEE: Barranquilla, Colombia, 2018; pp. 1–6. [CrossRef]
6. Martinez-Zamudio, P.; Gonzalez-Villela, V.J.; Lopez-Parra, M.; Ramirez-Reivich, A.C. Cinemática Diferencial de un Manipulador Paralelo Plano 3RRR-(RRR)v con Actuación Virtual Indirecta. *Ing. Mec. Tecnol. Desarro.* **2015**, *5*, 321–331.
7. Zubizarreta, A.; Cabanes, I.; Marcos, M.; Pinto, C.; Portillo, E. Redundant dynamic modelling of the 3RRR parallel robot for control error reduction. In Proceedings of the 2009 European Control Conference (ECC), Budapest, Hungary, 23–26 August 2009; pp. 2205–2210. [CrossRef]
8. Sheng, L.; Li, W. Optimization Design by Genetic Algorithm Controller for Trajectory Control of a 3-RRR Parallel Robot. *Algorithms* **2018**, *11*, 7. [CrossRef]
9. Rodelo, M.; Polo, S.; Duque, J.; Villa, J.; Yime, E. Robust Adaptive Control of a Planar 3RRR Parallel Robot for Trajectory-Tracking Applied to Crouch Gait Cycle in Children with Cerebral Palsy. In Proceedings of the 2019 IEEE 4th Colombian Conference on Automatic Control (CCAC), Medellin, Colombia, 15–18 October 2019; pp. 1–6. [CrossRef]
10. Oarcea, A.; Cobilean, V.; Stan, S.D. Trajectory planning of a 3-RRRRR planar parallel robot. In Proceedings of the 2021 9th International Conference on Modern Power Systems (MPS), Cluj-Napoca, Romania, 16–17 June 2021; pp. 1–6. [CrossRef]
11. Noshadi, A.; Mailah, M.; Zolfagharian, A. Active force control of 3-RRR planar parallel manipulator. In Proceedings of the 2010 International Conference on Mechanical and Electrical Technology, Singapore, 10–12 September 2010; pp. 77–81. [CrossRef]
12. Al-Mayyahi, A.; Aldair, A.A.; Chatwin, C. Control of a 3-RRR planar parallel robot using fractional order PID controller. *Int. J. Autom. Comput.* **2020**, *17*, 822–836. [CrossRef]
13. Saha, S.K. *Introducción a la Robótica*; McGraw-Hill: México City, Mexico, 2011.
14. Sierra, E.M.; Ordoñez-Avila, J.L. Mathematical Modeling of a Multi-Chamber Pneumatic Soft Actuator. *Actuators* **2022**, *11*, 221. [CrossRef]
15. Wang, Y.; Wang, Y.; Mushtaq, R.T.; Wei, Q. Advancements in Soft Robotics: A Comprehensive Review on Actuation Methods, Materials, and Applications. *Polymers* **2024**, *16*, 1087. [CrossRef]
16. Wong, C.C.; Tsai, C.Y.; Chen, R.J.; Chien, S.Y.; Yang, Y.H.; Wong, S.W.; Yeh, C.A. Generic Development of Bin Pick-and-Place System Based on Robot Operating System. *IEEE Access* **2022**, *10*, 65257–65270. [CrossRef]
17. Amador, L.D.F.; Castillo Castañeda, E.; Laribi, M.A.; Carbone, G. Design and Analysis of VARONE a Novel Passive Upper-Limb Exercising Device. *Robotics* **2024**, *13*, 29. [CrossRef]
18. Durango, S.; Restrepo, D.; Ruiz, O.; Restrepo-Giraldo, J.; Achiche, S. Symmetrical Observability of Kinematic Parameters in Symmetrical-Parallel Mechanisms. *Blucher Mech. Eng. Proc.* **2014**, *1*, 254–272. [CrossRef]
19. Chablat, D.; Wenger, P. The Kinematic Analysis of a Symmetrical Three-Degree-of-Freedom Planar Parallel Manipulator. In Proceedings of the 15th CISM-IFToMM Symposium on Robot Design, Dynamics and Control (Romansy), CISM-IFToMM, Montréal, QC, Canada, 14–18 June 2004; pp. 1–7.
20. Mejia, L.; Simas, H.; Martins, D. Force Capability Maximization Of a 3RRR Symmetric Parallel Manipulator by Topology Optimization. In Proceedings of the 22nd International Congress of Mechanical Engineering (COBEM 2013), Ribeirao Preto, Brasil, 3–7 November 2013.
21. Zhao, J.S.; Chu, F.; Feng, Z.J. Singularities within the workspace of spatial parallel mechanisms with symmetric structures. *Proc. Inst. Mech. Eng. Part C J. Mech. Eng. Sci.* **2010**, *224*, 459–472. [CrossRef]
22. Yang, G.; Chen, W.; Chen, I.-M. A geometrical method for the singularity analysis of 3-RRR planar parallel robots with different actuation schemes. In Proceedings of the IEEE/RSJ International Conference on Intelligent Robots and System, Las Vegas, NV, USA, 25–29 October 2020; IEEE: Lausanne, Switzerland, 2002; Volume 3, pp. 2055–2060. [CrossRef]
23. Lilge, S.; Nuelle, K.; Boettcher, G.; Spindeldreier, S.; Burgner-Kahrs, J. Tendon actuated continuous structures in planar parallel robots: A kinematic analysis. *J. Mech. Robot.* **2021**, *13*, 011025. [CrossRef]
24. Mauzé, B.; Dahmouche, R.; Laurent, G.J.; André, A.N.; Rougeot, P.; Sandoz, P.; Clévy, C. Nanometer precision with a planar parallel continuum robot. *IEEE Robot. Autom. Lett.* **2020**, *5*, 3806–3813. [CrossRef]

25. Nuelle, K.; Sterneck, T.; Lilge, S.; Xiong, D.; Burgner-Kahrs, J.; Ortmaier, T. Modeling, calibration, and evaluation of a tendon-actuated planar parallel continuum robot. *IEEE Robot. Autom. Lett.* **2020**, *5*, 5811–5818. [CrossRef]
26. Zaccaria, F.; Briot, S.; Chikhaoui, M.T.; Idà, E.; Carricato, M. An analytical formulation for the geometrico-static problem of continuum planar parallel robots. In Proceedings of the ROMANSY 23-Robot Design, Dynamics and Control: Proceedings of the 23rd CISM IFToMM Symposium 23, Sapporo, Japan, 20–24 September 2020; Springer: Berlin/Heidelberg, Germany, 2021; pp. 512–520.
27. Chen, G.; Kang, Y.; Liang, Z.; Zhang, Z.; Wang, H. Kinetostatics modeling and analysis of parallel continuum manipulators. *Mech. Mach. Theory* **2021**, *163*, 104380. [CrossRef]
28. Briot, S.; Goldsztejn, A. Singularity conditions for continuum parallel robots. *IEEE Trans. Robot.* **2021**, *38*, 507–525. [CrossRef]
29. Shao, Z.; Xie, G.; Zhang, Z.; Wang, L. Design and analysis of the cable-driven parallel robot for cleaning exterior wall of buildings. *Int. J. Adv. Robot. Syst.* **2021**, *18*, 1729881421990313. [CrossRef]
30. Mattioni, V.; Ida', E.; Carricato, M. Design of a planar cable-driven parallel robot for non-contact tasks. *Appl. Sci.* **2021**, *11*, 9491. [CrossRef]
31. Sun, C.; Gao, H.; Liu, Z.; Xiang, S.; Yu, H.; Li, N.; Deng, Z. Design and optimization of three-degree-of-freedom planar adaptive cable-driven parallel robots using the cable wrapping phenomenon. *Mech. Mach. Theory* **2021**, *166*, 104475. [CrossRef]
32. Ji, H.; Shang, W.; Cong, S. Adaptive synchronization control of cable-driven parallel robots with uncertain kinematics and dynamics. *IEEE Trans. Ind. Electron.* **2020**, *68*, 8444–8454. [CrossRef]
33. Yamine, J.; Prini, A.; Nicora, M.L.; Dinon, T.; Giberti, H.; Malosio, M. A planar parallel device for neurorehabilitation. *Robotics* **2020**, *9*, 104. [CrossRef]
34. Cardona Gutierrez, M.N. Dimensional Synthesis of 3RRR Planar Parallel Robots for Well-Conditioned Workspace. *IEEE Lat. Am. Trans.* **2015**, *13*, 409–415. [CrossRef]
35. Cero, I. Ejemplo de Aplicación de Cinemática Directa e Inversa a un Manipulador de 2 Grados de Libertad. 2016. Available online: https://www.researchgate.net/profile/Isra-Cero/publication/308102263_Ejemplo_de_aplicacion_de_cinematica_directa_e_inversa_a_un_manipulador_de_2_grados_de_libertad/links/57d9df0e08ae6399a39ae4e5/Ejemplo-de-aplicacion-de-cinematica-directa-e-inversa-a-un-manipulador-de-2-grados-de-libertad.pdf (accessed on 10 January 2024).
36. García-Samartín, J.F.; Barrientos, A. Kinematic Modelling of a 3RRR Planar Parallel Robot Using Genetic Algorithms and Neural Networks. *Machines* **2023**, *11*, 952. [CrossRef]
37. Apostolescu, T.C.; Cartal, L.A.; Udrea, I.; Ionascu, G.; Bogatu, L. Command and Control System of a Planar Parallel Robot for PCB Processing Operations. In Proceedings of the 2022 14th International Conference on Electronics, Computers and Artificial Intelligence (ECAI), Ploiesti, Romania, 30 June–1 July 2022; pp. 1–6. [CrossRef]
38. Acevedo, M.; Orvañanos-Guerrero, M.T.; Velázquez, R.; Arakelian, V. An Alternative Method for Shaking Force Balancing of the 3RRR PPM through Acceleration Control of the Center of Mass. *Appl. Sci.* **2020**, *10*, 1351. [CrossRef]
39. Shoaib, M.; Asadi, E.; Cheong, J.; Bab-Hadiashar, A. Cable Driven Rehabilitation Robots: Comparison of Applications and Control Strategies. *IEEE Access* **2021**, *9*, 110396–110420. [CrossRef]
40. Zou, Q.; Yi, B.J.; Zhang, D.; Shi, Y.; Huang, G. Design and Kinematic Analysis of a Novel Planar Parallel Robot With Pure Translations. *IEEE Access* **2024**, *12*, 9792–9809. [CrossRef]
41. Ordoñez-Avila, J.L.; Cardona, M.; Aguilar, D.A.; Ordoñez, M.; Garzón-Castro, C.L. A Novel Monitoring System for Contagious Diseases of Patients using a Parallel Planar Robot. In Proceedings of the 2022 IEEE International Conference on Machine Learning and Applied Network Technologies (ICMLANT), Soyapango, El Salvador, 15–16 December 2022; pp. 1–6. [CrossRef]

symmetry

MDPI

Article

Enhancing the Proportion of Sub-5 μm Atomized Droplet Size in Medical Air-Compression Nebulizer

Kun Xu [1], Zhongyou Lu [1], Li Wang [1] and Yunqing Tang [2,3,4,*]

1 College of Electrical Engineering, Henan University of Technology, Zhengzhou 450001, China; k.xu018@haut.edu.cn (K.X.); hautwangli@163.com (L.W.)
2 School of Mechanical Engineering, Shandong University, Jinan 250061, China
3 State Key Laboratory of Advanced Equipment and Technology for Metal Forming, Shandong University, Jinan 250061, China
4 Key Laboratory of High Efficiency and Clean Mechanical Manufacture, Ministry of Education, Jinan 250061, China
* Correspondence: tangyunqing@sdu.edu.cn

Abstract: Medical air-compression nebulizers deliver atomized medication to the lungs, providing rapid and painless treatment for respiratory diseases. However, the size of most atomized droplets is around 10 μm, limiting drug deposition in the lower airways and alveoli, with increasing the proportion of sub-5 μm droplets remaining challenging. In this work, finite element analysis was employed to model the effects of gas flow rate, liquid channel width, and broken baffle structure on droplet size distribution, aiming to optimize structure symmetrical parameters and operating conditions. A novel compression atomizer was developed and experimentally evaluated, incorporating an improved symmetrical structure for the crushing baffle. Following this modification, the proportion of sub-5 μm droplets increased from 54.6% to 59.25%, representing a 4.65% enhancement in the generation of sub-5 μm droplets. The effects of gas flow rate and liquid concentration on droplet size distribution were systematically investigated to further optimize the atomization performance. The results demonstrated a significant increase in the proportion of sub-5 μm droplets, thereby enhancing drug delivery efficiency to the lower respiratory tract and improving treatment efficacy for respiratory diseases.

Keywords: Venturi effect; finite element analysis; structure optimization; compression nebulizer; symmetry

1. Introduction

Respiratory diseases such as asthma and chronic obstructive pulmonary disease (COPD) are becoming increasingly prevalent and underscore the urgent need for improved therapeutic strategies. The incidence and mortality rates associated with these diseases are steadily increasing worldwide each year [1–3]. Nebulization therapy is a widely used physical treatment for respiratory diseases that aerosolizes medications into fine droplets to enhance therapeutic efficacy [4]. The droplet size distribution is critical, as smaller droplets can penetrate deeper into the respiratory tract, reaching distal lung areas. Droplets smaller than 5 μm can reach the lungs and alveoli [5]. Thus, optimizing nebulizer droplet size distribution to increase the proportion of these smaller droplets is essential for improving therapeutic efficacy in the lower respiratory tract.

In recent years, significant progress has been made in the simulation and experimental studies of medical nebulizers through the combination of computational fluid dynamics

(CFD) and simulation-driven design, particularly in the field of medical device optimization. For example, Schillaci et al. conducted a numerical study of liquid atomization regimes using conservative level-set simulations, offering valuable insights into the atomization process [6]. Similarly, Yu et al. explored the three-dimensional transient flow characteristics of gas–liquid twin-fluid atomizers, performing numerical simulations and analyses of the recirculating flow within the atomizer, which provided crucial theoretical data for optimization [7]. In the field of medical nebulization, Santati and his colleagues modified a small-volume jet nebulizer based on CFD simulations and evaluated its clinical outcomes in young children with asthma [8]. Additionally, Lelong et al. used ANSYS Workbench to design a model based on the latest airflow atomization and compared numerical simulations with experimental results, further advancing our understanding of atomization in jet nebulizers [9]. These studies collectively contribute to the ongoing optimization of atomizer designs for enhanced therapeutic efficacy.

Nebulizers can be classified based on aerosolization principles into compression nebulizers [5,10,11], ultrasonic nebulizers [12–14], and mesh nebulizers [15–17]. Compression nebulizers are among the most used devices for treating respiratory diseases. They provide several advantages, including high nebulization efficiency, which ensures effective medication delivery, and portability, which facilitates ease of use in various settings, from clinical environments to home care. Additionally, their reproducibility contributes to consistent therapeutic outcomes across treatments [5,18]. Consequently, compression nebulizers have emerged as a significant focus of recent research in the field of medical nebulization, with ongoing studies aimed at optimizing their performance and expanding their applications.

The nebulizing rate is a critical parameter affecting the efficacy of compression nebulizers. Using numerical simulations and experimental observations, researchers have explored the impact of nozzle design parameters, gas pressure, and liquid medication volume on nebulization rates [19]. The particle size distribution of nebulized droplets directly affects nebulization therapy efficacy. Researchers have utilized methods such as laser particle analyzers [20], phase Doppler anemometry (PDA) [21], laser diffraction [22], and aerosol measurement systems [23] to study the droplet size distribution of compression nebulizers. Although these studies have optimized nebulizer performance, targeted research on the proportion of small droplets for lower respiratory tract diseases, especially those under 5 μm, remains unreported.

In this work, we employed finite element analysis to investigate the impact of gas flow rate, liquid channel width, and broken baffle structure on droplet size distribution, to optimize symmetrical structural parameters and operating conditions. Based on these findings, we developed a novel compression nebulizer incorporating an improved symmetrical crushing baffle. Experimental evaluation demonstrated that the proportion of sub-5 μm droplets increased from 54.6% to 59.25%, marking a 4.65% enhancement in the generation of smaller droplets. Additionally, we systematically analyzed the effects of gas flow rate and liquid concentration on droplet size distribution to further refine the atomization performance. These results highlight the potential of this novel design to improve drug delivery efficiency to the lower respiratory tract, ultimately enhancing treatment outcomes for respiratory diseases.

2. Medical Compressed Nebulizer Based on the Venturi Effect

The Venturi effect is a fluid dynamics phenomenon where a region of low pressure is created in the section of a conduit or flow device where fluid velocity increases [24], while regions of higher pressure correspond to areas of lower velocity. According to Bernoulli's principle,

$$P_1 + \frac{1}{2}\rho v_1^2 + \rho g h_1 = P_2 + \frac{1}{2}\rho v_2^2 + \rho g h_2 \quad (1)$$

$$\Delta P = \frac{1}{2}\rho\left(v_2^2 - v_1^2\right) \tag{2}$$

In this context, P_1, v_1, and h_1 denote the fluid's pressure (Pa), velocity (m/s), and height (m) at one point, while P_2, v_2, and h_2 represent these variables at another point. ρ is the fluid density (kg/m^3), and g is the gravitational acceleration (m/s^2).

A symmetric Venturi tube comprises a converging section, a throat, and a diverging section, functioning according to Bernoulli's principle to facilitate fluid dynamics. As fluid flows through the converging section, its velocity increases while pressure decreases, reaching a minimum at the throat, where velocity is maximal. This pressure differential is integral to compression nebulizers, where the high-velocity airflow generates a low-pressure zone that draws liquid from a reservoir, atomizing it into fine droplets. This process enhances the mixing of the atomized liquid with air, thereby improving combustion efficiency in engines and optimizing delivery in various spray applications.

The structural model of a medical compressed atomizer based on the Venturi effect is depicted in Figure 1. The atomizer features a modular symmetric Venturi design comprising a gas channel (1), nozzle (the diameter is 0.5 mm) (2), liquid channel (3), broken baffle structure (4), liquid reservoir (5), pressure regulation channel (6), and mist outlet (7). In operation, compressed gas accelerates through the conical nozzle, reaching maximum velocity at the outlet, where it creates a negative pressure that draws the aqueous solution in as a liquid film. The pronounced velocity gradient between the liquid film and surrounding air generates shear forces that atomize the drug into fine droplets, facilitating their entry into the respiratory tract and lungs for deposition, thereby achieving a therapeutic effect.

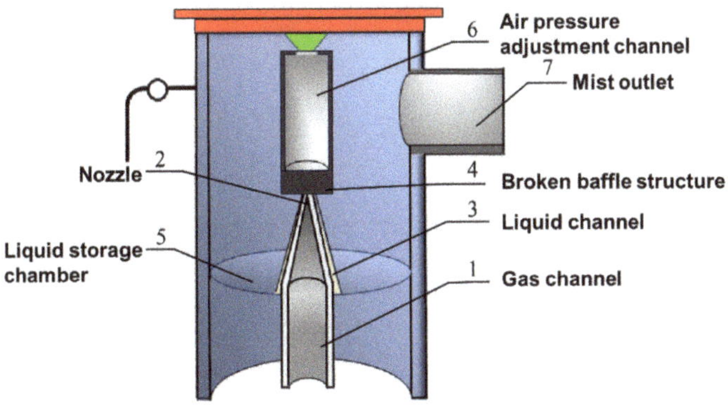

Figure 1. The structural model of a medical compression atomizer.

3. Model Building of Medical Compression Nebulizer

3.1. Model Assumption

Considering that the breakup of liquid droplets is a dynamic process, this study makes the following assumptions to enhance simulation efficiency: (1) The gas is treated as an ideal compressible fluid to account for its compressibility. (2) The influence of droplets on the gas-phase flow field is neglected. These assumptions are valid under conditions where the droplets are small, low in quantity, or have a low mass fraction, such that their momentum feedback on the gas flow can be considered negligible. While this assumption may lead to inaccuracies, especially under high velocities or in temperature-sensitive applications, it is commonly used to reduce computational complexity and still provides useful insights into the general behavior of the system. (3) The walls of the atomizer are assumed to be smooth. This condition refers to the assumption that the wall surface is perfectly smooth, without roughness or irregularities. Under this condition, the fluid velocity at the wall is assumed to be zero, adhering to the no-slip condition. (4) The effects of gravity on the gas are considered negligible.

However, these assumptions introduce certain limitations. For instance, treating the gas as an ideal fluid may not be accurate under high-pressure or low-temperature conditions, where intermolecular forces become significant and can affect the flow behavior. Similarly, the assumption of smooth walls overlooks surface roughness, which influences friction and the dynamics of flow near the wall. These simplifications, while reducing computational complexity, may lead to discrepancies between the simulated results and real-world behavior. It is important to consider these limitations when interpreting the simulation outcomes, particularly in systems where such factors play a significant role.

The computational workflow for solving the model is as follows: (1) The problem is reformulated into a format acceptable to the solver, including establishing the computational domain and mesh generation. (2) Files are imported into the solver interface for validation, followed by the configuration of the model and parameters, including the specification of boundary conditions for the computational domain, as outlined in Table 1. (3) Data visualization is conducted to enhance the clarity and accessibility of the computational results, facilitating effective communication of findings.

Table 1. Boundary conditions.

Parameters	Set Up
fluid type (kg·m^{-3})	liquid water
fluid density (kg·m·s^{-1})	998.2
operating temperature (°C)	300
speed inlet (m/s)	15
speed outlet (m/s)	0
working pressure (kPa)	120
wall condition	smooth wall

3.2. Model Simplification and Mesh Generation

In this study, numerical simulations were performed using ANSYS Fluent software 2022.R1. To improve computational efficiency, the atomizer's structural model was simplified to a two-dimensional planar representation, taking advantage of the symmetry of the Venturi structure. This simplification allowed for a focused examination of the effects of gas velocity and channel width on atomization performance.

To ensure the mesh independence of the numerical simulation results, simulations were conducted with three different mesh densities: 0.6 million, 1 million, and 1.5 million total mesh elements. Under consistent initial conditions, with the gas outlet velocity fixed at 15 m/s, the droplet size distributions at the atomizer outlet after 1 s of atomization were statistically analyzed for each mesh density. The statistical results, shown in Figure 2, indicate that variations in mesh density have a minimal impact on the droplet size distribution post-atomization. Specifically, the simulation results using a mesh density of 0.6 million showed slight discrepancies when compared to the results from the 1 million and 1.5 million mesh densities. However, the results for the 1 million and 1.5 million mesh densities were relatively stable. Based on these findings, and considering both computational accuracy and the reliable representation of atomization effects, a mesh density of 1.5 million elements was selected for the simulations in this study. This choice minimizes potential discrepancies caused by mesh resolution, thus ensuring the reliability and precision of the simulation results.

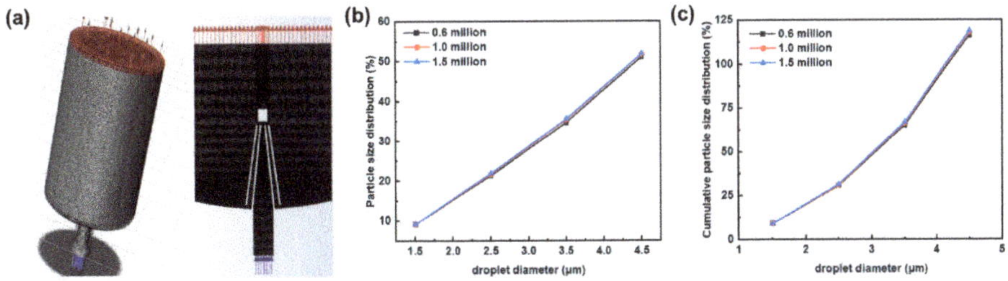

Figure 2. (**a**) Schematic diagram of mesh division. Percentage of droplets at varying mesh densities. (**b**) The distribution of droplet percentages by diameter, and (**c**) the cumulative percentage of droplets below 5 μm.

3.3. Gas–Liquid Flow Model

Mass conservation is a fundamental principle in fluid dynamics, and all fluid flow problems must adhere to it. Accordingly, the mass conservation equation is derived based on this principle. In the context of gas flow in a nebulizer, sharp outlet contractions can lead to very high gas velocities. Therefore, the gas's compressibility must be considered, often modeled as an ideal compressible fluid. As a result, the atomization process is simulated using both mass continuity and momentum conservation equations, which are essential for modeling compressible fluid behavior [25].

To describe the mixed-phase fluid flow (gas and liquid), the continuity equation and Navier–Stokes equations are used. These equations govern mass and momentum transport [26], incorporating assumptions for compressible fluid dynamics. The continuity equation ensures mass conservation, while the Navier–Stokes equation models momentum transport, accounting for pressure gradients, viscosity, and external forces, including surface tension and gravity.

For simulations, standard FLUENT models are used to represent fluid dynamics and gas–liquid interactions. The Volume of Fluid (VOF) method models the interface, and turbulence models are chosen based on expected flow characteristics. These models have been validated for accurate atomization dynamics simulations.

3.4. k-ε Turbulence Model

The k-ε turbulence model is a widely used model in CFD to predict turbulence motion and transport [27]. Proposed by Jones and Launder in 1972, it has been refined over time and is now one of the most common models. Based on the Reynolds-averaged Navier–Stokes equations, the model assumes turbulence is characterized by the turbulence kinetic energy (k) and dissipation rate (ε), with two equations used to represent these parameters.

$$\frac{\partial}{\partial t}(\rho_m k) + \nabla \cdot (\rho_m v_m k) = \nabla \cdot \left(\frac{\mu_{t,m}}{\sigma_k} \nabla k \right) + G_{k,m} - \rho_m \varepsilon \tag{3}$$

$$\frac{\partial}{\partial t}(\rho_m \varepsilon) + \nabla \cdot (\rho_m v_m \varepsilon) = \nabla \cdot \left(\frac{\mu_{t,m}}{\sigma_\varepsilon} \nabla \varepsilon \right) + \frac{\varepsilon}{k}(C_{1\varepsilon} G_{k,m} - C_{2\varepsilon} \rho_m \varepsilon) \tag{4}$$

Here, ρ_m represents the density of the mixed-phase fluid, v_m is the volume-averaged velocity of the mixed-phase fluid, and $\mu_{t,m}$ is the turbulent dynamic viscosity of the mixed phase. The expression for the mixed-phase density of the liquid and air is as follows:

$$\rho_m = \alpha_a \rho_a + \alpha_s \rho_s \tag{5}$$

The expression for the volume-averaged velocity of the mixed-phase fluid is as follows:

$$v_m = \frac{\alpha_a \rho_a v_a + \alpha_s \rho_s v_s}{\alpha_a \rho_a + \alpha_s \rho_s} \quad (6)$$

The expression for the turbulent dynamic viscosity of the mixed-phase fluid is given by the following:

$$\mu_{t,m} = \rho_m C_\mu \frac{k^2}{\varepsilon} \quad (7)$$

The rate of turbulent kinetic energy generation due to the velocity gradient is given by the following:

$$G_{k,m} = \mu_{t,m}\left(\nabla v_m + (\nabla v_m)^T\right)\nabla v_m \quad (8)$$

The standard values for the model constants are typically set as follows: $C_{1\varepsilon} = 1.44$, $C_{2\varepsilon} = 1.92$, $C_\mu = 0.09$, $\sigma_k = 1.0$, $\sigma_\varepsilon = 1.3$.

3.5. Breakup Model

The KH-RT (Kelvin–Helmholtz and Rayleigh–Taylor) model is used to predict droplet breakup [28]. During the atomization process, droplets are simultaneously subjected to both Kelvin–Helmholtz and Rayleigh–Taylor instabilities within the compressible airflow. This model assesses the breakup mechanism and timing of droplets based on disturbances caused by surface instability waves. The breakup model is expressed as follows:

$$\Omega_{RT} = \left(\frac{2}{3\sqrt{3\sigma_s}} \frac{[-g_t(\rho_l - \rho_a)]^{\frac{3}{2}}}{\rho_l + \rho_a}\right)^{\frac{1}{2}} \quad (9)$$

$$\tau_{RT} = \frac{C_\tau}{\Omega_{RT}} \quad (10)$$

Ω_{RT} is the Rayleigh–Taylor characteristic frequency (or breakup rate), $g_t = g n_j + a_{cc} n_j$, denoting the acceleration of the droplet in the trajectory direction, σ_s is the surface tension, ρ_l is the density of the liquid, ρ_a is the density of the gas, τ_{RT} denotes the droplet crushing time, and C_τ denotes the crushing time constant, which is usually taken as 1.

3.6. Model-Solving Calculations

The computational model utilizes the coupled VOF-DPM (Volume of Fluid–Discrete Droplet Model) approach [29]. The VOF model is used to track the continuous fluid phase and predict the phase boundary conditions. In the atomizer, the liquid is initially broken up as it collides with the baffle under high-velocity airflow. This preliminary breakup is followed by further movement of the liquid in the high-speed airflow, which generates significant surface instability. This instability causes tearing of the liquid film, leading to droplet formation. Once the liquid breaks into droplets, the equivalent diameter and sphericity criteria are applied to evaluate their shape. Large droplets that meet the transformation criteria are converted into spherical droplets. The resulting discrete droplets are then calculated using the DPM model.

3.7. Simulation Results and Analysis

The atomization performance of medical compression nebulizers primarily hinges on factors such as droplet size, atomization efficiency, and atomization rate. Notably, gas flow velocity and structural design exert significant influences on these outcomes. In this section, a focused simulation analysis is conducted to investigate the impact of gas flow velocity, liquid channel width, and the refined broken baffle structure (both before and after improvement) on the proportion of sub-5 μm atomized droplets.

3.7.1. Influence of Gas Flow Rate on Atomization (Simulation)

To investigate the influence of gas flow rate on atomization performance, simulations were conducted at gas exit velocities of 15 m/s, 18 m/s, 21 m/s, and 24 m/s for a duration of 1 s [30,31]. Figure 3 illustrates the velocity contour plots of the liquid phase at these varying gas flow rates, and the velocity cloud distribution shows symmetry. The comparisons indicate that higher gas exit velocities result in increased initial velocities of the drawn liquid, thereby enhancing the atomization rate of the liquid.

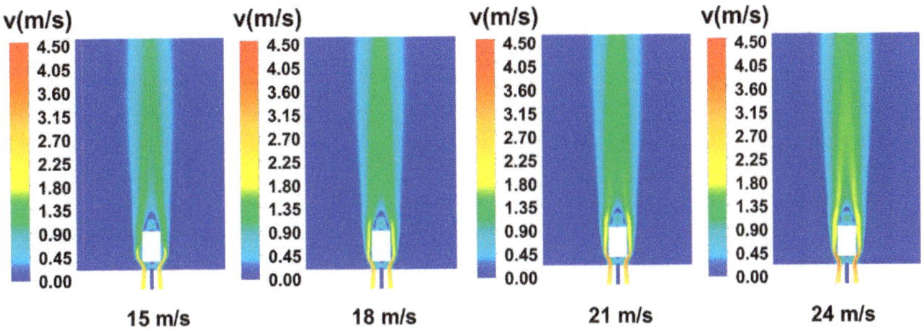

Figure 3. Velocity cloud for different gas flow rates (15 m/s, 18 m/s, 21 m/s, and 24 m/s).

Figure 4a,b present a statistical analysis of the distribution of droplet diameters within the range of 0–5 μm. At compressed air exit velocities of 15 m/s, 18 m/s, 21 m/s, and 24 m/s, the percentages of droplets with diameters below 5 μm were found to be 49.99%, 54.72%, 59.18%, and 63.7%, respectively. These data demonstrate a clear trend, indicating a strong correlation between increased gas flow velocities and a higher cumulative percentage of droplets under 5 μm at the atomizer outlet. This phenomenon can be attributed to two main factors. Firstly, an elevated gas outlet velocity augments the vacuum at the Venturi throat, thereby increasing the aspirated liquid medication volume. This higher initial liquid velocity, driven by the gas, facilitates the formation of smaller droplets. Secondly, increased gas flow velocity subjects the droplets within the atomization chamber to more vigorous vortical flows, centrifugal forces, and shear forces, enhancing atomization and elevating the proportion of droplets with diameters below 5 μm. The maximum exit gas velocity that the human throat can tolerate before it starts to be irritated is typically around 20–25 m/s. Gas velocities above this range can lead to discomfort, including throat dryness, a burning sensation, or coughing. This is due to the high-speed airflow impacting sensitive throat tissues. In our experiment, a gas velocity of around 21 m/s is considered optimal for medical nebulizers, as it ensures effective atomization of the medication while minimizing the risk of irritation.

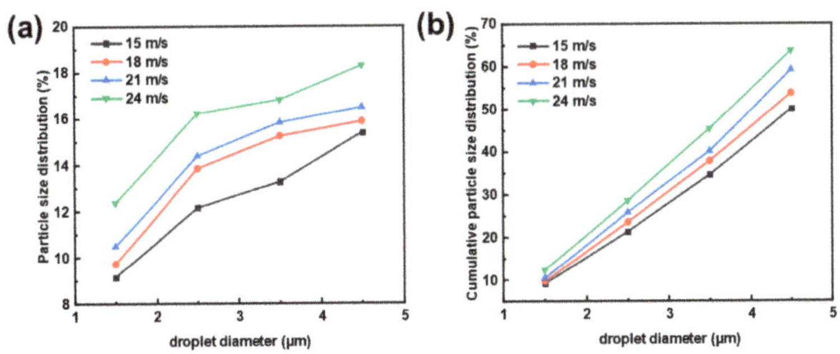

Figure 4. Percentage of droplets at varying exit velocities. (**a**) The distribution of droplet percentages by diameter, and (**b**) the cumulative percentage of droplets below 5 μm.

3.7.2. Influence of Channel Width on Atomization (Simulation)

To investigate the impact of channel width on atomization efficiency, simulations were performed at the same gas exit velocity, using liquid channel widths of 0.3 mm, 0.4 mm, 0.5 mm, 0.6 mm, and 0.7 mm. As shown in Figure 5, an increase in channel width correlates with a rise in both droplet velocity and exit rate. This is attributed to the larger volume of liquid drawn under negative pressure, increasing the quantity of smaller-diameter droplets. These smaller droplets exhibit higher velocities, enhancing the overall atomization rate. Additionally, the effect of channel width on the proportion of droplets below 5 μm was analyzed. Figure 6 illustrates the droplet diameter distribution at the outlet over one second of atomization. The results indicate that as channel width increases, the proportion of droplets with diameters less than 5 μm decreases, adversely affecting the atomization effect.

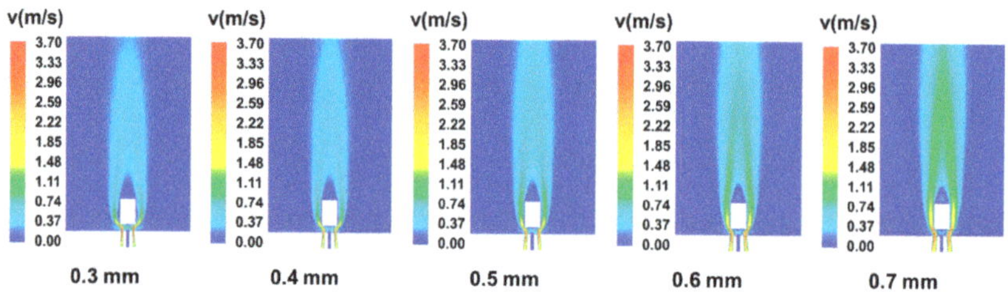

Figure 5. Velocity cloud for different channel widths (0.3 mm, 0.4 mm, 0.5 mm, 0.6 mm, and 0.7 mm).

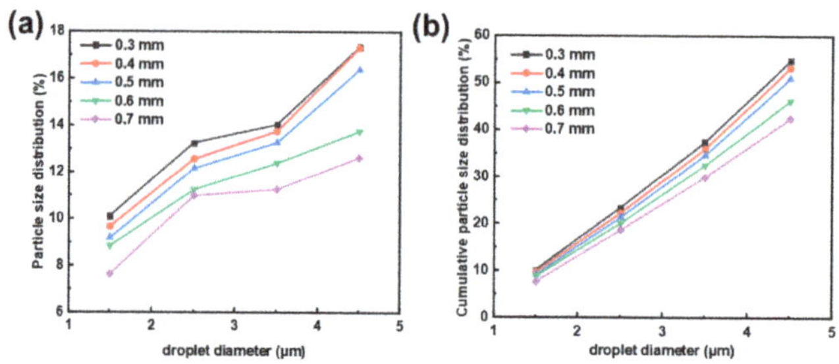

Figure 6. Percentage of droplets at varying channel width. (a) The distribution of droplet percentages by diameter, and (b) the cumulative percentage of droplets below 5 μm.

Analysis of the impact of liquid channel width on atomization reveals that, in terms of droplet kinetic energy, an increase in channel width leads to an increase in the kinetic energy of the droplets post-atomization. However, regarding the proportion of droplets with diameters below 5 μm, an increase in channel width results in a gradual decrease in this proportion. Considering both the kinetic energy of the droplets post-atomization and the proportion of droplets with diameters below 5 μm, the optimal liquid channel width for achieving the best atomization effect is determined to be 0.5 mm.

3.7.3. Influence of Broken Baffle Structure on Atomization (Simulation)

To assess the influence of symmetric broken baffle structures on atomization efficiency, the cross-sectional comparison of the symmetric broken baffle shape before and after modification is illustrated in Figure 7a. The baffle was transformed from a rectangular prism to a conical symmetric shape, allowing it to cover the liquid channel fully. This design ensures complete liquid impact while minimizing kinetic energy loss during collisions. Figure 7b illustrates the velocity contour of the liquid within the atomization chamber before and after the modification. The impact angle changed from 90° to 120°, leading to an increase in liquid velocity. This enhancement not only boosts the atomization rate but also increases the kinetic energy of the droplets, facilitating better delivery to targeted respiratory sites. Figure 7c,d show the droplet size distribution at the outlet at a gas exit velocity of 15 m/s. The proportion of droplets smaller than 5 μm increased from 49.99% to 56.02% following the baffle modification, representing a 6.03% improvement. These results indicate that the modified atomizer effectively enhances the proportion of smaller droplets, thereby improving atomization efficiency.

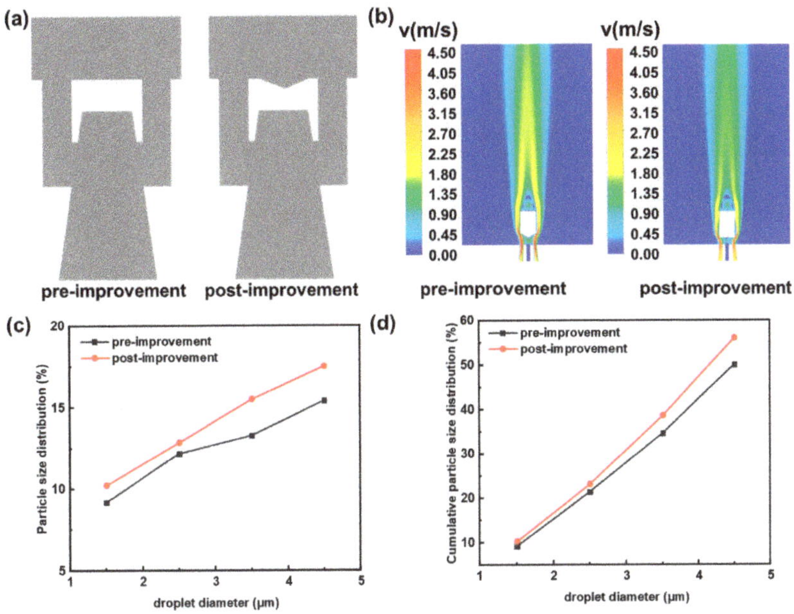

Figure 7. Comparison of pre-improvement and post-improvement. (**a**) Cross-sectional comparison of the broken baffle shape; (**b**) Velocity contour of the liquid in the atomizer. (**c**) The distribution of droplet percentages by diameter, and (**d**) the cumulative percentage of droplets below 5 μm.

The simulation results indicate that the improved broken baffle structure effectively enhances the proportion of sub-5 μm droplets due to its symmetry. Modifying the shape of the crushing baffle improves the contact area between the liquid and the baffle, leading to a more thorough impact and fragmentation of the liquid, further reducing the diameter of the atomized droplets. Additionally, the change in baffle shape alters the liquid's trajectory from a direct 90° impact to a 60° glancing blow. This reduction in impact angle minimizes kinetic energy loss upon collision with the baffle, thereby increasing the kinetic energy of the atomized droplets. The redesigned baffle significantly optimizes atomization dynamics, simultaneously enhancing both the kinetic energy of the atomized droplets and the proportion of small droplets necessary for effective respiratory delivery.

4. Characterization and Validation of Atomization Performance

Based on the simulation results presented above, this study conducted practical tests on the fabricated atomizer to investigate the effects of gas flow rate, broken baffle structure, and liquid concentration on atomization performance. Special emphasis was placed on examining the impact on the proportion of sub-5 μm droplets. The experimental system in this study consists of five key components: the computer, Microcontroller Unit (MCU), drive circuit, air compressor, gas flow sensor, and compression atomizer (Figure 8). The computer serves as both the user interface and the central data processing hub, communicating with the control unit to manage and monitor the system's overall operation. The MCU processes the control signals, driving the other components accordingly. The drive circuit supplies power to both the air compressor and the atomizer. The air compressor generates the pressurized air required for atomization, while the gas flow sensor continuously monitors and adjusts the airflow to maintain optimal conditions. The liquid atomizer then converts the liquid medicine into fine droplets suitable for inhalation, with its performance being influenced by the airflow dynamics and the atomizer's design. Before atomization, all experimental instruments were thoroughly calibrated, and necessary preparations were made, including the configuration of liquid solutions and adjustment of key parameters, such as the gas flow rate. Once the system reached a stable operational state, data collection commenced. This allowed for precise monitoring of changes in liquid medicine and facilitated an in-depth analysis of its behavior following atomization.

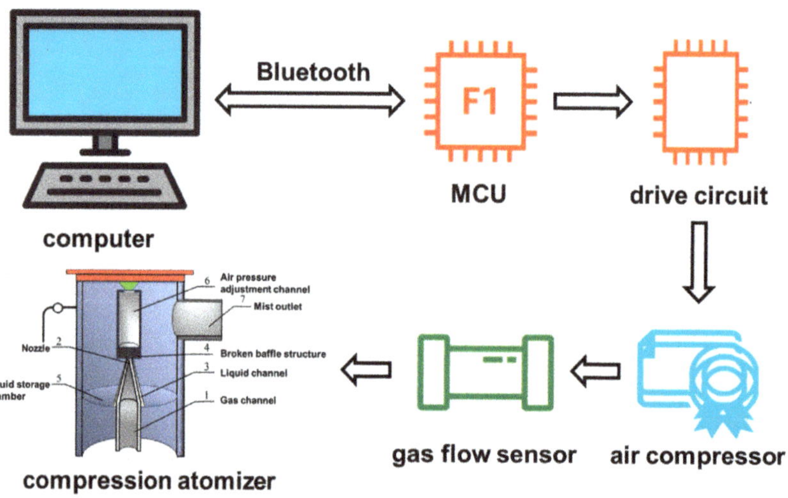

Figure 8. The experimental system of compression atomizer.

4.1. *Influence of Gas Flow Rate on Atomization (Experiment)*

As presented in Table 2, the current experiment investigated gas flow rates of 4 L/min, 6 L/min, 8 L/min, and 10 L/min. With the increase in gas flow rate, a significant reduction in atomization time was observed, accompanied by a gradual decrease in the residual liquid volume and a corresponding enhancement in atomization rate. Furthermore, the projection distance of the atomized droplets expanded, indicating an augmentation in droplet kinetic energy as the gas flow rate increased.

Table 2. Experimental results at different gas flow rates.

Gas Flow Rates	Times	Residual Liquid Volume	Atomization Rate	Range Distance
4 L/min	9.47 min	0.7 mL	0.348 mL/min	115 mm
6 L/min	8.09 min	0.55 mL	0.426 mL/min	152 mm
8 L/min	6.32 min	0.47 mL	0.558 mL/min	195 mm
10 L/min	5.36 min	0.43 mL	0.657 mL/min	237 mm

As shown in Figure 9a,b, under identical testing conditions and after one minute of atomization, the proportion of droplets with diameters below 5 μm increased as the gas flow rate was elevated. Specifically, the cumulative percentages of droplets with diameters under 5 μm at gas flow rates of 4 L/min, 6 L/min, 8 L/min, and 10 L/min were 46.56%, 50.74%, 54.6%, and 59.52%, respectively. This indicates a clear trend where higher gas flow rates promote the generation of smaller droplets, enhancing atomization efficiency. The data obtained from the experiment closely align with the trends observed in the numerical simulations, reinforcing the reliability of the results. The increased gas flow rate not only reduced the atomization time but also contributed to a finer droplet size distribution. This finding is crucial for optimizing medical nebulizers, as smaller droplets (under 5 μm) are more effectively inhaled into the lower respiratory tract, improving the overall drug delivery efficiency. In practical applications, as the gas flow rate increases, the droplet size decreases, and excessive gas flow rates can lead to discomfort such as dryness or irritation. Therefore, to optimize the design of a nebulizer, a balance must be struck between nebulization efficiency and throat tolerance to ensure effective drug delivery without irritation.

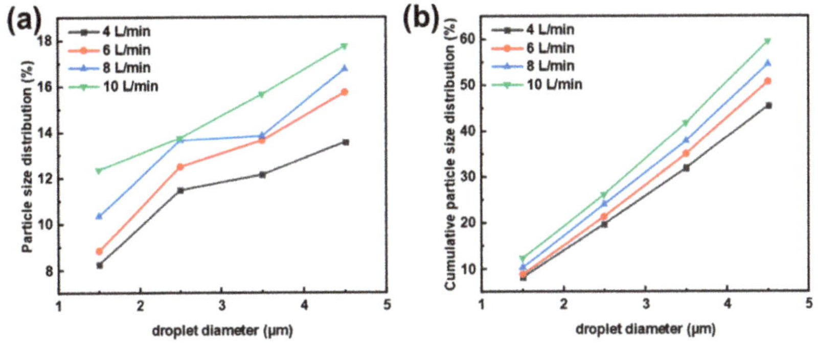

Figure 9. Practical tests at different flow rates. (**a**) the percentage distribution of droplet diameters, and (**b**) the cumulative percentage of droplets below 5 μm.

4.2. Influence of Broken Baffle Structure on Atomization (Experiment)

Table 3 presents the performance of the atomizer before and after the modification of the broken baffle. The results indicate that under identical testing conditions, the modified atomizer exhibited a shorter time of atomization and a reduced amount of residual liquid. By calculating the atomization rates at each velocity before and after the improvement of the liquid-crushing baffle, it is evident that the atomization rate increased following the modification. Specifically, at a gas flow rate of 8 L/min, the spread of atomized droplets increased from 195 mm before the modification to 214 mm after. These findings suggest an increase in the kinetic energy of the atomized droplets due to the enhancement of the symmetric baffle, which aligns with the results from the numerical simulations.

Table 3. Pre-improvement and post-improvement experimental results.

Broken Baffle Structure	Gas Flow Rates	Times	Residual Liquid Volume	Atomization Rate
Pre-improvement	4 L/min	9.47 min	0.7 mL	0.348 mL/min
	6 L/min	8.09 min	0.55 mL	0.426 mL/min
	8 L/min	6.32 min	0.47 mL	0.558 mL/min
	10 L/min	5.36 min	0.43 mL	0.657 mL/min
Post-improvement	4 L/min	9.07 min	0.63 mL	0.371 mL/min
	6 L/min	7.89 min	0.51 mL	0.442 mL/min
	8 L/min	6.02 min	0.43 mL	0.593 mL/min
	10 L/min	4.96 min	0.35 mL	0.735 mL/min

As shown in Figure 10a,b, comprehensive measurements were performed under consistent conditions to determine the droplet size distribution at the exit point after 1 min of atomization, both before and after the modification of the conical symmetric broken baffle structure. The results indicate a significant improvement in the droplet size distribution, with the percentage of droplets with diameters below 5 μm increasing from 54.6% to 59.25% following the modification of the broken baffle. This represents a 4.65% increase in the proportion of sub-5 μm droplets, underscoring enhanced performance due to the optimized symmetric conical structure. These findings align with the trends predicted by the numerical simulations, confirming that the modification of the broken baffle effectively promotes the generation of finer droplets, improving the atomizer's overall efficiency.

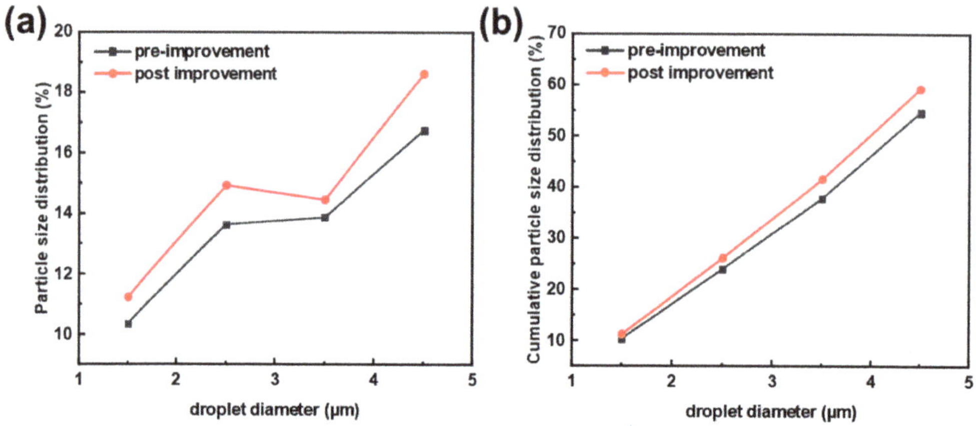

Figure 10. Practical tests of pre-improvement and post-improvement parameters. (**a**) The percentage distribution of droplet diameters, and (**b**) the cumulative percentage of droplets below 5 μm.

4.3. Influence of Liquid Concentration on Atomization (Experiment)

In Figure 11a–c, a detailed analysis of atomization parameters is conducted across varying liquid concentrations and gas flow rates. Specifically, Figure 11a demonstrates that atomization time decreases with increasing gas flow rate at a constant liquid concentration but increases with higher liquid concentrations at a fixed gas flow rate. Conversely, Figure 11b shows a reduction in residual liquid volume in the atomizer as either the gas flow rate or liquid concentration increases, indicating enhanced atomization efficiency. Figure 11c further shows that, with a constant liquid concentration, the atomization rate increases as the gas flow rate increases. However, at a fixed flow rate, the atomization rate decreases with higher liquid concentrations, emphasizing the complex interaction between these two factors on atomization performance. Additionally, Figure 11d,e provide insights into droplet size distribution at the atomizer exit after one minute of atomization at a gas flow rate of 8 L/min. The figures illustrate the distribution, which quantifies the decline in the proportion of droplets with diameters below 5 μm as liquid concentration increases, reinforcing the critical role of liquid concentration in controlling droplet size. The data offer valuable information for optimizing atomizer performance and understanding the operational dynamics of the atomization process.

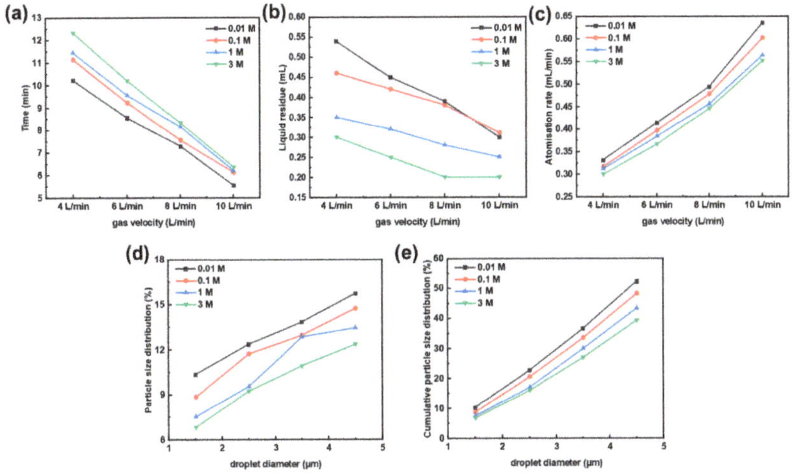

Figure 11. (**a**) Atomization time, (**b**) residual liquid volume, and (**c**) atomization rate at varying liquid concentrations and gas flow rates. Practical tests of different liquid concentrations, (**d**) the percentage distribution of droplet diameters, and (**e**) the cumulative percentage of droplets below 5 μm.

5. Conclusions

This study focuses on the atomization process of the nebulizer, analyzing the atomization mechanism and the formation of droplets. A physical model of the nebulizer and a mathematical model of the atomization process were established. Using the finite element analysis software Fluent, the effects of gas flow rate, liquid channel width, and broken baffle structure on atomization performance were investigated. Simulation results indicated that after optimizing the broken baffle structure, a symmetrical conical structure not only enhanced the droplet kinetic energy but also increased the proportion of droplets with diameters below 5 μm at the outlet by approximately 6%. Increasing the gas flow rate and reducing the channel width were found to improve the proportion of droplets smaller than 5 μm at the outlet to some extent. The results demonstrated a significant increase in the proportion of droplets under 5 μm. This improvement enhances drug delivery to the lower respiratory system and improves treatment outcomes for respiratory diseases.

Additionally, an experimental study was conducted using a system that integrated the master computer, the main control unit, the drive circuit, the air compressor, and the gas flow sensor with the compression atomizer. The effects of different gas flow rates, the broken baffle before and after modification, and liquid concentration on atomization were examined. The results showed that by increasing the gas flow rate, the atomization rate of the designed compression nebulizer increased from 0.348 mL/min to 0.657 mL/min, and the range of atomized droplets increased from 115 mm to 237 mm. The proportion of sub-5 μm droplets at the outlet increased from 46.56% to 59.52%. When modifying the broken baffle structure under the same gas flow rate conditions, the proportion of droplets smaller than 5 μm at the outlet increased by approximately 5%, and the atomization rate also increased. However, for different liquid concentrations, the atomization rate decreased at the same gas flow rate as the concentration increased, and the proportion of droplets under 5 μm also decreased.

This study highlights the optimization of the atomization process in medical air-compressing nebulizers, demonstrating significant improvements in droplet size distribution and drug delivery to the lower respiratory system. While the results are promising, further research is needed to explore a wider range of operating conditions and assess long-term clinical efficacy. Nonetheless, this work lays the foundation for the development of more efficient nebulizers, offering the potential for improved treatments of respiratory diseases.

Author Contributions: Conceptualization, Y.T. and K.X.; methodology, Z.L.; software, Z.L.; validation, Z.L. and K.X.; formal analysis, K.X.; investigation, L.W.; data curation, Y.T.; writing—original draft preparation, Z.L.; writing—review and editing, K.X.; visualization, Z.L.; supervision, Y.T.; project administration, L.W.; funding acquisition, Y.T. All authors have read and agreed to the published version of the manuscript.

Funding: This research was funded by National Natural Science Foundation of China (No. 52305203, 62303094), "Qilu Young Scholar" program of Shandong University (No. 31360082363167), Shandong Excellent Young Scientists Fund Program (Overseas) (No. 2024HWYQ-003), and Young Taishan Scholars Program of Shan-dong Province (No. tsqn202306087), during the course of this work.

Data Availability Statement: Data are contained within the article.

Acknowledgments: The authors would like to express their gratitude to the editors and reviewers for handling and reviewing our paper.

Conflicts of Interest: The authors declare no conflicts of interest.

References

1. Barnes, P.J. Asthma-COPD Coexistence. *J. Allergy Clin. Immunol.* **2024**, *154*, 275–277. [CrossRef] [PubMed]
2. McIntyre, A.; Busse, W.W. Asthma Exacerbations: The Achilles Heel of Asthma Care. *Trends Mol. Med.* **2022**, *28*, 1112–1127. [CrossRef]
3. Rhee, C.K.; Chau, N.Q.; Yunus, F.; Matsunaga, K.; Perng, D.; on behalf the COPD Assembly of the APSR. Management of COPD in Asia: A Position Statement of the Asian Pacific Society of Respirology. *Respirology* **2019**, *24*, 1018–1025. [CrossRef] [PubMed]
4. Barjaktarevic, I.Z.; Milstone, A.P. Nebulized Therapies in COPD: Past, Present, and the Future. *Int. J. Chron. Obstruct. Pulmon. Dis.* **2020**, *15*, 1665–1677. [CrossRef] [PubMed]
5. Yu, M.; Wang, R.; Yao, J.; Liu, K.; Han, Y. Transient Atomization Modeling and Optimal Design of a Medical Air-Compressing Nebulizer. *IEEE Access* **2020**, *8*, 211553–211561. [CrossRef]
6. Schillaci, E.; Antepara, O.; Balcázar, N.; Rigola, J.; Oliva, A. A Numerical Study of Liquid Atomization Regimes by Means of Conservative Level-Set Simulations. *Comput. Fluids* **2019**, *179*, 137–149. [CrossRef]
7. Minghao, Y.; Rui, W.; Kai, L.; Jiafeng, Y. Numerical Simulation of Three-Dimensional Transient Flow Characteristics for a Dual-Fluid Atomizer. *Eng. Appl. Comput. Fluid Mech.* **2019**, *13*, 1144–1152. [CrossRef]
8. Santati, S.; Thongsri, J.; Sarntima, P. Modified Small-Volume Jet Nebulizer Based on CFD Simulation and Its Clinical Outcomes in Small Asthmatic Children. *J. Healthc. Eng.* **2019**, *2019*, 2524583. [CrossRef] [PubMed]
9. Lelong, N.; Vecellio, L.; Sommer De Gélicourt, Y.; Tanguy, C.; Diot, P.; Junqua-Moullet, A. Comparison of Numerical Simulations to Experiments for Atomization in a Jet Nebulizer. *PLoS ONE* **2013**, *8*, e78659. [CrossRef] [PubMed]
10. Li, T. One-Step Synthesis of Magnetic Carboxymethyl Chitosan-Copper Microspheres with Hollow Porous Structure Using a Compression Nebulizer for Adsorption Studies of Bovine Hemoglobin. *Sep. Purif. Technol.* **2024**, *340*, 126755. [CrossRef]
11. Yang, T.; Guan, Y.; Ren, B.; Sun, Y.; Jiang, X.; Li, T. One-Step Synthesis of Superparamagnetic Copper Alginate Gel Microspheres (Cu^{2+}-Alg@Fe_3O_4) Using a Compression Nebulizer. *Chem. Eng. Process.-Process Intensif.* **2024**, *204*, 109950. [CrossRef]
12. Henriques, T.M.; Rito, B.; Proença, D.N.; Morais, P.V. Application of an Ultrasonic Nebulizer Closet in the Disinfection of Textiles and Footwear. *Int. J. Environ. Res. Public Health* **2022**, *19*, 10472. [CrossRef] [PubMed]
13. Lee, J.; Han, C.H.; Oh, I.H.; Allu, S.; Kim, H.J.; Kim, J.; Kim, W.-S.; Park, B.J. Fabrication and Evaluation of Stable Amorphous Polymer-Drug Composite Particles via a Nozzle-Free Ultrasonic Nebulizer. *Int. J. Pharm.* **2024**, *657*, 124177. [CrossRef]
14. Al-Mogahed, W.; Voigt, S.; Mehner, P.J.; Paschew, G.; Richter, A.; Mehner, J. Multi Fourier Horn Ultrasonic Nebulizer Based on Laser Structuring of Silicon. In Proceedings of the 2024 Symposium on Design, Test, Integration and Packaging of MEMS/MOEMS (DTIP), Dresden, Germany, 2–5 June 2024; pp. 1–5.
15. Pritchard, J.N.; Hatley, R.H.; Denyer, J.; Hollen, D.V. Mesh Nebulizers Have Become the First Choice for New Nebulized Pharmaceutical Drug Developments. *Ther. Deliv.* **2018**, *9*, 121–136. [CrossRef]
16. Carrigy, N.B.; Chang, R.Y.; Leung, S.S.Y.; Harrison, M.; Petrova, Z.; Pope, W.H.; Hatfull, G.F.; Britton, W.J.; Chan, H.-K.; Sauvageau, D.; et al. Anti-Tuberculosis Bacteriophage D29 Delivery with a Vibrating Mesh Nebulizer, Jet Nebulizer, and Soft Mist Inhaler. *Pharm. Res.* **2017**, *34*, 2084–2096. [CrossRef]

17. Moon, S.-H.; Chang, K.H.; Park, H.M.; Park, B.J.; Yoo, S.K.; Nam, K.C. Effects of Driving Frequency and Voltage on the Performances of Vibrating Mesh Nebulizers. *Appl. Sci.* **2021**, *11*, 1296. [CrossRef]
18. Collins, M.S.; Murray, T.S. Repeated Hot Water and Steam Disinfection of Pari LC Plus® Nebulizers Alters Nebulizer Output. *J. Cyst. Fibros.* **2019**, *18*, e17–e18. [CrossRef] [PubMed]
19. Soni, H.; Beaudry, B.; O'Neil, J.; Villasmil-Urdaneta, L. Dissecting the Performance of a Jet Nebulizer: The Effect of Changing Design Parameters and Operating conditions. *Aerosol Sci. Technol.* **2024**, *58*, 1–12. [CrossRef]
20. Li, W.; Qian, L.; Song, S.; Zhong, X. Numerical Study on the Influence of Shaping Air Holes on Atomization Performance in Pneumatic Atomizers. *Coatings* **2019**, *9*, 410. [CrossRef]
21. Kapulla, R.; Najera, S.B. Operation Conditions of a Phase Doppler Anemometer: Droplet Size Measurements with Laser Beam Power, Photomultiplier Voltage, Signal Gain and Signal-to-Noise Ratio as Parameters. *Meas. Sci. Technol.* **2006**, *17*, 221–227. [CrossRef]
22. Aladwani, G.; Momin, M.A.M.; Spence, B.; Farkas, D.R.; Bonasera, S.; Hassan, A.; Hindle, M.; Longest, W. Effects of Different Mesh Nebulizer Sources on the Dispersion of Powder Formulations Produced with a New Small-Particle Spray Dryer. *Int. J. Pharm.* **2023**, *642*, 123138. [CrossRef]
23. Mišík, O.; Kejíková, J.; Cejpek, O.; Malý, M.; Jugl, A.; Bělka, M.; Mravec, F.; Lízal, F. Nebulization and In Vitro Upper Airway Deposition of Liposomal Carrier Systems. *Mol. Pharm.* **2024**, *21*, 1848–1860. [CrossRef]
24. Jia, B.; Zhang, S.; Yan, L.; He, S.; Liu, B.; Shen, H.; Yang, P. Self-aspiration Sampling Extractive Electrospray Ionization Mass Spectrometry (EESI-MS) for High-throughput Analysis of Liquid Samples. *Rapid Commun. Mass Spectrom.* **2016**, *30*, 56–61. [CrossRef]
25. Yang, J.; Wang, S. Convergence of Compressible Navier-Stokes-Maxwell Equations to Incompressible Navier-Stokes Equations. *Sci. China Math.* **2014**, *57*, 2153–2162. [CrossRef]
26. Monreal, J. General Navier–Stokes-like Momentum and Mass-Energy Equations. *Ann. Phys.* **2015**, *354*, 565–569. [CrossRef]
27. Salvador, F.J.; Gimeno, J.; Pastor, J.M.; Martí-Aldaraví, P. Effect of Turbulence Model and Inlet Boundary Condition on the Diesel Spray Behavior Simulated by an Eulerian Spray Atomization (ESA) Model. *Int. J. Multiph. Flow* **2014**, *65*, 108–116. [CrossRef]
28. Li, Y.; Huang, Y.; Luo, K.; Liang, M.; Lei, B. Development and Validation of an Improved Atomization Model for GDI Spray Simulations: Coupling Effects of Nozzle-Generated Turbulence and Aerodynamic Force. *Fuel* **2021**, *299*, 120871. [CrossRef]
29. Nan, X.; Shen, Z.; Li, G.; Hou, J. Numerical Modeling of Structural Body Deformation under Free Surface Flow Based on Volume of Fluid–Discrete Element Method Coupling. *Phys. Fluids* **2024**, *36*, 033321. [CrossRef]
30. Zhang, X.; Zhao, L.; Ma, C.; Jiao, G.; Xu, K. Development and Performance of a Photoelectric Salt Concentration Sensor. *IEEE Sens. J.* **2018**, *18*, 1694–1702. [CrossRef]
31. Zhao, L.; Zhang, X. Medical Salt Powder Atomizer and Atomization Effect Experimental Research. Master's Thesis, Jiangsu University, Zhenjiang, China, 2019.

symmetry

MDPI

Article

Galerkin-Type Solution of the Föppl–von Kármán Equations for Square Plates

Sergey Lychev [1,*], Alexander Digilov [1] and Nikolay Djuzhev [2]

1 Ishlinsky Institute for Problems in Mechanics RAS, 119526 Moscow, Russia; digilov@ipmnet.ru
2 MEMSEC R&D Center, National Research University of Electronic Technology (MIET), 124498 Moscow, Russia; dyuzhev@ckp-miet.ru
* Correspondence: lychevsa@mail.ru

Abstract: The solution of the non-linear Föppl–von Kármán equations for square plates in the form of expansion over a system of eigenfunctions, generated by a linear self-adjoint operator, is obtained. The coefficients of the expansion are determined via the reduction method from the infinite-dimensional system of cubic equations. This allows the proposed solution to be considered as a non-linear generalization of the classical Galerkin approach. The novelty of the study is in the strict formulation of the auxiliary boundary problem, which makes it possible to take into account a rigid fixation against any displacements along the boundary. To verify the proposed solution, it is compared with experimental data. The latter is obtained by the holographic interferometry of small deflection increments superimposed on the large deflection caused by initial pressure. Experiment and theory show a good agreement.

Keywords: Föppl–von Kármán equations; finite deformations; clamped square plates; non-linear PDEs; expansion over a basis in space of square-integrable functions; Galerkin method; infinite-dimensional system of cubic equations; reduction; experimental verification

1. Introduction

The Föppl–von Kármán equations, which describe the stress-strain state of flexible plates, have attracted the attention of physicists and mathematicians since August Föppl introduced their prototype in the treatise "Vorlesungen über technische Mechanik" [1]. A. Föppl assumed that in the case of finite deflections, both in-plane linear strains and the rotations are small with respect to unity, while the rotations and squares of in-plane strains are infinitesimals of the same order. Furthermore, he supposed that the bending strains can be thought of infinitesimally small with respect to membrane ones. With these assumptions he derived the following system of equations:

$$hL(w, \phi) + p = 0, \quad \nabla^2\nabla^2\phi + \frac{E}{2}L(w, w) = 0. \tag{1}$$

Here, h is plate thickness, E stands for Young's modulus of the material, and p is the function of transversal load applied to the plate surface. The function $w(x, y)$ represents the deflections of the plate middle-plane, while $\phi(x, y)$ is the Airy stress function related to the components of the stress tensor within it:

$$\sigma_x = \frac{\partial^2\phi}{\partial y^2}, \quad \sigma_y = \frac{\partial^2\phi}{\partial x^2}, \quad \sigma_{xy} = \sigma_{yx} = -\frac{\partial^2\phi}{\partial x \partial y}. \tag{2}$$

Symbols $\nabla^2\nabla^2$ and L stand for the biharmonic operator and the Monge–Ampère form [2], respectively. In Cartesian coordinates, they can be represented as follows:

$$\nabla^2\nabla^2 f = \frac{\partial^4 f}{\partial x^4} + 2\frac{\partial^4 f}{\partial x^2 \partial y^2} + \frac{\partial^4 f}{\partial y^4}, \qquad L(f, g) = \frac{\partial^2 f}{\partial x^2}\frac{\partial^2 g}{\partial y^2} + \frac{\partial^2 f}{\partial y^2}\frac{\partial^2 g}{\partial x^2} - 2\frac{\partial^2 f}{\partial x \partial y}\frac{\partial^2 g}{\partial x \partial y}. \quad (3)$$

Theodor von Kármán improved the first equation in (1) by adding an item that accounts for the influence of in-plane stresses on bending and derived the system of equations in a well-known, almost symmetric form [3]:

$$D\nabla^2\nabla^2 w - hL(w, \phi) = p, \quad \nabla^2\nabla^2\phi + \frac{E}{2}L(w, w) = 0. \quad (4)$$

Here, D denotes the bending stiffness of the plate, which is expressed through h, E, and the Poisson's ratio ν:

$$D = \frac{Eh^3}{12(1-\nu^2)}.$$

The Föppl–von Kármán equations were initially developed as a mathematical tool for buckling analysis of ship hull parts [4]. However, wide application of the structural elements that were able to bend sufficiently without undergoing plastic deformations calls for further development and improvement of the non-linear theories of plates, subjected to various loads, including transverse ones. The Föppl–von Kármán equations proved to be a good approach, taking into account non-linear effects but not having the intimidating bulky form of the non-linear elasticity equations. At present these equations are widely used to describe deformations of various elements of modern technical systems, particularly MEMS, for which the problem of accurate mathematical modeling of stress-strain state in flexible parts is extremely relevant [5,6].

First solutions of the non-linear bending problem were obtained for circular plates subjected to axially symmetric loading, which allowed one to significantly simplify the Föppl–von Kármán equations. Thus, the first attempt to solve them for a clamped circular plate under uniform pressure was undertaken by Nádai [7] in 1925. Another variant of the approximate solution was obtained by Timoshenko, who used the Rayleigh–Ritz method [8], and by Way through the power series expansions [9]. In the first part of their study [10], Friedrichs and Stoker presented numerical solutions for a simply supported circular plate under compressive forces at the boundary, obtained by three distinct algorithms. Furthermore, in [11], they introduced a change in variables, which reduced the order of the governing system of equations from 8th to 4th. In addition, in the second part of [10], they presented a proof of the existence theorem and demonstrated that, apart from a trivial solution, there is at most another one (accurate to a sign). In the present paper, only square plates will be discussed, so the part of the review focused on circular plates is exhausted by a few references. A detailed survey can be found in [12].

One of the first approximate solutions of the Föppl–von Kármán equations for a simply supported square plate under uniform pressure was derived by Kaiser [13] using the finite-difference method. An alternative approach to approximate solutions was proposed by Way [14]. He obtained a solution for clamped rectangular plates with various aspect ratios by the Ritz method. The Galerkin method was also successfully applied for this problem by Panov [15]. Levy first derived exact solutions in terms of double Fourier series for a simply supported rectangular plate [16] and a clamped square one [17]. Later, he and Greenman solved the problem for a clamped rectangular plate having an aspect ratio of 3/2 [18]. This method has also been applied in [19,20] for long clamped and simply supported plates under combined load. It is important to note that in works [16–19], it is assumed that clamped edges are movable, i.e., they are rigidly clamped against normal displacements and rotations while allowing for movement in plane. A numerical solution for a clamped rectangular plate with immovable edges was obtained by Wang [21]. He also considered the important case of a riveted panel [22]. Lastly, it is worth noting the article by Green and Southwell on the application of the relaxation method to solve the Föppl–von Kármán system [23]. A detailed description of the majority of the aforementioned results, as well as many others, can be found in the remarkable books [4,24].

Briefly discuss the recent works on this topic. In the works [25,26], the solution for a simply supported plate was obtained by the "accelerated" iterative method. The perturbation method was successfully applied in the works [27–30]. In addition, the work [29] is notable because of the application of the spline function method for solving a set of linear equations. A number of authors successfully applied the finite-difference method [31–33]. A modification of this approach was proposed by Petit [34], who employed the finite-difference method in combination with the relaxation one. Furthermore, this method was also used in the work [35] for a plate under combined load. The finite element method was used extensively [36–39]. Also, a few attempts to apply the boundary element method were made [40–44]. The Ritz method was widely used as well. Boresi and Terner [45] applied it to obtain the solution for a simply supported rectangular plate with stress-free edges. In the work [46], the rectangular plates with various boundary conditions were considered using the automated Ritz method procedure. This procedure was also applied to calculate large deflections of orthotropic plates [47] and functionally graded material ones [48]. Enem [49] introduced the Ritz method solutions for rectangular plates with twelve different boundary conditions. Recently, the homotopy analysis method has begun to be successfully applied to solve the Föppl–von Kármán system [50–53]. A sufficiently detailed review of works managed before the 1970s is presented in the article [54]. Also, a comprehensive list of works managed before the 2000s is given in the book [55], while a detailed review containing more recent works can be found, for example, in the PhD thesis by Khoa [56].

An original approach to solve the Föppl–von Kármán system was proposed by H. Berger in his PhD thesis [57]. This approach was based on the hypothesis that the strain tensor second invariant makes a small contribution to the full potential energy in comparison with the contribution made by the first invariant. Although this hypothesis lacked a physical basis, it allowed one to decouple the Föppl–von Kármán system. Berger obtained solutions for circular and rectangular plates with various boundary conditions using this method. The further development of Berger's technique was continued by a number of other scientists [58–61]. A similar method for decoupling the Föppl–von Kármán equations was employed in the recent works [62–64]. An alternative approach to decoupling of this system was proposed by Banerjee and Datta [65].

A large amount of results were obtained by the Galerkin method. Yamaki [66] used a modification of the one-term Galerkin method to solve the dynamic analogue of the Föppl–von Kármán equations. He also gave a classification of flexible plate boundary conditions. In their work [67], Iyengar and Naqvi applied biorthogonal expansions of the displacements and the Airy function to obtain solutions for a simply supported plate and a clamped one. For each boundary condition type, two options were considered: all edges are stress-free, and all edges are immovable. In the papers [68,69], the authors obtained the Galerkin method-based system of cubic equations for Kármán plates in an explicit form. In the work [70], this result was supplemented with the explicit expression of the inverted Galerkin matrix. This procedure allowed us to obtain the solution more effectively due to avoiding a laborious calculation of the inverted Galerkin matrix. Recently, the wavelet-Galerkin method has been widely used [71,72]. This method leads to simplification of calculations due to the choice of wavelet functions as the basis.

It is advisable to consider some works devoted to the analysis of the Föppl–von Kármán equations in terms of pure math. Initially, they did not have a rigorous mathematical justification and were derived through an engineering approach significantly based on a few auxiliary hypotheses [2]. The justification of this system has been provided by Ciarlet, who has demonstrated [2,73] that this system can be obtained using the asymptotic expansion of the equations of non-linear elasticity theory by a small parameter, assuming certain hypotheses about the asymptotic orders of displacement vector components. This approach was also applied in the recent work [74] to construct the family of Kármán-like plate models.

The existence theorem for the Föppl–von Kármán boundary-value problem with sufficiently smooth boundary and homogeneous boundary conditions was proved by Morozov [75] using the Leray–Schauder method [76]. The conditions for the existence of a non-trivial equilibrium state of a Kármán plate under compression forces were formulated by M. Berger and Fife [77]. Knightly [78] demonstrated the existence of a solution for a plate subjected to combined loading, and moreover, he proven that if the magnitude of the loading is sufficiently small, then this problem has a unique solution. Further, Knightly and Sather [79] have modified this uniqueness result. Hlaváček and Naumann have obtained similar results for the various types of inhomogeneous boundary conditions [80,81]. It is important to highlight that results of Knightly, Hlaváček, and Naumann are applicable to plates with the boundary that consists of curves from C^3 and may have corners at junctions of the arcs, i.e., they are valid for the majority of plates used in engineering.

From this brief analysis, it is observed that, to date, there are no long-standing methods to solve Föppl–von Kármán equations for square and rectangular plates fixed at the boundary with respect to all displacement components. Moreover, direct 3D FEM modeling of thin plates results in an ill-conditioned algebraic system due to the small parameter, which is the plate thickness. However, it is the full fixation that is usually implemented in experiment and practice. In this context, the development of Föppl–von Kármán equations solution methods for square and rectangular plates, fully fixed at the boundary, is an urgent problem in computational mechanics. This is the subject of the present paper. A new solution is proposed in it via the form of decomposition over a basis in functional space of square-integrable functions. The coefficients of expansion are determined from the infinite-dimensional cubic system via the reduction method. This approach could provide a non-linear counterpart to the Galerkin method. The principal difference from similar studies [67,70] consists in a strict formulation of the auxiliary boundary problem, allowing for accounting natural ways of plate fixing.

2. Problem Statement

Consider a clamped square plate of side $2a$ and thickness h, made from a linear-elastic material with moduli E and ν and subjected to an arbitrary transversal load $p(x, y)$ (see Figure 1). For a closed formulation of the initial boundary problem, we should complement Equation (4) with suitable boundary conditions, which we will now deal with. The boundary value problem will be formulated in the region:

$$\Omega = (-a, a) \times (-a, a) \subset \mathbb{R}^2,$$

with the boundary

$$\partial\Omega = \mathcal{E}_1 \cup \mathcal{E}_2,$$

where the parts of the boundary $\mathcal{E}_1$, $\mathcal{E}_2$ correspond to pairs of opposite edges, i.e.,

$$\mathcal{E}_1 = (\{-a\} \times [-a, a]) \cup (\{a\} \times [-a, a]), \qquad \mathcal{E}_2 = ([-a, a] \times \{-a\}) \cup ([-a, a] \times \{a\}).$$

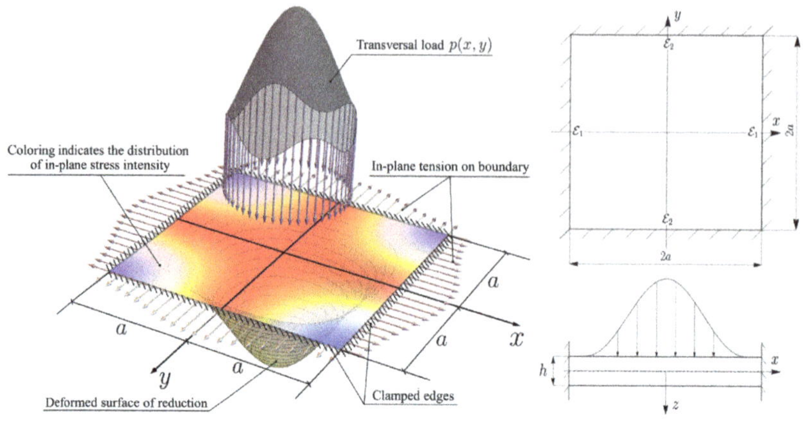

Figure 1. Problem statement.

The boundary conditions for the flexible plates usually are gathered into two sequences associated with plate bending and with plane stress-strain state, respectively. Consider the first one. In the case of clamped plates, it can be written as

$$\left.\frac{\partial w}{\partial x}\right|_{\mathcal{E}_1} = 0, \quad \left.\frac{\partial w}{\partial y}\right|_{\mathcal{E}_2} = 0, \quad w|_{\partial\Omega} = 0. \tag{5}$$

Other variants of the boundary conditions related to bending are well known [24], and we will not comment on them, as (5) are the ones that will be stated hereinafter. The boundary conditions associated with the in-plane stress-strain state are not so widely covered in literature, and we will discuss their formulation in more detail. Following [17,21], we consider two cases:

– movable edges:

$$u|_{\mathcal{E}_1} = 0, \quad v|_{\mathcal{E}_2} = 0, \quad \left.\frac{\partial^2 \phi}{\partial x \partial y}\right|_{\partial\Omega} = 0; \tag{6}$$

– immovable edges:

$$u|_{\partial\Omega} = 0, \quad v|_{\partial\Omega} = 0. \tag{7}$$

Note that despite the designation, moving edges are not free from fixing at all. The points on them can only be moved along the edges, while their displacement in a perpendicular direction is constrained. In Equations (6) and (7), $u(x, y)$ and $v(x, y)$ denote the in-plane plate displacements (the in-plane displacements of the points on the plate middle plane). These functions are not explicitly included in the governing system (4), so in order to solve the boundary-value problem, it is necessary to express them through the desired functions w, ϕ. The sought relations can be obtained using the Cesàro formula, which expresses the displacement vector $\boldsymbol{u}$ and the rotation vector $\boldsymbol{\omega}$ via the infinitesimal strain tensor $\boldsymbol{\varepsilon}$ [82]:

$$\boldsymbol{u}_M = \boldsymbol{u}_0 + \boldsymbol{\omega}_0 \times (\boldsymbol{x} - \boldsymbol{x}_0) + \int_{M_0}^{M} [\boldsymbol{\varepsilon} + (\boldsymbol{x} - \boldsymbol{x}_M) \times (\nabla \times \boldsymbol{\varepsilon})] \cdot d\boldsymbol{x}. \tag{8}$$

Note that the displacement vector $\boldsymbol{u}$ is defined with account of the kinematic hypothesis of the Kirchhoff-like plate theory:

$$\boldsymbol{u} = \left(u - z\frac{\partial w}{\partial x}\right)\boldsymbol{i} + \left(v - z\frac{\partial w}{\partial y}\right)\boldsymbol{j} + w\boldsymbol{k}.$$

Therefore, to relate the displacements $\boldsymbol{u}$ to the Airy stress function ϕ, it is necessary to express the infinitesimal strain tensor $\boldsymbol{\varepsilon}$ through the stress tensor $\boldsymbol{\sigma}$. The essence of the Kármán theory is based on a number of simplifications that allow only the use of the Saint-Venant–Kirchhoff constitutive equation [73,83]:

$$\boldsymbol{\sigma} = \frac{E}{1+\nu}\boldsymbol{E}^* + \frac{\nu E}{1-\nu^2}(\mathrm{tr}\boldsymbol{E}^*)\boldsymbol{I}, \tag{9}$$

where $\boldsymbol{E}^*$ is the reduced Green–Saint-Venant strain tensor (its components are also known as the Kármán strains):

$$\boldsymbol{E}^* = \boldsymbol{\varepsilon} + \frac{1}{2}(\nabla w) \otimes (\nabla w) = \frac{1}{2}\Big[\nabla \boldsymbol{u}^{\mathrm{T}} + \nabla \boldsymbol{u} + (\nabla w) \otimes (\nabla w)\Big]. \tag{10}$$

Finally, substituting the relation (10) into the Cesàro formula (8), one can obtain:

$$\begin{aligned} \boldsymbol{u}_M &= \boldsymbol{u}_0 + \boldsymbol{\omega}_0 \times (\boldsymbol{x} - \boldsymbol{x}_0) \\ &+ \int\limits_{M_0}^{M} \left\{\boldsymbol{E}^* - \frac{1}{2}(\nabla w) \otimes (\nabla w) + (\boldsymbol{x} - \boldsymbol{x}_M) \times \left[\nabla \times \left(\boldsymbol{E}^* - \frac{1}{2}(\nabla w) \otimes (\nabla w)\right)\right]\right\} \cdot d\boldsymbol{x}. \end{aligned} \tag{11}$$

Remark 1. *This representation of displacements does not reproduce the accurate non-linear analogues of the Cesàro formula [84–87]. However, since the displacements are calculated only in the plane of reduction and the reduced finite strain measure* (10) *takes into account quadratic terms only for the transversal component of displacements, such an inaccurate formulation is usually acceptable for engineering calculations [4,21,66].*

Formula (11) can be significantly simplified by choosing an appropriate path of integration. Taking into account that displacements and rotations of the whole plate are defined up to an additive constants any point of it can be fixed. Apart from that, if the transversal load function is odd or even by both coordinates, then the center of a plate will be displaced vertically without changing its in-plane position. For the sake of simplicity, assume that the load function p and, as a consequence, sought functions w, ϕ are even by both coordinates (three other cases of symmetry are derived from reasoning completely similar to the following):

$$\{w,\, \phi,\, p\} \subset \mathcal{F}_s = \Big\{f \in \mathfrak{L}^2(\Omega) : f(x,y) = f(-x,y),\, f(x,y) = f(y,x)\Big\}. \tag{12}$$

Here, we also restrict the class of considered functions p, w, ϕ only to square-integrable ones. This allows us to decompose them over basis and reduce the boundary-value problem to a system of cubic equations.

With account of the transversal loading function symmetry (12), the displacement on $\mathcal{E}_1$ can be calculated as follows:

$$u\big|_{x=\pm a} = \int\limits_0^{\pm a} \left(E^*_{xx} - \frac{1}{2}\left(\frac{\partial w}{\partial x}\right)^2\right) dx, \qquad v\big|_{x=\pm a} = \int\limits_0^{y} \left(E^*_{yy} - \frac{1}{2}\left(\frac{\partial w}{\partial y}\right)^2\right) dy\Bigg|_{x=\pm a}.$$

Substitution of relations (9) and (2) into the above formula results in

$$u\big|_{x=\pm a} = \int_0^{\pm a} \left[\frac{\partial^2 \phi}{\partial y^2} - \nu \frac{\partial^2 \phi}{\partial x^2} - \frac{E}{2}\left(\frac{\partial w}{\partial x}\right)^2 \right] dx = 0, \tag{13}$$

$$v\big|_{x=\pm a} = \int_0^{y} \left[\frac{\partial^2 \phi}{\partial x^2} - \nu \frac{\partial^2 \phi}{\partial y^2} - \frac{E}{2}\left(\frac{\partial w}{\partial y}\right)^2 \right] dy \Bigg|_{x=\pm a} = 0.$$

However, taking into account that a function on a simply connected region is identically zero iff it takes value zero at least at one point and its first derivative is zero also, the latter integral condition can be reduced to a differential relation:

$$\left[\frac{\partial^2 \phi}{\partial x^2} - \nu \frac{\partial^2 \phi}{\partial y^2} - \frac{E}{2}\left(\frac{\partial w}{\partial y}\right)^2 \right] \Bigg|_{x=\pm a} = 0.$$

The last terms in the left-hand side of this relation can be omitted due to the fact that the deflection function is identically zero on the edges (5):

$$\left(\frac{\partial^2 \phi}{\partial x^2} - \nu \frac{\partial^2 \phi}{\partial y^2} \right) \Bigg|_{x=\pm a} = 0.$$

Thus, the above relation, together with the first integral equality (13), fully describes the boundary conditions for the in-plane displacements of the plate on $\mathcal{E}_1$. The formula for the boundary conditions on $\mathcal{E}_2$ can be written in a similar way up to the replacement of the variable x by y and function u by v, respectively.

Now we can move on to the formulation of a boundary value problem. To this end, introduce the following dimensionless variables:

$$\tilde{x} := \frac{x}{a}, \quad \tilde{y} := \frac{y}{a}, \quad \tilde{z} := \frac{z}{h}, \quad \tilde{w} := \frac{w}{h}, \quad \tilde{\phi} := \frac{\phi h}{D}, \quad \tilde{p} := \frac{pa^4}{Dh}.$$

In the new variables the Föppl–von Kármán Equation (4) and corresponding boundary conditions can be written as

$$\tilde{\nabla}^2 \tilde{\nabla}^2 \tilde{w} - \tilde{L}(\tilde{w}, \tilde{\phi}) = \tilde{p}, \qquad \tilde{\nabla}^2 \tilde{\nabla}^2 \tilde{\phi} + \kappa \tilde{L}(\tilde{w}, \tilde{w}) = 0, \tag{14}$$

$$\frac{\partial \tilde{w}}{\partial \tilde{x}}\bigg|_{\tilde{x}=\pm 1} = 0, \quad \frac{\partial \tilde{w}}{\partial \tilde{y}}\bigg|_{\tilde{y}=\pm 1} = 0, \quad \tilde{w}\big|_{\tilde{x}=\pm 1} = 0, \quad \tilde{w}\big|_{\tilde{y}=\pm 1} = 0, \tag{15}$$

$$\int_0^{\pm 1} \left[\frac{\partial^2 \tilde{\phi}}{\partial \tilde{y}^2} - \nu \frac{\partial^2 \tilde{\phi}}{\partial \tilde{x}^2} - \kappa \left(\frac{\partial \tilde{w}}{\partial \tilde{x}}\right)^2 \right] d\tilde{x} = 0, \quad \int_0^{\pm 1} \left[\frac{\partial^2 \tilde{\phi}}{\partial \tilde{x}^2} - \nu \frac{\partial^2 \tilde{\phi}}{\partial \tilde{y}^2} - \kappa \left(\frac{\partial \tilde{w}}{\partial \tilde{y}}\right)^2 \right] d\tilde{y} = 0; \tag{16}$$

- movable edges:

$$\frac{\partial^2 \tilde{\phi}}{\partial \tilde{x} \partial \tilde{y}}\bigg|_{\tilde{x}=\pm 1} = 0, \quad \frac{\partial^2 \tilde{\phi}}{\partial \tilde{x} \partial \tilde{y}}\bigg|_{\tilde{y}=\pm 1} = 0; \tag{17}$$

- immovable edges:

$$\left(\frac{\partial^2 \tilde{\phi}}{\partial \tilde{y}^2} - \nu \frac{\partial^2 \tilde{\phi}}{\partial \tilde{x}^2} \right) \Bigg|_{\tilde{y}=\pm 1} = 0, \quad \left(\frac{\partial^2 \tilde{\phi}}{\partial \tilde{x}^2} - \nu \frac{\partial^2 \tilde{\phi}}{\partial \tilde{y}^2} \right) \Bigg|_{\tilde{x}=\pm 1} = 0, \tag{18}$$

where $\kappa := 6(1 - \nu^2)$, $\tilde{\nabla}^2 := a^2 \nabla^2$, $\tilde{L}(\cdot, \cdot) := a^4 L(\cdot, \cdot)$. In what follows, we will omit the tildes above these operators for brevity. The Equations (14)–(18) constitute two different boundary-value problems for a clamped square plate.

To obtain an approximate solution of the problems (14)–(17) or (18), a recursive approach is often used. Briefly describe its essence for comparison with the proposed solution below. The basis of the recursion $\tilde{w}_0$ can be taken as a solution of the Sophie Germain–Lagrange equation:

$$\tilde{w}_0 := \mathcal{M}_w^{-1}\tilde{p},$$

while the step of the recursion is defined as follows (here subscript indicates recursion step number):

$$\tilde{w}_k = \mathcal{M}_w^{-1}\Big(\tilde{p} - \kappa L\Big(\tilde{w}_{k-1}, \mathcal{M}_\phi^{-1}L(\tilde{w}_{k-1}, \tilde{w}_{k-1})\Big)\Big).$$

Stress function $\tilde{\phi}_k$ can be obtained on each step from the relation

$$\tilde{\phi}_k := -\kappa\mathcal{M}_\phi^{-1}L(\tilde{w}_k, \tilde{w}_k).$$

Here, $\mathcal{M}_w, \mathcal{M}_\phi$ stand for differential operators, defined by the same biharmonic differential expression (3), but by different boundary conditions, corresponding to deflection (15) and stress function (16)–(18). Computational efficiency of recursive processes depends on the way to represent inverse linear operators $\mathcal{M}_w$ and $\mathcal{M}_\phi$. It can be performed differently, but for square plates, the usage of the complex potentials method (Goursat–Muskhelishvili–Kalandiya) [88] seems to be most appropriate. Within this method one can define the solution of linear problem $\mathcal{M}_w\tilde{w} = \tilde{p}$ as the sum of some particular solution $\tilde{w}_*$ that satisfies the non-homogeneous biharmonic differential equation but does not meet boundary conditions and the boundary solution of the homogeneous equation $\tilde{w}_0$, such that $\tilde{w} = \tilde{w}_* + \tilde{w}_0$ satisfies the conditions on the boundary.

Particular solution $\tilde{w}_*$ can be obtained as

$$\tilde{w}_*(z) = \frac{1}{16}\int_0^z\int_0^{\bar{z}}(z-\beta)(\bar{z}-\alpha)\,\tilde{p}\left(\frac{\alpha+\beta}{2}, \frac{\beta-\alpha}{2i}\right)d\alpha\,d\beta,$$

or

$$\begin{aligned}\tilde{w}_*(x, y) = \frac{1}{16}\Bigg[&\int_0^x\int_0^x(x-iy-\alpha)(x+iy-\beta)\,\tilde{p}\left(\frac{\alpha+\beta}{2}, \frac{\beta-\alpha}{2i}\right)d\alpha\,d\beta\\ &+\int_0^x\int_0^y(y-\alpha)(x+iy-\beta)\,\tilde{p}\left(\frac{x-i\alpha+\beta}{2}, \frac{-x+i\alpha+\beta}{2i}\right)d\alpha\,d\beta\\ &-\int_0^x\int_0^y(y-\alpha)(x-iy-\beta)\,\tilde{p}\left(\frac{x+i\alpha+\beta}{2}, \frac{x+i\alpha-\beta}{2i}\right)d\alpha\,d\beta\\ &+\int_0^y\int_0^y(y-\alpha)(y-\beta)\,\tilde{p}\left(\frac{2x-i\alpha+i\beta}{2}, \frac{\alpha+\beta}{2}\right)d\alpha\,d\beta\Bigg].\end{aligned}$$

while the boundary solution w_0 can be expressed in terms of complex-valued series with respect to $z = x+iy$ as follows:

$$\tilde{w}_0 = 2\mathrm{Re}(\gamma\overline{\psi}+\chi).$$

Here,

$$\chi = \int\psi\frac{\partial\gamma}{\partial z}dz, \qquad \phi = \sum a_k z^k, \qquad \psi = \sum b_k z^k,$$

where a_k, b_k can be obtained from the solution of the algebraic counterpart for the boundary problem (see [88,89] for details). In the above formula, $\bar{z}$ indicates the complex conjugate of z, while γ stands for the conformal map between the unit disk and rectangular domain.

This mapping can be exactly expressed via hypergeometric function, but for computations it is more convenient to use polynomial approximation of order n:

$$\gamma = e^{-\frac{i\pi}{4}} \frac{\sqrt{2}\Gamma\left(\frac{3}{4}\right)}{\sqrt{\pi}\Gamma\left(\frac{5}{4}\right)} \sum_{k=0}^{n} \frac{z^{4k+1}}{(4k+1)k!} \prod_{s=0}^{k-1}\left(s+\frac{1}{2}\right).$$

The inverse operator M_{ϕ}^{-1} can be obtained in a similar way. Note, that the advantage of complex potential method is that the integral boundary conditions can be expressed directly via potentials with Love relations:

$$u = \frac{1}{2\mu}\mathrm{Re}\left(\frac{3-\nu}{1+\nu}\varphi - \gamma\frac{\partial\overline{\phi}}{\partial\gamma} - \overline{\psi}\right), \qquad v = \frac{1}{2\mu}\mathrm{Im}\left(\frac{3-\nu}{1+\nu}\varphi - \gamma\frac{\partial\overline{\phi}}{\partial\gamma} - \overline{\psi}\right).$$

Before embarking on the general procedure of problems (14)–(18), let us try to approach it with the recursive method mentioned above. In doing so, keep in mind that in order to identify the characteristic features of non-linear deformation of the plate, it is sufficient to perform at least one step of recursion. The calculation results for the model problem ($a = 1$, $D = 1$, $p = 1$, $\nu = 1/3$, $h = 1/150$) are shown in Figure 2. In it, the subfigure (a) shows the distribution of deflections $\tilde{w}_0$ obtained without accounting for non-linear effects (it is the base of recursion, i.e., the solution of the Sophie Germain–Lagrange problem). Subfigure (b) shows the distribution of plain deformation incompatibility, induced by bending, namely $L(\tilde{w}_0, \tilde{w}_0)$. In subfigure (c), one can see stress function $\tilde{\phi}$, corresponding to such incompatibility. In the second line of figures, the distributions for normal and tangential in-plane stresses are shown, while the last subfigure (f) demonstrates dummy transverse loading $\tilde{q} := \kappa L\left(\tilde{w}_0, \mathcal{M}_{\phi}^{-1}L(\tilde{w}_0, \tilde{w}_0)\right)$ that has to be subtracted from actual loading on the first recursive step.

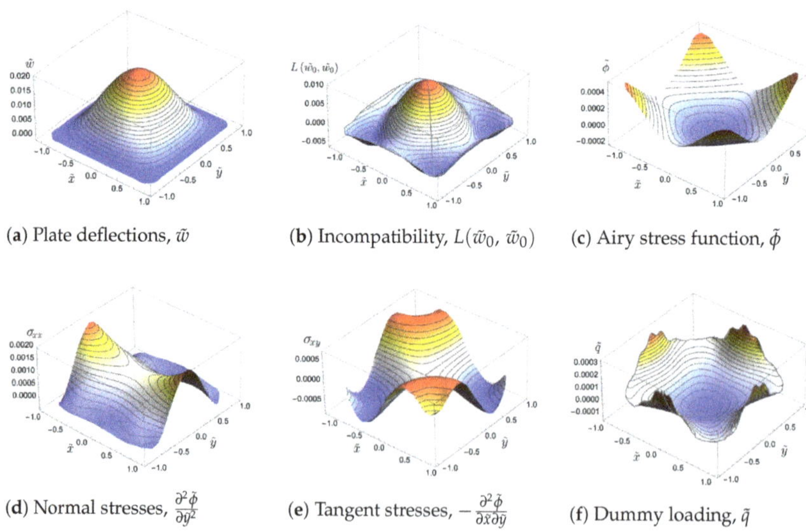

(**a**) Plate deflections, $\tilde{w}$ (**b**) Incompatibility, $L(\tilde{w}_0, \tilde{w}_0)$ (**c**) Airy stress function, $\tilde{\phi}$

(**d**) Normal stresses, $\frac{\partial^2\tilde{\phi}}{\partial\tilde{y}^2}$ (**e**) Tangent stresses, $-\frac{\partial^2\tilde{\phi}}{\partial\tilde{x}\partial\tilde{y}}$ (**f**) Dummy loading, $\tilde{q}$

Figure 2. Results for the model problem.

Already from the analysis of the results obtained in the first step of recursion, it is possible to give a qualitative estimate of the effect of the dummy load and, accordingly, of the non-linear effects depending on the expected value of the maximal deflections. In particular, with evenly distributed load, the maximal value of dummy load (according to the distribution shown in Figure 2f for some load $\alpha\tilde{p}$ with dimensionless intensity α)

can be estimated as $4 \times 10^{-4}\kappa\alpha^3$. Therefore, the given and dummy loads will be of the same order when $\alpha \approx 20$, which, with account of distribution for deflections (Figure 2a), is approximately equivalent to maximal deflections, obtained in a linear approach, of 20% from the edge length of the plate. This value is a very rough estimate from above, because in fact the displacements will be much lower if we take into account non-linear effects, calculating the following recursion steps. Also, the following can be noticed:

- The influence of the dummy load increases proportionally to the cube of maximum deflection;
- The geometric shape of the bending surface at the final deformation significantly differs from the corresponding shape for small deflections because the dummy load changes the distribution of RHS of the bending equation on successive recurrent steps. We will see this again later when we consider the general procedure of solution.

The mentioned strategy for the solution of the non-linear bending problem has computational stability for moderate bending only. When the amount of deflection increases, the dummy load is no longer a small adjustment and may even exceed the given transverse load. In this case, the sequence of recursively constructed approximations tends to diverge. Although special regularization procedures [5] can be used, the diverging cannot be avoided. For this reason, below a different strategy is proposed, implementing the full algebraization of the problem and transforming it to the infinite-dimensional system of cubic equations. Reducing this system to a finite-dimensional one and using any variation of the successive approximation method to find the solution of the reduced system, it is possible to avoid computational instability even for deflections that are comparable with the size of the plate in the plane. The rest of the article is devoted to the construction and analysis of such algebraization.

3. Solution of the Linear Problem

Since the solution of the non-linear problem will be obtained in the form of decomposition over some basis in $\mathfrak{L}^2(\Omega)$, generated by a linear self-adjoint operator, we will start with the linear problem. It is advisable to consider the linear bending of a square plate as such a prototype for this problem because, first, it generates a basis in terms of well-known Krylov–Duncan functions, and second, the problem itself can be used for testing a non-linear solution at zero approach.

The boundary-value problem for clamped Sophie Germain's plate can be stated as follows:

$$\nabla^2\nabla^2\tilde{w} = \tilde{p}, \tag{19}$$

$$\left.\frac{\partial\tilde{w}}{\partial\tilde{x}}\right|_{\tilde{x}=\pm1} = 0, \quad \left.\frac{\partial\tilde{w}}{\partial\tilde{y}}\right|_{\tilde{y}=\pm1} = 0, \quad \tilde{w}\big|_{\tilde{x}=\pm1} = 0, \quad \tilde{w}\big|_{\tilde{y}=\pm1} = 0.$$

In order to expand the deflection function $\tilde{w}$ into a Fourier series, it is necessary to choose a basis in $\mathfrak{L}^2(\Omega)$. According to the Hilbert–Schmidt theorem [90], a system of eigenfunctions of any compact self-adjoint operator constitutes a basis of $\mathfrak{L}^2(\Omega)$. However, the biharmonic operator has a sufficiently complicated system of eigenfunctions even with homogeneous boundary conditions, so it is more preferable to use another self-adjoint operator. It is appropriate to use the following operator:

$$\mathcal{L} = \frac{\partial^4}{\partial\tilde{x}^4} + \frac{\partial^4}{\partial\tilde{y}^4}.$$

The sequence of eigenfunctions for this operator can be obtained as a combination of solutions of an auxiliary Sturm–Liouville problem.

3.1. Auxiliary Sturm–Liouville Problem

Let us consider a Sturm–Liouville problem:

$$\frac{\partial^4 f}{\partial v^4} - \eta^4 f = 0, \qquad f\big|_{v=\pm 1} = 0, \qquad \frac{\partial f}{\partial v}\bigg|_{v=\pm 1} = 0.$$

A fundamental system of solutions for this equation is well known:

$$\Psi = \{\sin(\eta v), \cos(\eta v), \sinh(\eta v), \cosh(\eta v)\}.$$

By substituting a linear combination of these functions into the boundary conditions, we obtain the spectral equation with respect to eigenvalues η:

$$[\sinh(\eta)\cos(\eta) - \sin(\eta)\cosh(\eta)][\sinh(\eta)\cos(\eta) + \sin(\eta)\cosh(\eta)] = 0.$$

Thus, a spectrum of the operator consists of two sequences of eigenvalues (in the strict sense, λ_m, μ_m are not eigenvalues; they are fourth power roots of eigenvalues; however, for brevity, we will not specify this further):

$$\{\eta_m\}_{m=1}^{\infty} = \{\lambda_m\}_{m=1}^{\infty} \cup \{\mu_m\}_{m=1}^{\infty},$$

which are defined as roots of transcendental equations:

$$\tanh(\lambda_m) + \tan(\lambda_m) = 0, \qquad \tanh(\mu_m) - \tan(\mu_m) = 0.$$

These roots can be found numerically with arbitrary precision, but for sufficiently large values of m, very precise approximations can be obtained from asymptotic formula:

$$\lambda_m = \pi m - \frac{\pi}{4}, \qquad \mu_m = \pi m + \frac{\pi}{4}. \tag{20}$$

The question arises: what does it mean to be large enough? The following computation (Tables 1 and 2) shows that for standard machine precision, it is enough to refine only the five first values, while all the rest can be calculated directly by Formula (20). Of course, for calculations with enhanced accuracy, the number of refined roots should be increased.

Table 1. Eigenvalues λ.

Eigenvalue Number	Numerical Solution	Approximation Formula	Relative Error
1	2.36502	2.35619	0.00372
2	5.49780	5.49779	$3.05126 \cdot 10^{-6}$
3	8.63938	8.63938	$3.62616 \cdot 10^{-6}$
4	11.7810	11.7810	$4.96600 \cdot 10^{-12}$
5	14.9226	14.9226	$7.38037 \cdot 10^{-15}$

Table 2. Eigenvalues μ.

Eigenvalue Number	Numerical Solution	Approximation Formula	Relative Error
1	3.92660	3.92699	0.00010
2	7.06858	7.06858	$1.02559 \cdot 10^{-7}$
3	10.2102	10.2102	$1.32593 \cdot 10^{-10}$
4	13.3518	13.3518	$1.89320 \cdot 10^{-13}$
5	16.4934	16.4934	$2.15403 \cdot 10^{-16}$

Each sequence of the eigenvalues is associated with a sequence of the auxiliary eigenfunctions:

$$\begin{aligned} \mathfrak{u}_m(v) &= \cos(\lambda_m v)\cosh(\lambda_m) - \cos(\lambda_m)\cosh(\lambda_m v), \\ \mathfrak{v}_m(v) &= \sin(\mu_m v)\cosh(\mu_m) - \cos(\mu_m)\sinh(\mu_m v). \end{aligned} \quad (21)$$

Such sequences of functions are convenient for derivations due to their compact view; however, for calculations, it is more convenient to use normalized functions:

$$\tilde{\mathfrak{u}}_m(v) = \frac{\mathfrak{u}_m}{\sqrt{\cosh(\lambda_m)^2 + \cos(\lambda_m)^2}}, \qquad \tilde{\mathfrak{v}}_m(v) = \frac{\mathfrak{v}_m}{\sqrt{\cosh(\mu_m)^2 - \cos(\mu_m)^2}}.$$

Curves of the first several normalized eigenfunctions are shown in Figure 3.

It is evident that each pairwise combination of functions (21) presents an eigenfunction of two-dimensional operator $\mathcal{L}$. Thus, we arrive at four families of eigenfunctions for $\mathcal{L}$:

$$\left\{\mathfrak{u}_i(\tilde{x})\mathfrak{u}_j(\tilde{y}),\ \mathfrak{u}_i(\tilde{x})\mathfrak{v}_j(\tilde{y}),\ \mathfrak{v}_i(\tilde{x})\mathfrak{u}_j(\tilde{y}),\ \mathfrak{v}_i(\tilde{x})\mathfrak{v}_j(\tilde{y})\right\}. \quad (22)$$

Although each term of these families is marked by two indexes, it is preferable to use a one-indexed numeration system. Such a system can be provided by any bijection $\mathbb{Z}^+ \times \mathbb{Z}^+ \to \mathbb{Z}^+$. The optimal numeration system is well known from classical courses of calculus:

$$k = \frac{1}{2}(i+j)(i+j-1) - j + 1. \quad (23)$$

This bijection is named as the Cantor pairing function for the positive integers and was proposed by him in the article [91]. The correspondence between two systems of numeration can be illustrated by a diagram as shown in Figure 3c. This way of numeration is the most preferable for square regions because it allows one to construct the reduced sums symmetrically with respect to both indices. In order to achieve this, it is sufficient to include complete diagonals in the sum.

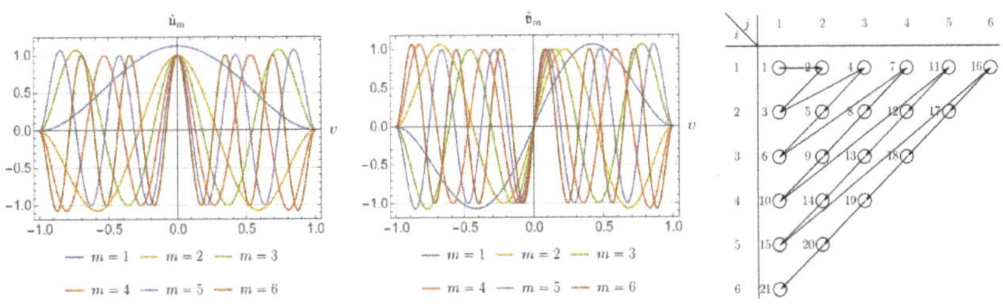

(a) First sequence, $\mathfrak{u}$ (b) Second sequence, $\mathfrak{v}$ (c) Cantor function

Figure 3. Normalized auxiliary eigenfunctions and Cantor pairing function diagram.

Thus the two-indexed system of basis functions (22) can be replaced by the following one:

$$\{f_{1k}(\tilde{x}, \tilde{y}),\ f_{2k}(\tilde{x}, \tilde{y}),\ f_{3k}(\tilde{x}, \tilde{y}),\ f_{4k}(\tilde{x}, \tilde{y})\}. \quad (24)$$

One can go back to the functions (22) with the reverse function to (23). In order to construct this function, it is necessary to relate a number of the current term k with the number of diagonal N on which this term is located:

$$N = \left\lfloor \frac{1 + \sqrt{1 + 8(k-1)}}{2} \right\rfloor,$$

where the symbols $\lfloor \dots \rfloor$ denote the floor. From here, it is simple to obtain the inverse function:

$$i = k - \frac{N(N-1)}{2} =: S(k), \qquad j = \frac{N(N+1)}{2} - k + 1 =: C(k).$$

Hereinafter, S and C are brief notations for components of the inverse function.

3.2. Solution of the Linear Boundary-Value Problem

Now it is time to recover the boundary-value problem (19). We will seek the solution of this problem in the form of expansion in terms of a series of normalized basis functions (24):

$$\tilde{w} = \sum_{r=1}^{4}\sum_{k=1}^{\infty} g_{rk}\frac{f_{rk}}{\|f_{rk}\|} = \sum_{r=1}^{4}\sum_{k=1}^{\infty} g_{rk}\tilde{f}_{rk}. \tag{25}$$

This solution is guaranteed to satisfy the boundary conditions (15) due to the appropriate choice of the basis. Consequently, in order to solve the boundary-value problem, it is sufficient to find the Fourier coefficients g_{rk}, with which (25) satisfies the inhomogeneous Sophie Germain–Lagrange Equation (19):

$$\sum_{r=1}^{4}\sum_{k=1}^{\infty} g_{rk}\nabla^2\nabla^2\tilde{f}_{rk} = \sum_{r=1}^{4}\sum_{k=1}^{\infty} g_{rk}\left(\xi_{rk}\tilde{f}_{rk} + 2\frac{\partial^4\tilde{f}_{rk}}{\partial\tilde{x}^2\partial\tilde{y}^2}\right) = \tilde{p}. \tag{26}$$

Here with ξ_{rk}, we denote the eigenvalues of the operator $\mathcal{L}$:

$$\xi_{1k} = \lambda^4_{S(k)} + \lambda^4_{C(k)}, \quad \xi_{2k} = \lambda^4_{S(k)} + \mu^4_{C(k)},$$
$$\xi_{3k} = \mu^4_{S(k)} + \lambda^4_{C(k)}, \quad \xi_{4k} = \mu^4_{S(k)} + \mu^4_{C(k)}.$$

One can find g_{rk} by acting the projection operators:

$$\mathcal{P}_{sq}(F(\tilde{x},\tilde{y})) = \int_{-1}^{1}\int_{-1}^{1} F\tilde{f}_{sq}\,d\tilde{x}\,d\tilde{y}, \quad q = 1,\dots,\infty, \tag{27}$$

on both sides of Equation (26):

$$g_{sq}\xi_{sq} + \sum_{r=1}^{4}\sum_{k=1}^{\infty} g_{rk}\int_{-1}^{1}\int_{-1}^{1} 2\frac{\partial^4\tilde{f}_{rk}}{\partial\tilde{x}^2\partial\tilde{y}^2}\tilde{f}_{sq}\,d\tilde{x}\,d\tilde{y} = \int_{-1}^{1}\int_{-1}^{1}\tilde{p}\tilde{f}_{sq}\,d\tilde{x}\,d\tilde{y}.$$

It is essential to note that the basis functions lying in different sequences have a different parity. Consequently, due to the fact that the second derivative of any function has the same parity as the function itself, the double sum in the previous relation can be reduced to the sum:

$$g_{sq}\xi_{sq} + \sum_{k=1}^{\infty} g_{sk}\int_{-1}^{1}\int_{-1}^{1} 2\frac{\partial^4\tilde{f}_{sk}}{\partial\tilde{x}^2\partial\tilde{y}^2}\tilde{f}_{sq}\,d\tilde{x}\,d\tilde{y} = \int_{-1}^{1}\int_{-1}^{1}\tilde{p}\tilde{f}_{sq}\,d\tilde{x}\,d\tilde{y}.$$

Thus, the differential equation is reduced to an infinite-dimensional system of algebraic equations. By applying the reduction method [92,93], we arrive at the sequence of finite-dimensional systems of linear algebraic equations, which can be written in matrix form:

$$\boldsymbol{G}^s\boldsymbol{g}^s = \boldsymbol{b}^s, \quad s = \{1,\dots,4\}, \tag{28}$$

where $\boldsymbol{G}^s$ are matrices $n \times n$ known as Galerkin matrices. Their components, G^s_{qk}, can be obtained via formula (here, δ_{qk} is the Kronecker delta):

$$[\boldsymbol{G}^s]_{qk} = G^s_{qk} = \xi_{sq}\delta_{qk} + \int\limits_{-1}^{1}\int\limits_{-1}^{1} 2\frac{\partial^4 \tilde{f}_{sk}}{\partial\tilde{x}^2\partial\tilde{y}^2}\tilde{f}_{sq}\, d\tilde{x}\, d\tilde{y}.$$

Vector $\boldsymbol{b}^s$ contains Fourier coefficients of transverse load function b^s_q (for brevity, we will refer to it as the load vector):

$$[\boldsymbol{b}^s]_q = b^s_q = \int\limits_{-1}^{1}\int\limits_{-1}^{1} \tilde{p}\tilde{f}_{sq}\, d\tilde{x}\, d\tilde{y}. \tag{29}$$

Recall now that we consider functions only from $\mathcal{F}$ (12), so it is sufficient to calculate only the first Galerkin matrix $\boldsymbol{G}^1$ and the first load vector $\boldsymbol{b}^1$, which correspond to even by both variables functions. In this regard, hereinafter we will omit the first index and adopt the following notations:

$$f_k := f_{1k}, \quad \tilde{f}_k := \tilde{f}_{1k}, \quad \mathcal{P}_q := \mathcal{P}_{1q}, \quad \xi_k := \xi_{1k}, \quad \boldsymbol{G} := \boldsymbol{G}^1, \quad G_{qk} := G^1_{qk}, \tag{30}$$
$$\boldsymbol{b} := \boldsymbol{b}^1, \quad b_q := b^1_q, \quad \boldsymbol{g} := \boldsymbol{g}^1, \quad g_k := g^1_k.$$

Thus, the problem has been reduced to one matrix equation. In the case of an arbitrary non-symmetric load function, further steps would be similar; however, they will require finding of all matrices $\boldsymbol{G}^s$ and vectors $\boldsymbol{b}^s$, $\boldsymbol{g}^s$.

Proceed now to calculating the Galerkin matrix and load vector components. It is convenient to represent an expression for them using an auxiliary integral I_G (the auxiliary integrals are provided in the Appendix A):

$$G_{qk} = \xi_q\delta_{qk} + \frac{2\lambda^2_{S(k)}\lambda^2_{C(k)}}{\|f_k\|\,\|f_q\|} I_G\left(\lambda_{S(q)}, \lambda_{S(k)}\right) I_G\left(\lambda_{C(q)}, \lambda_{C(k)}\right) \tag{31}$$

where $\|f_k\|$ denote norms of the basis functions:

$$\|f_k\| = \sqrt{\cosh^2\left(\lambda_{S(k)}\right) + \cos^2\left(\lambda_{S(k)}\right)}\sqrt{\cosh^2\left(\lambda_{C(k)}\right) + \cos^2\left(\lambda_{C(k)}\right)}.$$

The components of the load vector are calculated by Formula (29) if we assume $s = 1$.

Now, the Fourier coefficients g_k can be calculated from the first matrix equation system (28) using, for example, Cramer's rule:

$$g_k = \frac{\det \boldsymbol{G}_k}{\det \boldsymbol{G}}, \qquad \boldsymbol{G}_k = \boldsymbol{G}\cdot(\boldsymbol{E} - \boldsymbol{E}_{kk}) + \sum_{i=1}^{n} b_i \boldsymbol{E}_{ik}. \tag{32}$$

Here, $\boldsymbol{E}$ is the identity matrix, and $\boldsymbol{E}_{ij}$ are so-called matrix units which have only one unit component with indexes ij, while other components are zero.

Finally, the solution can be obtained by summing the basis functions with the calculated Fourier coefficients g_k:

$$\tilde{w}_n = \sum_{k=1}^{n} \frac{\det \boldsymbol{G}_k}{\det \boldsymbol{G}} \tilde{f}_k(\tilde{x}, \tilde{y}), \quad w_n = \frac{pa^4}{D}\sum_{k=1}^{n} \frac{\det \boldsymbol{G}_k}{\det \boldsymbol{G}} \tilde{f}_k\left(\frac{x}{a}, \frac{x}{a}\right), \quad w = \lim_{n\to\infty} w_n. \tag{33}$$

The above limit exists because the original infinite system is fully regular (see [93]). It should be noted that the Formula (33), based on Cramer's rule, are merely of academic interest, while in numerical realization it is convenient to use more efficient algorithms of matrix inverting.

To illustrate the obtained solution model, calculations were performed for two types of load function: uniform load, $\tilde{p} \equiv 1$, and bump-function, which is defined within the region Ω as polynomial function $\tilde{p} = \left(1 - \tilde{x}^2\right)^2\left(1 - \tilde{y}^2\right)^2$ and is zero outside of it. The trial load function and its approximations by reduced Fourier sums are shown in Figure 4. The purpose of these calculations is to show how the convergence rate of the sequence of partial sums $\tilde{w}_n$ depends on the smoothness of the load function and its belonging to the differential operator's domain (obviously the first function does not belong to the domain, whereas the second one belongs to it).

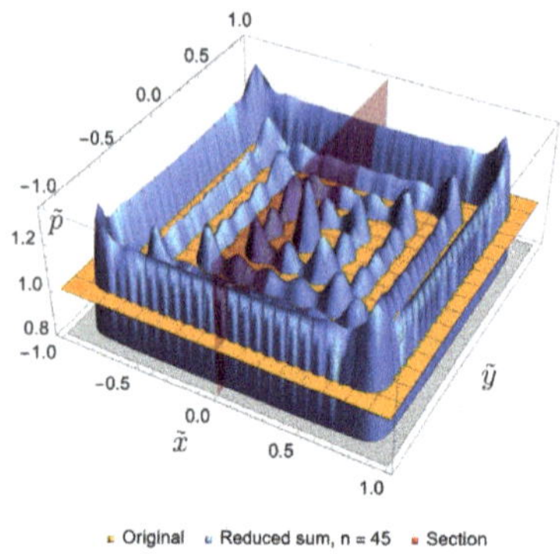

(**a**) Uniform load function, $\tilde{p} = 1$

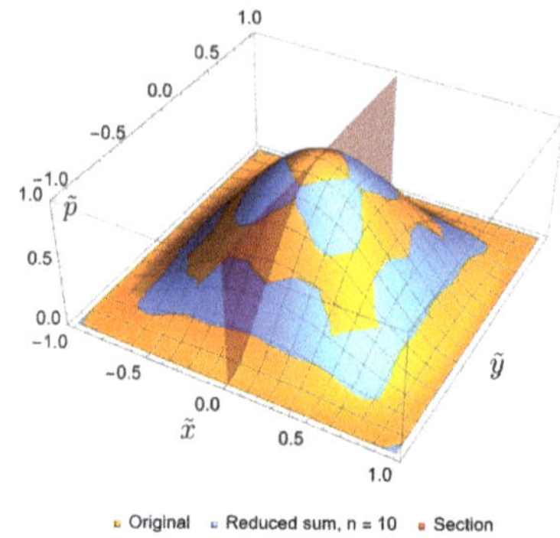

(**b**) Bump-function, $\tilde{p} = \left(1 - \tilde{x}^2\right)^2\left(1 - \tilde{y}^2\right)^2$

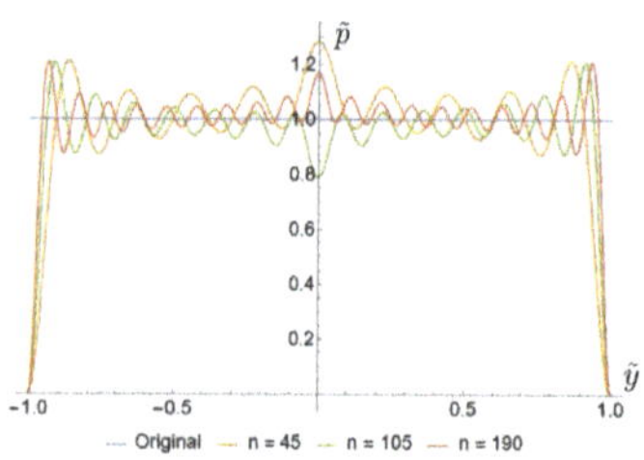

(**c**) Uniform load function, section $\tilde{x} = 0$

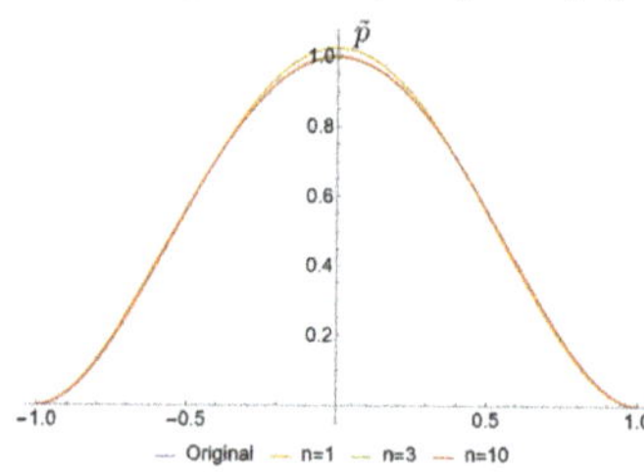

(**d**) Bump-function, section $\tilde{x} = 0$

Figure 4. Trial load functions and their reduced Fourier sums.

Results of these calculations are presented in Figures 5 and 6. Subfigures (a) contain deflected shapes of plate. In subfigures (b), the inaccuracy functions for the corresponding load functions are shown, which are defined as distributions of unbalanced transversal forces absolute values:

$$\delta_n = \left| \tilde{p} - \sum_{k=1}^{n} g_k \nabla^2 \nabla^2 \tilde{f}_k \right|.$$

Subfigures (c) give integral estimations of the inaccuracy functions versus the number of accounted terms in reduced sums:

$$\varepsilon_n = \left| 1 - \sum_{k=1}^{n} g_k \frac{\int_{-1}^{1} \int_{-1}^{1} \nabla^2 \nabla^2 \tilde{f}_k \, d\tilde{x} \, d\tilde{y}}{\int_{-1}^{1} \int_{-1}^{1} \tilde{p} \, d\tilde{x} \, d\tilde{y}} \right|.$$

In these subfigures, the orange points correspond to sums that are symmetric with respect to the coordinates, i.e., with any term $g_k \tilde{f}_k = g_k \tilde{u}_i(\tilde{x}) \tilde{u}_j(\tilde{y})$, they contain a symmetric term $g_q \tilde{u}_j(\tilde{x}) \tilde{u}_i(\tilde{y})$.

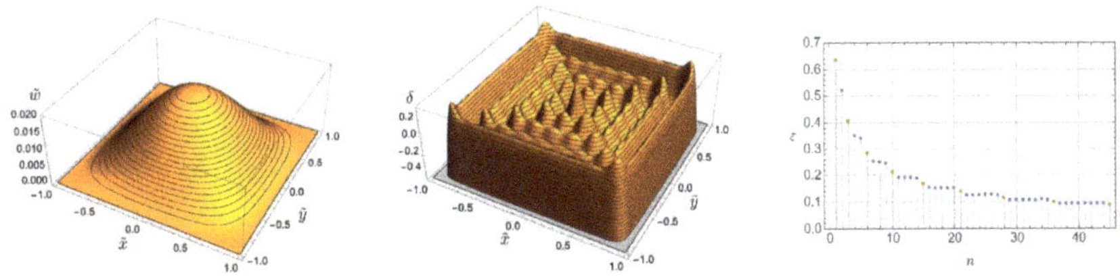

(**a**) Plate deflections, $\tilde{w}$ (**b**) Inaccuracy, δ_n; n = 45 (**c**) Convergence diagram, ε_n

Figure 5. Test solutions for plate under uniform load.

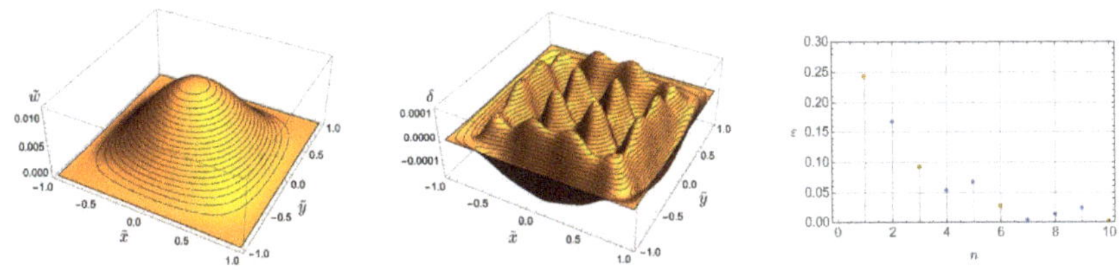

(**a**) Plate deflections, $\tilde{w}$ (**b**) Inaccuracy, δ_n; n = 10 (**c**) Convergence diagram, ε_n

Figure 6. Test solutions for plate under bump-function load.

4. Solution of the Föppl–von Kármán System

Proceed to solving the Föppl-von Kármán boundary-value problems (14)–(18). As above, we will represent the unknown functions $\tilde{w}$ and $\tilde{\phi}$ as decompositions over basis functions (24). However, in this case such expansions do not allow the boundary conditions to be satisfied automatically. In order to overcome this difficulty, we augment the decompositions by the solution of the specific boundary problem.

4.1. Auxiliary Boundary Problem

Consider a homogenous biharmonic equation with the following boundary conditions:

$$\nabla^2\nabla^2\tilde{\phi} = 0, \tag{34}$$

$$\frac{\partial^2\tilde{\phi}}{\partial\tilde{x}^2}|_{\tilde{y}=\pm 1} = \psi_1(\tilde{x}), \quad \frac{\partial^2\tilde{\phi}}{\partial\tilde{y}^2}|_{\tilde{x}=\pm 1} = \psi_2(\tilde{y}), \tag{35}$$

$$\frac{\partial^2\tilde{\phi}}{\partial\tilde{x}\partial\tilde{y}}|_{\tilde{y}=1} = -\frac{\partial^2\tilde{\phi}}{\partial\tilde{x}\partial\tilde{y}}|_{\tilde{y}=-1} = \chi_1(\tilde{x}), \quad \frac{\partial^2\tilde{\phi}}{\partial\tilde{x}\partial\tilde{y}}|_{\tilde{x}=1} = -\frac{\partial^2\tilde{\phi}}{\partial\tilde{x}\partial\tilde{y}}|_{\tilde{x}=-1} = \chi_2(\tilde{y}),$$

where ψ_1, ψ_2, χ_1, χ_2 are arbitrary square-integrable functions. The solution of this problem can be found in the form of a sum:

$$\tilde{\phi} = \tilde{\phi}_0 + \tilde{\phi}^*, \tag{36}$$

where $\tilde{\phi}^*$ meets inhomogeneous boundary conditions (35) but does not satisfy the homogeneous biharmonic Equation (34). Another summand $\tilde{\phi}_0$, instead, meets the homogeneous boundary condition and inhomogeneous equation:

$$\nabla^2\nabla^2\tilde{\phi}_0 = -\nabla^2\nabla^2\tilde{\phi}^*.$$

Thus, the sum (36) will represent the solution of the boundary problem (34) and (35).

Taking into account symmetry relation (12), it is possible to simplify boundary conditions (35):

$$\frac{\partial^2\tilde{\phi}}{\partial\tilde{x}^2}|_{\tilde{y}=\pm1} = \psi(\tilde{x}), \quad \frac{\partial^2\tilde{\phi}}{\partial\tilde{y}^2}|_{\tilde{x}=\pm1} = \psi(\tilde{y}),$$

$$\frac{\partial^2\tilde{\phi}}{\partial\tilde{x}\partial\tilde{y}}|_{\tilde{y}=1} = -\frac{\partial^2\tilde{\phi}}{\partial\tilde{x}\partial\tilde{y}}|_{\tilde{y}=-1} = \chi(\tilde{x}), \quad \frac{\partial^2\tilde{\phi}}{\partial\tilde{x}\partial\tilde{y}}|_{\tilde{x}=1} = -\frac{\partial^2\tilde{\phi}}{\partial\tilde{x}\partial\tilde{y}}|_{\tilde{x}=-1} = \chi(\tilde{y}),$$

where ψ is an even function, while χ is an odd one. In this case, the function $\tilde{\phi}^*$ may be presented as follows:

$$\tilde{\phi}^* = \alpha_0\left(\frac{\tilde{x}^2}{4}+\frac{\tilde{y}^2}{4}\right) - \sum_{k=1}^{\infty}\frac{\alpha_k}{\pi^2k^2}[\cos(\pi k\tilde{x}) + \cos(\pi k\tilde{y})] \tag{37}$$
$$+\sum_{k=1}^{\infty}\frac{2\gamma_k}{\pi^2 k}\left\{\cos\left(\frac{\pi\tilde{y}}{2}\right)\left[\cos(\pi k\tilde{x}) - (-1)^k\right] + \cos\left(\frac{\pi\tilde{x}}{2}\right)\left[\cos(\pi k\tilde{y}) - (-1)^k\right]\right\}.$$

Substitution of this function in (35) results in expansions of functions ψ and χ into the Fourier series:

$$\frac{\partial^2\tilde{\phi}}{\partial\tilde{x}^2}|_{\tilde{y}=\pm1} = \frac{\alpha_0}{2} + \sum_{k=1}^{\infty}\alpha_k\cos(\pi k\tilde{y}), \quad \frac{\partial^2\tilde{\phi}}{\partial\tilde{y}^2}|_{\tilde{x}=\pm1} = \frac{\alpha_0}{2} + \sum_{k=1}^{\infty}\alpha_k\cos(\pi k\tilde{x}),$$

$$\frac{\partial^2\tilde{\phi}}{\partial\tilde{x}\partial\tilde{y}}|_{\tilde{y}=\pm1} = \pm\sum_{k=1}^{\infty}\gamma_k\sin(\pi k\tilde{x}), \quad \frac{\partial^2\tilde{\phi}}{\partial\tilde{x}\partial\tilde{y}}|_{\tilde{x}=\pm1} = \pm\sum_{k=1}^{\infty}\gamma_k\sin(\pi k\tilde{y}).$$

Thus, this solution allows one to satisfy any boundary conditions that may occur in the case of the loading function from (12). In (37), the sums are represented in the most natural form; however, for further derivation, it is convenient to separate the first term of the second sum and renumber the coefficients:

$$\tilde{\phi}^* = \alpha_0\left(\frac{\tilde{x}^2}{4}+\frac{\tilde{y}^2}{4}\right) - \sum_{k=1}^{\infty}\frac{\alpha_k}{\pi^2k^2}[\cos(\pi k\tilde{x}) + \cos(\pi k\tilde{y})]$$
$$+\frac{2\gamma_0}{\pi^2}\left\{\cos\left(\frac{\pi\tilde{y}}{2}\right)[\cos(\pi\tilde{x})+1] + \cos\left(\frac{\pi\tilde{x}}{2}\right)[\cos(\pi\tilde{y})+1]\right\}$$
$$+\sum_{k=1}^{\infty}\frac{2\gamma_k}{\pi^2(k+1)}\left\{\cos\left(\frac{\pi\tilde{y}}{2}\right)\left[\cos(\pi(k+1)\tilde{x}) + (-1)^k\right] + \cos\left(\frac{\pi\tilde{x}}{2}\right)\left[\cos(\pi(k+1)\tilde{y}) + (-1)^k\right]\right\}.$$

It is now necessary to define the function $\tilde{\phi}_0$ in such a way that the full solution (36) satisfies the homogeneous biharmonic equation. In order to do this, this function will be expressed as a series over basis functions $\tilde{f}_k$:

$$\tilde{\phi}_0 = \sum_{k=1}^{\infty}\zeta_k\tilde{f}_k.$$

Substitution of this expression and the boundary solution (37) into the biharmonic equation leads to the following relation:

$$\sum_{k=1}^{\infty}\zeta_k\nabla^2\nabla^2\tilde{f}_k=\sum_{k=1}^{\infty}\alpha_k\pi^2k^2[\cos(\pi k\tilde{x})+\cos(\pi k\tilde{y})]$$
$$-\gamma_0\frac{\pi^2}{8}\left[\cos\left(\frac{\pi y}{2}\right)(25\cos\pi x+1)+\cos\left(\frac{\pi x}{2}\right)(25\cos\pi y+1)\right]$$
$$-\sum_{k=1}^{\infty}\gamma_k\frac{\pi^2}{8(k+1)}\cos\left(\frac{\pi y}{2}\right)\left[\left(4(k+1)^2+1\right)^2\cos(\pi(k+1)x)+(-1)^k\right]$$
$$-\sum_{k=1}^{\infty}\gamma_k\frac{\pi^2}{8(k+1)}\cos\left(\frac{\pi x}{2}\right)\left[\left(4(k+1)^2+1\right)^2\cos(\pi(k+1)y)+(-1)^k\right].$$

As before, with projector operator $\mathcal{P}_q$ (27) and (30), one can transform the above relation to an infinite-dimensional system of algebraic equations with respect to ζ_k, α_k and γ_k:

$$\sum_{k=1}^{\infty}\zeta_k\left(\xi_q\delta_{qk}+\int_{-1}^{1}\int_{-1}^{1}2\frac{\partial^4\tilde{f}_k}{\partial\tilde{x}^2\partial\tilde{y}^2}\tilde{f}_q\,d\tilde{x}\,d\tilde{y}\right)=\sum_{k=1}^{\infty}\alpha_k\pi^2k^2\int_{-1}^{1}\int_{-1}^{1}[\cos(\pi k\tilde{x})+\cos(\pi k\tilde{y})]\tilde{f}_q\,d\tilde{x}\,d\tilde{y}$$
$$-\gamma_0\frac{\pi^2}{8}\int_{-1}^{1}\int_{-1}^{1}\left[\cos\left(\frac{\pi y}{2}\right)(25\cos\pi x+1)+\cos\left(\frac{\pi x}{2}\right)(25\cos\pi y+1)\right]\tilde{f}_q\,d\tilde{x}\,d\tilde{y}$$
$$-\sum_{k=1}^{\infty}\gamma_k\frac{\pi^2}{8(k+1)}\int_{-1}^{1}\int_{-1}^{1}\cos\left(\frac{\pi y}{2}\right)\left[\left(4(k+1)^2+1\right)^2\cos(\pi(k+1)x)+(-1)^k\right]\tilde{f}_q\,d\tilde{x}\,d\tilde{y}$$
$$-\sum_{k=1}^{\infty}\gamma_k\frac{\pi^2}{8(k+1)}\int_{-1}^{1}\int_{-1}^{1}\cos\left(\frac{\pi x}{2}\right)\left[\left(4(k+1)^2+1\right)^2\cos(\pi(k+1)y)+(-1)^k\right]\tilde{f}_q\,d\tilde{x}\,d\tilde{y},$$

and reduce it to a finite-dimensional one [92,93]. It is convenient to write the reduced system in a matrix form:

$$\boldsymbol{\zeta}=\boldsymbol{G}^{-1}\cdot\left[(\boldsymbol{A}+\boldsymbol{B})\cdot\boldsymbol{\alpha}+\left(\boldsymbol{C}^0+\boldsymbol{D}^0\right)\gamma_0+(\boldsymbol{C}+\boldsymbol{D})\cdot\boldsymbol{\gamma}\right], \tag{38}$$

where $\boldsymbol{\zeta}$, $\boldsymbol{\alpha}$, $\boldsymbol{\gamma}$ are vectors of corresponding Fourier coefficients, $\boldsymbol{C}^0$, $\boldsymbol{D}^0$ are auxiliary vectors and $\boldsymbol{A}$, $\boldsymbol{B}$, $\boldsymbol{C}$, $\boldsymbol{D}$ are auxiliary matrices. Recall that the components of the Galerkin matrix, $\boldsymbol{G}$, are presented by the relation (31). The components of other matrices can be tersely expressed through the corresponding auxiliary integrals $I_0(\ldots),\ldots,I_5(\ldots)$ (see Appendix A):

$$A_{qk}=\frac{\pi^2k^2}{\|f_q\|}\left[I_4\left(\lambda_{S(q)},k\right)-I_5\left(\lambda_{S(q)},k\right)\right]I_0\left(\lambda_{C(q)}\right),$$
$$C_{qk}=\frac{\pi^2}{8(k+1)\cosh\left(\frac{\pi}{2}\right)\|f_q\|}\left[I_2\left(\lambda_{C(q)},\frac{\pi}{2}\right)-I_3\left(\lambda_{C(q)},\frac{\pi}{2}\right)\right]$$
$$\times\left[\left(I_4\left(\lambda_{S(q)},k+1\right)-I_5\left(\lambda_{S(q)},k+1\right)\right)\left(4(k+1)^2+1\right)^2+(-1)^kI_0\left(\lambda_{S(q)}\right)\right].$$

Here, the components of only two matrices are shown. Matrices $\boldsymbol{B}$ and $\boldsymbol{D}$ can be written in a similar way to $\boldsymbol{A}$ and $\boldsymbol{C}$, respectively, using the replacement of eigenvalues $\lambda_{S(q)}$ by $\lambda_{C(q)}$ and vice versa. The components of vectors $\boldsymbol{C}^0$ and $\boldsymbol{D}^0$ can be obtained from the corresponding components of $\boldsymbol{C}$ and $\boldsymbol{D}$ by assuming k is equal to zero:

$$C_q^0=C_{q0},\quad D_q^0=D_{q0}.$$

Thus, the solution (36) is fully defined. Further, for brevity, we will use the following notations:

$$\tilde{\phi}^*_{\alpha k} := -\frac{1}{\pi^2 k^2}\cos(\pi k \tilde{x}), \qquad \tilde{\phi}^*_{\beta k} := -\frac{1}{\pi^2 k^2}\cos(\pi k \tilde{y}),$$
$$\tilde{\phi}^*_{\gamma 0} := \frac{2}{\pi^2}\cos\left(\frac{\pi\tilde{y}}{2}\right)[\cos(\pi\tilde{x})+1], \qquad \tilde{\phi}^*_{\delta 0} := \frac{2}{\pi^2}\cos\left(\frac{\pi\tilde{x}}{2}\right)[\cos(\pi\tilde{y})+1],$$
$$\tilde{\phi}^*_{\gamma k} := \frac{2}{\pi^2(k+1)}\cos\left(\frac{\pi\tilde{y}}{2}\right)\left[\cos(\pi(k+1)\tilde{x}) + (-1)^k\right],$$
$$\tilde{\phi}^*_{\delta k} := \frac{2}{\pi^2(k+1)}\cos\left(\frac{\pi\tilde{x}}{2}\right)\left[\cos(\pi(k+1)\tilde{y}) + (-1)^k\right].$$

With this notation, the solution (37) takes the following form:

$$\tilde{\phi}^* = \alpha_0\left(\frac{\tilde{x}^2}{4}+\frac{\tilde{y}^2}{4}\right) + \gamma_0\left(\tilde{\phi}^*_{\gamma 0}+\tilde{\phi}^*_{\delta 0}\right) + \sum_{k=1}^{\infty}\left[\alpha_k\left(\tilde{\phi}^*_{\alpha k}+\tilde{\phi}^*_{\beta k}\right) + \gamma_k\left(\tilde{\phi}^*_{\gamma k}+\tilde{\phi}^*_{\delta k}\right)\right].$$

As an example, consider boundary-value problem (35) and the equation above (37) with following boundary values (see Figure 7):

$$\psi(t) = \left(1-t^2\right)^2, \qquad \chi(t) = t\left(1-t^2\right)^2.$$

The corresponding Airy function and stress distributions are shown in Figure 8.

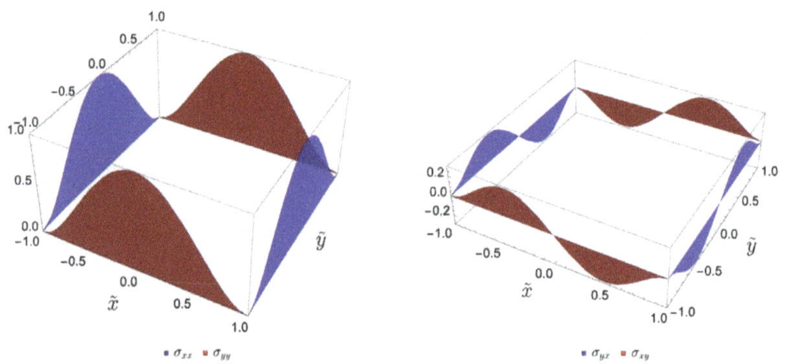

(**a**) Normal stresses, $\sigma_{\tilde{x}\tilde{x}}$, $\sigma_{\tilde{y}\tilde{y}}$ (**b**) Tangent stresses, $\sigma_{\tilde{x}\tilde{y}}$, $\sigma_{\tilde{y}\tilde{x}}$

Figure 7. Stress functions.

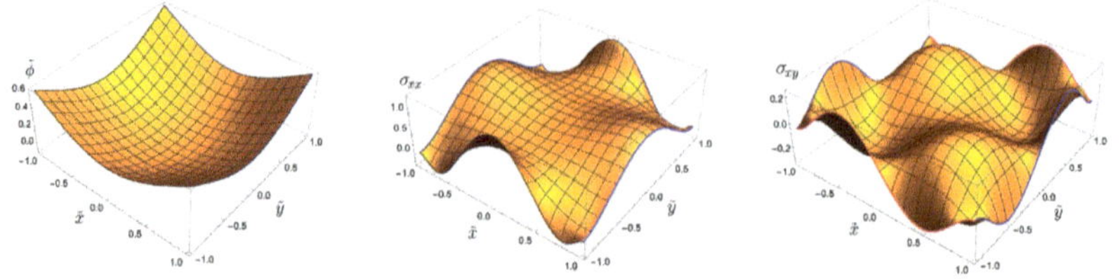

(**a**) Airy stress function, $\tilde{\phi}$ (**b**) Normal stresses, σ_{xx} (**c**) Tangent stresses, σ_{xy}, σ_{yx}

Figure 8. The Airy function and corresponding stress distributions.

4.2. Plate with Movable Edges

Consider now the first main boundary-value problem (14)–(17) for a square plate. We will seek the solution in the following form:

$$\tilde{w} = \sum_{k=1}^{\infty} g_k \tilde{f}_k, \qquad \tilde{\phi} = \sum_{k=1}^{\infty} (s_k + \zeta_k)\tilde{f}_k + \tilde{\phi}^*. \tag{39}$$

Due to the specificity of boundary conditions (17), the solution of the auxiliary boundary problem can be significantly simplified:

$$\tilde{\phi}^* = \alpha_0\left(\frac{\tilde{x}^2}{4} + \frac{\tilde{y}^2}{4}\right) + \sum_{k=1}^{\infty} \alpha_k\left(\tilde{\phi}^*_{\alpha k} + \tilde{\phi}^*_{\beta k}\right). \tag{40}$$

Expressions for the coefficients of the other part of the auxiliary solution (38) can be written in the following form:

$$\zeta = G^{-1}\cdot(A + B)\cdot\boldsymbol{\alpha}. \tag{41}$$

Although Equation (14) are non-linear, the bilinearity of the Monge–Ampère form (3) allows us to use the expansions (39) and (40) for reducing them to the system of algebraic equations:

$$\begin{gathered}\sum_{k=1}^{\infty} g_k\left\{\nabla^2\nabla^2\tilde{f}_k - \frac{\alpha_0}{2}\left(\frac{\partial^2\tilde{f}_k}{\partial\tilde{y}^2} + \frac{\partial^2\tilde{f}_k}{\partial\tilde{x}^2}\right) - \sum_{r=1}^{\infty}\left[(s_r + \zeta_r)L(\tilde{f}_k, \tilde{f}_r) + \alpha_r L\left(\tilde{f}_k, \tilde{\phi}^*_{\alpha r} + \tilde{\phi}^*_{\beta r}\right)\right]\right\} = \tilde{p}, \\ \sum_{k=1}^{\infty}\left\{s_k\nabla^2\nabla^2\tilde{f}_k + \kappa g_k\sum_{r=1}^{\infty} g_r L(\tilde{f}_k, \tilde{f}_r)\right\} = 0.\end{gathered} \tag{42}$$

Boundary conditions (15) and (17) are satisfied automatically due to an appropriate choice of basis functions in (39) and (40). Apart from that, the symmetry of the problem allows us to consider boundary conditions (16) satisfied if at least one of them is met. Then, substituting (39) and (40) into, for example, the first condition from (16), we have

$$\sum_{k=1}^{\infty}\left\{\int_0^1\left[\frac{\partial^2}{\partial\tilde{y}^2}\left(s_k\tilde{f}_k + \zeta_k\tilde{f}_k + \alpha_k\tilde{\phi}^*_{\beta k}\right) - \kappa g_k\sum_{r=1}^{\infty} g_r\frac{\partial\tilde{f}_k}{\partial\tilde{x}}\frac{\partial\tilde{f}_r}{\partial\tilde{x}}\right]d\tilde{x}\right\} + \frac{\alpha_0}{2}(1-\nu) = 0. \tag{43}$$

As before, one can act on these equations by projector operator and reduce them to an infinite-dimensional system of algebraic equations. For Equation (42) it is possible to use operator (27):

$$\begin{gathered}\sum_{k=1}^{\infty} g_k\left\{\zeta_k\delta_{qk} + \int_{-1}^{1}\int_{-1}^{1} 2\frac{\partial^4\tilde{f}_k}{\partial\tilde{x}^2\partial\tilde{y}^2}\tilde{f}_q\,d\tilde{x}\,d\tilde{y} - \frac{\alpha_0}{2}\int_{-1}^{1}\int_{-1}^{1}\left(\frac{\partial^2\tilde{f}_k}{\partial\tilde{y}^2} + \frac{\partial^2\tilde{f}_k}{\partial\tilde{x}^2}\right)\tilde{f}_q\,d\tilde{x}\,d\tilde{y}\right. \\ \left.-\sum_{r=1}^{\infty}\left[(s_r + \zeta_r)\int_{-1}^{1}\int_{-1}^{1} L(\tilde{f}_k, \tilde{f}_r)\tilde{f}_q\,d\tilde{x}\,d\tilde{y} + \alpha_r\int_{-1}^{1}\int_{-1}^{1} L\left(\tilde{f}_k, \tilde{\phi}^*_{\alpha r} + \tilde{\phi}^*_{\beta r}\right)\tilde{f}_q\,d\tilde{x}\,d\tilde{y}\right]\right\} = \int_{-1}^{1}\int_{-1}^{1}\tilde{p}\tilde{f}_q\,d\tilde{x}\,d\tilde{y}, \\ \sum_{k=1}^{\infty}\left\{s_k\left[\zeta_k\delta_{qk} + \int_{-1}^{1}\int_{-1}^{1} 2\frac{\partial^4\tilde{f}_k}{\partial\tilde{x}^2\partial\tilde{y}^2}\tilde{f}_q\,d\tilde{x}\,d\tilde{y}\right] + \kappa g_k\sum_{r=1}^{\infty} g_r\int_{-1}^{1}\int_{-1}^{1} L(\tilde{f}_k, \tilde{f}_r)\tilde{f}_q\,d\tilde{x}\,d\tilde{y}\right\} = 0.\end{gathered} \tag{44}$$

while boundary relation (43) should be subjected by the action of a one-dimensional projector operator. Since an auxiliary part of the Airy function is presented through cosines, it is convenient to use the projector operators with a cosine kernel:

$$\mathcal{F}_0(F(v)) = \int_{-1}^{1} F\,dv, \qquad \mathcal{F}_q^c(F(v)) = \int_{-1}^{1} F\cos(\pi q v)\,dv, \quad q = 1\ldots\infty.$$

The action of these operators on the boundary relation results in

$$-\kappa\sum_{k=1}^{\infty}\sum_{r=1}^{\infty}\int_{-1}^{1}\int_{0}^{1} g_k g_r \frac{\partial\tilde{f}_k}{\partial\tilde{x}}\frac{\partial\tilde{f}_r}{\partial\tilde{x}}\,d\tilde{x}\,d\tilde{y} + \alpha_0(1-\nu) = 0, \tag{45}$$

$$\sum_{k=1}^{\infty}\left\{\int_{-1}^{1}\int_{0}^{1}\left[(s_k+\zeta_k)\frac{\partial^2\tilde{f}_k}{\partial\tilde{y}^2}\cos(\pi q\tilde{y}) - \kappa g_k\sum_{r=1}^{\infty} g_r\frac{\partial\tilde{f}_k}{\partial\tilde{x}}\frac{\partial\tilde{f}_r}{\partial\tilde{x}}\cos(\pi q\tilde{y})\right]d\tilde{x}\,d\tilde{y} + \alpha_k\delta_{qk}\right\} = 0.$$

For solving systems (44) and (45), it is necessary to reduce it to a finite-dimensional one. The application of the reduction method to an infinite system of linear equations is validated by well-developed theory rooted in early works of Koyalovich [92,93]. However, there is no rigorous foundation for its application to non-linear systems in the present. In spite of this fact, the reduction method is widely used [16,67,70] and provides results in good agreement with practice. "Why should I refuse a good dinner simply because I do not understand the digestive processes involved?" (Oliver Heaviside).

So, applying the reduction method, one can transform (44) and (45) to the following system:

$$\begin{gathered}\left[\boldsymbol{G} - \frac{\alpha_0}{2}\left(\boldsymbol{M}^{\alpha 0} + \boldsymbol{M}^{\beta 0}\right) - \boldsymbol{M}\cdot(\boldsymbol{s}+\boldsymbol{\zeta}) - \left(\boldsymbol{M}^{\alpha} + \boldsymbol{M}^{\beta}\right)\cdot\boldsymbol{\alpha}\right]\cdot\boldsymbol{g} = \boldsymbol{b},\\ \boldsymbol{G}\cdot\boldsymbol{s} + \kappa(\boldsymbol{M}\cdot\boldsymbol{g})\cdot\boldsymbol{g} = \boldsymbol{0},\\ -\kappa\left(\boldsymbol{U}^{w0}\cdot\boldsymbol{g}\right)\cdot\boldsymbol{g} + \alpha_0(1-\nu) = 0, \qquad \boldsymbol{U}\cdot(\boldsymbol{s}+\boldsymbol{\zeta}) + \boldsymbol{E}\cdot\boldsymbol{\alpha} - \kappa(\boldsymbol{U}^{w}\cdot\boldsymbol{g})\cdot\boldsymbol{g} = \boldsymbol{0},\end{gathered} \tag{46}$$

where $\boldsymbol{M}^{\alpha 0}$, $\boldsymbol{M}^{\beta 0}$, $\boldsymbol{U}^{w0}$, $\boldsymbol{U}$ are matrices $n\times n$ and $\boldsymbol{M}^{\alpha}$, $\boldsymbol{M}^{\beta}$, $\boldsymbol{U}^{w}$, $\boldsymbol{M}$ are cubic matrices $n\times n\times n$, the expression for the components of which are presented in Table 3.

Table 3. Components of auxiliary structures (movable edges case).

Structure	Components Expression
matrix $\boldsymbol{M}^{\alpha 0}$	$M^{\alpha 0}_{qk} = -\frac{\lambda^2_{C(k)} I_G\left(\lambda_{C(q)},\lambda_{C(k)}\right)}{\lVert f_k\rVert\lVert f_q\rVert}\left(\cos^2\left(\lambda_{S(k)}\right) + \cosh^2\left(\lambda_{S(k)}\right)\right)\delta_{S(q)S(k)}$
matrix $\boldsymbol{M}^{\beta 0}$	$M^{\beta 0}_{qk} = -\frac{\lambda^2_{S(k)} I_G\left(\lambda_{S(q)},\lambda_{S(k)}\right)}{\lVert f_k\rVert\lVert f_q\rVert}\left(\cos^2\left(\lambda_{C(k)}\right) + \cosh^2\left(\lambda_{C(k)}\right)\right)\delta_{C(q)C(k)}$
matrix $\boldsymbol{U}^{w0}$	$U^{w0}_{qk} = \frac{\lambda_{S(k)}\lambda_{S(q)} I_U\left(\lambda_{S(q)},\lambda_{S(k)}\right)}{2\lVert f_k\rVert\lVert f_q\rVert}\left(\cos^2\left(\lambda_{C(k)}\right) + \cosh^2\left(\lambda_{C(k)}\right)\right)\delta_{C(q)C(k)}$
matrix $\boldsymbol{U}$	$U_{qk} = -\frac{\lambda^2_{C(k)}\left(I_4\left(\lambda_{C(k)},q\right) + I_5\left(\lambda_{C(k)},q\right)\right) I_0\left(\lambda_{S(k)}\right)}{2\lVert f_k\rVert}$

Table 3. *Cont.*

Structure	Components Expression
cubic matrix $\boldsymbol{M}$	$M_{qkr} = \frac{I_M\left(\lambda_{S(q)}, \lambda_{S(k)}, \lambda_{S(r)}, \lambda_{C(q)}, \lambda_{C(k)}, \lambda_{C(r)}\right)}{\\|f_k\\|\\|f_r\\|\\|f_q\\|}$
cubic matrix $\boldsymbol{M}^{\alpha}$	$M^{\alpha}_{qkr} = -\frac{\lambda^2_{C(k)} I_{M3}\left(\lambda_{S(q)}, \lambda_{S(k)}, \pi r\right) I_G\left(\lambda_{C(q)}, \lambda_{C(k)}\right)}{\\|f_k\\|\\|f_q\\| \cosh(\pi r)}$
cubic matrix $\boldsymbol{M}^{\beta}$	$M^{\beta}_{qkr} = -\frac{\lambda^2_{S(k)} I_{M3}\left(\lambda_{C(q)}, \lambda_{C(k)}, \pi r\right) I_G\left(\lambda_{S(q)}, \lambda_{S(k)}\right)}{\\|f_k\\|\\|f_q\\| \cosh(\pi r)}$
cubic matrix $\boldsymbol{U}^{w}$	$U^{w}_{qkr} = \frac{\lambda_{S(k)}\lambda_{S(r)} I_{M3}\left(\lambda_{C(k)}, \lambda_{C(r)}, \pi q\right) I_U\left(\lambda_{S(k)}, \lambda_{S(r)}\right)}{2\\|f_k\\|\\|f_r\\| \cosh(\pi q)}$

Turning now to the solution of the constructed finite-dimensional algebraic system. With account of (41), exclude the auxiliary vector $\boldsymbol{\zeta}$ from the last equation of (46):

$$\boldsymbol{U}\cdot\boldsymbol{s} + \left[\boldsymbol{U}\cdot\boldsymbol{G}^{-1}\cdot(\boldsymbol{A}+\boldsymbol{B}) + \boldsymbol{E}\right]\cdot\boldsymbol{\alpha} - \kappa(\boldsymbol{U}^{w}\cdot\boldsymbol{g})\cdot\boldsymbol{g} = \boldsymbol{0}. \tag{47}$$

Moreover, it is possible to express in turn vectors $\boldsymbol{s}$, $\boldsymbol{\alpha}$ and constant α_0 from system (46)–(47):

$$\boldsymbol{s} = -\kappa\left(\boldsymbol{G}^{-1}\cdot\boldsymbol{M}\cdot\boldsymbol{g}\right)\cdot\boldsymbol{g}, \qquad \alpha_0 = \frac{\kappa}{1-\nu}\left(\boldsymbol{U}^{w0}\cdot\boldsymbol{g}\right)\cdot\boldsymbol{g},$$
$$\boldsymbol{\alpha} = \kappa\left[\boldsymbol{U}\cdot\boldsymbol{G}^{-1}\cdot(\boldsymbol{A}+\boldsymbol{B}) + \boldsymbol{E}\right]^{-1}\cdot\left[\left(\boldsymbol{U}\cdot\boldsymbol{G}^{-1}\cdot\boldsymbol{M} + \boldsymbol{U}^{w}\right)\cdot\boldsymbol{g}\right]\cdot\boldsymbol{g},$$

and substitute them into the first equation of (46), then the whole system can be reduced to the following system of cubic equations:

$$\boldsymbol{G}\cdot\boldsymbol{g} - \kappa\left[(\boldsymbol{S}\cdot\boldsymbol{g})\cdot\boldsymbol{g} + \frac{1}{2(1-\nu)}\left(\left(\boldsymbol{U}^{w0}\cdot\boldsymbol{g}\right)\cdot\boldsymbol{g}\right)\left(\boldsymbol{M}^{\alpha 0} + \boldsymbol{M}^{\beta 0}\right)\right]\cdot\boldsymbol{g} = \boldsymbol{b}. \tag{48}$$

Here, $\boldsymbol{S}$ is the quadratic matrix with known components:

$$\boldsymbol{S} = -\boldsymbol{M}\cdot\boldsymbol{G}^{-1}\cdot\boldsymbol{M} + \left[\boldsymbol{M}\cdot\boldsymbol{G}^{-1}\cdot(\boldsymbol{A}+\boldsymbol{B}) + \boldsymbol{M}^{\alpha} + \boldsymbol{M}^{\beta}\right]\cdot\boldsymbol{R},$$

where $\boldsymbol{R}$ is an auxiliary cubic matrix deriving through known matrices:

$$\boldsymbol{R} = \left[\boldsymbol{U}\cdot\boldsymbol{G}^{-1}\cdot(\boldsymbol{A}+\boldsymbol{B}) + \boldsymbol{E}\right]^{-1}\cdot\left(\boldsymbol{U}\cdot\boldsymbol{G}^{-1}\cdot\boldsymbol{M} + \boldsymbol{U}^{w}\right).$$

Although calculation of even finite numbers of components of cubic matrix $\boldsymbol{S}$ requires some rather cumbersome calculations, it does not present any insurmountable difficulties.

Remark 2. *In general, the system of cubic equations allows multiple solutions, so it is usually necessary to analyze the branching of them [94]. However, if the in-plane stresses are of tensile type, the solution is unique [78,79].*

To solve the cubic Equation (48), it is possible to use Newton's method for linearizing the equation at each iteration. Denote a non-linear algebraic operator in LHS of (48) as $\mathcal{A}$:

$$\mathcal{A}(\boldsymbol{g}) := \boldsymbol{G}\cdot\boldsymbol{g} - \kappa\left[(\boldsymbol{S}\cdot\boldsymbol{g})\cdot\boldsymbol{g} + \frac{1}{2(1-\nu)}\left(\left(\boldsymbol{U}^{w0}\cdot\boldsymbol{g}\right)\cdot\boldsymbol{g}\right)\left(\boldsymbol{M}^{\alpha 0} + \boldsymbol{M}^{\beta 0}\right)\right]\cdot\boldsymbol{g}.$$

Let g_0 be the initial value of the solution obtained from the linear problem (32) or as an expansion of some approximate solution. Then, the cubic equation can be reduced to a linear one in the neighbour of g_0 as follows:

$$\mathcal{A}(g) = \mathcal{A}(g_0) + \frac{\partial \mathcal{A}}{\partial g}\bigg|_{g=g_0} \cdot (g - g_0) + o(g - g_0). \tag{49}$$

where

$$\begin{aligned}
\frac{\partial \mathcal{A}}{\partial g}\bigg|_{g=g_0} &= \sum_{k=1}^{n} G \cdot E_{kk} - \kappa \sum_{k=1}^{n} \{[(S \cdot E_{kk}) \cdot g_0] \cdot g_0 + [(S \cdot g_0) \cdot E_{kk}] \cdot g_0 + [(S \cdot g_0) \cdot g_0] \cdot E_{kk}\} \\
&- \frac{\kappa}{2(1-\nu)} \sum_{k=1}^{n} \Big\{ \Big[\Big(U^{w0} \cdot E_{kk} \Big) \cdot g_0 + \Big(U^{w0} \cdot g_0 \Big) \cdot E_{kk} \Big] \Big(M^{\alpha 0} + M^{\beta 0} \Big) \cdot g_0 \\
&+ \Big(\Big(U^{w0} \cdot g_0 \Big) \cdot g_0 \Big) \Big(M^{\alpha 0} + M^{\beta 0} \Big) \cdot E_{kk} \Big\}.
\end{aligned}$$

4.3. Plate with Immovable Edges

Consider the second main boundary-value problem, defined by boundary conditions (18) In this case the additional term appears in the right-hand side of (40):

$$\tilde{\Phi}^* = [\ldots] + \gamma_0 \Big(\tilde{\Phi}^*_{\gamma 0} + \tilde{\Phi}^*_{\delta 0} \Big) + \sum_{k=1}^{\infty} \gamma_k \Big(\tilde{\Phi}^*_{\gamma k} + \tilde{\Phi}^*_{\delta k} \Big).$$

Hereafter, we denote terms contained in corresponding equations in the previous case by the symbol $[\ldots]$. The first and the third equations of system (42) are modified as follows:

$$[\ldots] - \sum_{k=1}^{\infty} g_k \left[\gamma_0 L\Big(f_k, \tilde{\Phi}^*_{\gamma 0} + \tilde{\Phi}^*_{\delta 0}\Big) + \sum_{r=1}^{\infty} \gamma_r L\Big(f_k, \tilde{\Phi}^*_{\gamma r} + \tilde{\Phi}^*_{\delta r}\Big) \right] = \tilde{p},$$

$$[\ldots] + \int_0^1 \left\{ \gamma_0 \left[\frac{\partial^2}{\partial \tilde{y}^2} \Big(\tilde{\Phi}^*_{\gamma 0} + \tilde{\Phi}^*_{\delta 0} \Big) - \nu \frac{\partial^2}{\partial \tilde{x}^2} \tilde{\Phi}^*_{\delta 0} \right] + \sum_{k=1}^{\infty} \gamma_k \left[\frac{\partial^2}{\partial \tilde{y}^2} \Big(\tilde{\Phi}^*_{\gamma k} + \tilde{\Phi}^*_{\delta k} \Big) - \nu \frac{\partial^2}{\partial \tilde{x}^2} \tilde{\Phi}^*_{\delta k} \right] \right\} d\tilde{x} = 0,$$

while the second equation remains unchangeable. Apart from that, this system must be added by one more boundary condition equation:

$$\left[\frac{\alpha_0}{2}(1-\nu) + \gamma_0 \frac{\partial^2 \tilde{\Phi}^*_{\gamma 0}}{\partial \tilde{x}^2} + \sum_{k=1}^{\infty} \left((s_k + \zeta_k) \frac{\partial^2 \tilde{f}_k}{\partial \tilde{x}^2} + \alpha_k \frac{\partial^2 \tilde{\Phi}^*_{\alpha k}}{\partial \tilde{x}^2} + \gamma_k \frac{\partial^2 \tilde{\Phi}^*_{\gamma k}}{\partial \tilde{x}^2} - \nu \alpha_k \frac{\partial^2 \tilde{\Phi}^*_{\beta k}}{\partial \tilde{y}^2} \right) \right] \Bigg|_{\tilde{x}=1} = 0.$$

As in the previous case, projecting into basis functions with subsequent reduction allows us to obtain a finite system of algebraic equations:

$$\begin{aligned}
\Big[G - \frac{\alpha_0}{2} \Big(M^{\alpha 0} + M^{\beta 0} \Big) - \gamma_0 \Big(M^{\gamma 0} + M^{\delta 0} \Big) - M \cdot (s + \zeta) \\
- \Big(M^{\alpha} + M^{\beta} \Big) \cdot \alpha - \Big(M^{\gamma} + M^{\delta} \Big) \cdot \gamma \Big] \cdot g &= b, \\
G \cdot s + \kappa (M \cdot g) \cdot g &= 0, \\
U \cdot (s + \zeta) + E \cdot \alpha + \Big(U^{\gamma} + U^{\delta 1} - \nu U^{\delta 2} \Big) \cdot \gamma + \gamma_0 \Big(u^{\gamma} + u^{\delta 1} - \nu u^{\delta 2} \Big) - \kappa (U^{w} \cdot g) \cdot g &= 0, \\
V^{\gamma} \cdot \gamma + V \cdot (s + \zeta) + \gamma_0 v^{\gamma} - \nu E \cdot \alpha &= 0, \\
\Big(u^{\gamma 0} - \nu u^{\delta 0} \Big) \cdot \gamma - \kappa \Big(U^{w0} \cdot g \Big) \cdot g + \alpha_0 (1-\nu) - \frac{2\gamma_0}{\pi}(1-\nu) &= 0, \\
v^{\gamma 0} \cdot \gamma + v^{0} \cdot (s + \zeta) + v^{\alpha 0} \cdot \alpha + \alpha_0 (1-\nu) + \frac{8\gamma_0}{\pi} &= 0.
\end{aligned}$$

Here, the additional vectors, matrices, and cubic matrices are used. The expressions for components of some of it are presented in Table 4. Components of vectors $\boldsymbol{u}^{\gamma}$, $\boldsymbol{u}^{\delta 1}$, $\boldsymbol{u}^{\delta 2}$, $\boldsymbol{v}^{\gamma}$ and matrices $\boldsymbol{M}^{\gamma 0}$, $\boldsymbol{M}^{\delta 0}$ can be found from corresponding matrices $\boldsymbol{U}^{\gamma}$, $\boldsymbol{U}^{\delta 1}$, $\boldsymbol{U}^{\delta 2}$, $\boldsymbol{V}^{\gamma}$ and cubic matrices $\boldsymbol{M}^{\gamma}$, $\boldsymbol{M}^{\delta}$ if we assume their last indexes as zero.

Table 4. Components of auxiliary structures (immovable edges case).

Structure	Components Expression
vectors $\boldsymbol{u}^{\gamma 0}, \boldsymbol{u}^{\delta 0}$	$u_q^{\gamma 0} = u_q^{\delta 0} = -\dfrac{2(-1)^q}{\pi(q+1)}$
vector $\boldsymbol{v}^{\gamma 0}$	$v_q^{\gamma 0} = \dfrac{8(-1)^q(q+1)}{\pi}$
vector $\boldsymbol{v}^{\alpha 0}$	$v_q^{\alpha 0} = 2(-1)^q$
vector $\boldsymbol{v}^{0}$	$v_q^{0} = -\dfrac{2\lambda_{S(q)}^2 \cos\left(\lambda_{S(q)}\right)\cosh\left(\lambda_{S(q)}\right) I_0\left(\lambda_{C(q)}\right)}{\Vert f_q \Vert}$
matrix $\boldsymbol{U}^{\gamma}$	$U_{qk}^{\gamma} = \dfrac{2(-1)^{q+k}}{\pi(k+1)(4q^2-1)}$
matrix $\boldsymbol{U}^{\delta 1}$	$U_{qk}^{\delta 1} = -4q\pi^{-1}\delta_{q(k+1)}$
matrix $\boldsymbol{U}^{\delta 2}$	$U_{qk}^{\delta 2} = -(\pi q)^{-1}\delta_{q(k+1)}$
matrix $\boldsymbol{V}^{\gamma}$	$V_{qk}^{\gamma} = -\dfrac{8(k+1)(-1)^{q+k}}{\pi(4q^2-1)}$
matrix $\boldsymbol{V}$	$V_{qk} = -\dfrac{2\lambda_{S(k)}^2 \cos\left(\lambda_{S(k)}\right)\cosh\left(\lambda_{S(k)}\right)\left(I_4\left(\lambda_{C(k)}, q\right) - I_5\left(\lambda_{C(k)}, q\right)\right)}{\Vert f_k \Vert}$
cubic matrix $\boldsymbol{M}^{\gamma}$	$M_{qkr}^{\gamma} = \dfrac{I_B\left(\lambda_{S(q)}, \lambda_{S(k)}, \lambda_{C(q)}, \lambda_{C(k)}, r+1\right)}{\Vert f_k \Vert \Vert f_q \Vert}$
cubic matrix $\boldsymbol{M}^{\delta}$	$M_{qkr}^{\delta} = \dfrac{I_B\left(\lambda_{C(q)}, \lambda_{C(k)}, \lambda_{S(q)}, \lambda_{S(k)}, r+1\right)}{\Vert f_k \Vert \Vert f_q \Vert}$

Turn to solve this system. First, scalars γ^0 can be expressed through other variables as follows:

$$\begin{aligned}\gamma_0 = &-\Big\{\left[\boldsymbol{v}^{\alpha 0} + \boldsymbol{v}^0\cdot\boldsymbol{G}^{-1}\cdot(\boldsymbol{A}+\boldsymbol{B})\right]\cdot\boldsymbol{\alpha} + \left[\boldsymbol{v}^{\gamma 0} + \boldsymbol{v}^0\cdot\boldsymbol{G}^{-1}\cdot(\boldsymbol{C}+\boldsymbol{D}) - \boldsymbol{u}^{\gamma 0} + \nu\boldsymbol{u}^{\delta 0}\right]\cdot\boldsymbol{\gamma} \\ &+\kappa\left[\left(\boldsymbol{U}^{w0} - \boldsymbol{v}^0\cdot\boldsymbol{G}^{-1}\cdot\boldsymbol{M}\right)\cdot\boldsymbol{g}\right]\cdot\boldsymbol{g}\Big\}\left[\frac{2(5-\nu)}{\pi} + \boldsymbol{v}^0\cdot\boldsymbol{G}^{-1}\cdot\left(\boldsymbol{C}^0+\boldsymbol{D}^0\right)\right]^{-1} \\ =:&\ \mathcal{S}_{\gamma\alpha}^0\cdot\boldsymbol{\alpha} + \mathcal{S}_{\gamma\gamma}^0\cdot\boldsymbol{\gamma} + \left(\mathcal{S}_{\gamma g}^0\cdot\boldsymbol{g}\right)\cdot\boldsymbol{g}.\end{aligned}$$

Further, one can express vector γ:

$$\begin{aligned}\boldsymbol{\gamma} = &-\left[\boldsymbol{V}^{\gamma} + \boldsymbol{V}\cdot\boldsymbol{G}^{-1}\cdot(\boldsymbol{C}+\boldsymbol{D}) + \left(\boldsymbol{v}^{\gamma} + \boldsymbol{V}\cdot\boldsymbol{G}^{-1}\cdot\left(\boldsymbol{C}^0+\boldsymbol{D}^0\right)\right)\otimes\mathcal{S}_{\gamma\gamma}^0\right]^{-1} \\ &\cdot\Big\{\left[\left(\boldsymbol{v}^{\gamma} + \boldsymbol{V}\cdot\boldsymbol{G}^{-1}\cdot\left(\boldsymbol{C}^0+\boldsymbol{D}^0\right)\right)\otimes\mathcal{S}_{\gamma\alpha}^0 + \boldsymbol{V}\cdot\boldsymbol{G}^{-1}\cdot(\boldsymbol{A}+\boldsymbol{B}) - \nu\boldsymbol{E}\right]\cdot\boldsymbol{\alpha} \\ &+\left(\boldsymbol{v}^{\gamma} + \boldsymbol{V}\cdot\boldsymbol{G}^{-1}\cdot\left(\boldsymbol{C}^0+\boldsymbol{D}^0\right)\right)\left(\left(\mathcal{S}_{\gamma g}^0\cdot\boldsymbol{g}\right)\cdot\boldsymbol{g}\right) - \kappa\left(\boldsymbol{V}\cdot\boldsymbol{G}^{-1}\cdot\boldsymbol{M}\cdot\boldsymbol{g}\right)\cdot\boldsymbol{g}\Big\} \\ =:&\ \mathcal{S}_{\gamma\alpha}\cdot\boldsymbol{\alpha} + \left(\mathcal{S}_{\gamma g}\cdot\boldsymbol{g}\right)\cdot\boldsymbol{g} + \mathcal{T}_{\gamma g}\left(\mathcal{S}_{\gamma g}^0\cdot\boldsymbol{g}\right)\cdot\boldsymbol{g},\end{aligned}$$

where $\otimes$ is the Kronecker product for vectors, which assigns a matrix to a pair of vectors by the following rule:

$$\boldsymbol{a} \otimes \boldsymbol{b} := \boldsymbol{M}, \quad \boldsymbol{M}_{ij} = a_i b_j.$$

Then $\alpha 0$, $\boldsymbol{\alpha}$:

$$\begin{aligned}
\boldsymbol{\alpha} =& \Big\{ \boldsymbol{U}\cdot\boldsymbol{G}^{-1}\cdot\Big[(\boldsymbol{A}+\boldsymbol{B}) + (\boldsymbol{C}+\boldsymbol{D})\cdot\mathcal{S}_{\gamma\alpha} + \left(\boldsymbol{C}^0+\boldsymbol{D}^0\right)\otimes\left(\mathcal{S}^0_{\gamma\gamma}\cdot\mathcal{S}_{\gamma\alpha}+\mathcal{S}^0_{\gamma\alpha}\right)\Big] + \boldsymbol{E} \\
&+ \left(\boldsymbol{U}^\gamma+\boldsymbol{U}^{\delta 1}-\nu\boldsymbol{U}^{\delta 2}\right)\cdot\mathcal{S}_{\gamma\alpha} + \left(\boldsymbol{u}^\gamma+\boldsymbol{u}^{\delta 1}-\nu\boldsymbol{u}^{\delta 2}\right)\otimes\left(\mathcal{S}^0_{\gamma\gamma}\cdot\mathcal{S}_{\gamma\alpha}+\mathcal{S}^0_{\gamma\alpha}\right)\Big\}^{-1} \\
&\cdot\Big\{\Big[\Big(\kappa\boldsymbol{U}\cdot\boldsymbol{G}^{-1}\cdot\boldsymbol{M} - \boldsymbol{U}\cdot\boldsymbol{G}^{-1}\cdot(\boldsymbol{C}+\boldsymbol{D})\cdot\mathcal{S}_{\gamma g} - \left(\boldsymbol{U}^\gamma+\boldsymbol{U}^{\delta 1}-\nu\boldsymbol{U}^{\delta 2}\right)\cdot\mathcal{S}_{\gamma g} + \kappa\boldsymbol{U}^w \\
&- \left(\boldsymbol{U}\cdot\boldsymbol{G}^{-1}\cdot\left(\boldsymbol{C}^0+\boldsymbol{D}^0\right) + \left(\boldsymbol{u}^\gamma+\boldsymbol{u}^{\delta 1}-\nu\boldsymbol{u}^{\delta 2}\right)\right)\otimes\mathcal{S}^0_{\gamma\gamma}\cdot\mathcal{S}_{\gamma g}\Big)\cdot\boldsymbol{g}\Big]\cdot\boldsymbol{g} \\
&- \Big[\left(\boldsymbol{U}\cdot\boldsymbol{G}^{-1}\cdot(\boldsymbol{C}+\boldsymbol{D}) + \left(\boldsymbol{U}^\gamma+\boldsymbol{U}^{\delta 1}-\nu\boldsymbol{U}^{\delta 2}\right)\right)\cdot\mathcal{T}_{\gamma g} \\
&- \left(\boldsymbol{U}\cdot\boldsymbol{G}^{-1}\cdot\left(\boldsymbol{C}^0+\boldsymbol{D}^0\right) + \left(\boldsymbol{u}^\gamma+\boldsymbol{u}^{\delta 1}-\nu\boldsymbol{u}^{\delta 2}\right)\right)\left(1+\mathcal{S}^0_{\gamma\gamma}\cdot\mathcal{T}_{\gamma g}\right)\Big]\left(\mathcal{S}^0_{\gamma g}\cdot\boldsymbol{g}\right)\cdot\boldsymbol{g}\Big\} \\
=:& \left(\mathcal{S}_{\alpha g}\cdot\boldsymbol{g}\right)\cdot\boldsymbol{g} + \mathcal{T}_{\alpha g}\left(\mathcal{S}^0_{\gamma g}\cdot\boldsymbol{g}\right)\cdot\boldsymbol{g},
\end{aligned}$$

$$\begin{aligned}
\boldsymbol{\alpha}_0 =& \frac{1}{1-\nu}\Big\{\Big[\Big(\frac{2}{\pi}(1-\nu)\left(\mathcal{S}^0_{\gamma\alpha}\cdot\mathcal{S}_{\alpha g}+\mathcal{S}^0_{\gamma\gamma}\cdot\mathcal{S}_{\gamma\alpha}\cdot\mathcal{S}_{\alpha g}+\mathcal{S}^0_{\gamma\gamma}\cdot\mathcal{S}_{\gamma g}+\mathcal{S}^0_{\gamma g}\right) \\
&- \left(\boldsymbol{u}^{\gamma 0}-\nu\boldsymbol{u}^{\delta 0}\right)\cdot\left(\mathcal{S}_{\gamma\alpha}\cdot\mathcal{S}_{\alpha g}+\mathcal{S}_{\gamma g}\right) + \kappa\boldsymbol{U}^{w0}\Big)\cdot\boldsymbol{g}\Big]\cdot\boldsymbol{g} \\
&+ \Big[\frac{2}{\pi}(1-\nu)\left(\mathcal{S}^0_{\gamma\alpha}\cdot\mathcal{T}_{\alpha g}+\mathcal{S}^0_{\gamma\gamma}\cdot\mathcal{S}_{\gamma\alpha}\cdot\mathcal{T}_{\alpha g}+\mathcal{S}^0_{\gamma\gamma}\cdot\mathcal{T}_{\gamma g}\right) \\
&- \left(\boldsymbol{u}^{\gamma 0}-\nu\boldsymbol{u}^{\delta 0}\right)\cdot\left(\mathcal{S}_{\gamma\alpha}\cdot\mathcal{T}_{\alpha g}+\mathcal{T}_{\gamma g}\right)\Big)\Big]\cdot\left(\mathcal{S}^0_{\gamma g}\cdot\boldsymbol{g}\right)\cdot\boldsymbol{g}\Big\} =: \left(\mathcal{S}^0_{\alpha g}\cdot\boldsymbol{g}\right)\cdot\boldsymbol{g} + \mathcal{T}^0_{\alpha g}\left(\mathcal{S}^0_{\gamma g}\cdot\boldsymbol{g}\right)\cdot\boldsymbol{g}.
\end{aligned}$$

After introducing a set of auxiliary structures $\mathcal{S}$ and $\mathcal{T}$, one can simply reduce the full system to the cubic system:

$$\boldsymbol{G}\cdot\boldsymbol{g} - \left[\left(\tilde{\mathbf{S}}\cdot\boldsymbol{g}\right)\cdot\boldsymbol{g} + \left(\left(\mathcal{S}^0_{\gamma g}\cdot\boldsymbol{g}\right)\cdot\boldsymbol{g}\right)\tilde{\boldsymbol{T}}\right]\cdot\boldsymbol{g} = \boldsymbol{b},$$

where $\tilde{\mathbf{S}}$ is a quadric matrix:

$$\begin{aligned}
\tilde{\mathbf{S}} =& -\kappa\boldsymbol{M}\cdot\boldsymbol{G}^{-1}\cdot\boldsymbol{M} + \frac{1}{2}\left(\boldsymbol{M}^{\alpha 0}+\boldsymbol{M}^{\beta 0}\right)\cdot\mathcal{S}^0_{\alpha g} + \left[\boldsymbol{M}\cdot\boldsymbol{G}^{-1}\cdot(\boldsymbol{A}+\boldsymbol{B}) + \boldsymbol{M}^\alpha + \boldsymbol{M}^\beta\right]\cdot\mathcal{S}_{\alpha g} \\
&+ \left[\boldsymbol{M}\cdot\boldsymbol{G}^{-1}\cdot(\boldsymbol{C}+\boldsymbol{D}) + \boldsymbol{M}^\gamma + \boldsymbol{M}^\delta\right]\cdot\left(\mathcal{S}_{\gamma\alpha}\cdot\mathcal{S}_{\alpha g}+\mathcal{S}_{\gamma g}\right) \\
&+ \left[\boldsymbol{M}\cdot\boldsymbol{G}^{-1}\cdot\left(\boldsymbol{C}^0+\boldsymbol{D}^0\right) + \boldsymbol{M}^{\gamma 0} + \boldsymbol{M}^{\delta 0}\right]\cdot\left(\mathcal{S}^0_{\gamma\alpha}\cdot\mathcal{S}_{\alpha g}+\mathcal{S}^0_{\gamma\gamma}\cdot\mathcal{S}_{\gamma\alpha}\cdot\mathcal{S}_{\alpha g}+\mathcal{S}^0_{\gamma\gamma}\cdot\mathcal{S}_{\gamma g}\right)
\end{aligned}$$

and $\tilde{\boldsymbol{T}}$ is a vector:

$$\begin{aligned}
\tilde{\boldsymbol{T}} =& \frac{1}{2}\mathcal{T}^0_{\alpha g}\left(\boldsymbol{M}^{\alpha 0}+\boldsymbol{M}^{\beta 0}\right) + \left[\boldsymbol{M}\cdot\boldsymbol{G}^{-1}\cdot(\boldsymbol{A}+\boldsymbol{B}) + \boldsymbol{M}^\alpha + \boldsymbol{M}^\beta\right]\cdot\mathcal{T}_{\alpha g} \\
&+ \left[\boldsymbol{M}\cdot\boldsymbol{G}^{-1}\cdot(\boldsymbol{C}+\boldsymbol{D}) + \boldsymbol{M}^\gamma + \boldsymbol{M}^\delta\right]\cdot\left(\mathcal{S}_{\gamma\alpha}\cdot\mathcal{T}_{\alpha g}+\mathcal{T}_{\gamma g}\right) \\
&+ \left[\boldsymbol{M}\cdot\boldsymbol{G}^{-1}\cdot\left(\boldsymbol{C}^0+\boldsymbol{D}^0\right) + \boldsymbol{M}^{\gamma 0} + \boldsymbol{M}^{\delta 0}\right]\left(\mathcal{S}^0_{\gamma\alpha}\cdot\mathcal{T}_{\alpha g}+\mathcal{S}^0_{\gamma\gamma}\cdot\mathcal{S}_{\gamma\alpha}\cdot\mathcal{T}_{\alpha g}+\mathcal{S}^0_{\gamma\gamma}\cdot\mathcal{T}_{\gamma g}+1\right)
\end{aligned}$$

As in the previous case, the obtained cubic equation can be linearized by a formula similar to (49) and solved by Newton's method.

5. Numerical Results and Discussion

In order to verify the proposed solution, mechanical tests with thin, square, completely clamped plates were carried out. The essence of tests is as follows. A plate initially undergoes finite strain under given pressure p_0 and then reloads by small increments of additional pressure δp. Corresponding increments of normal deflections δw are determined with holographic interferometry [95]. To this end, the plates, fixed in a square, thick-walled steel chamber, were employed. The initial deformed state was considered a reference and recorded on a reference hologram. Superimposing on it another hologram, recorded after the increase in pressure, results in a fringe pattern that identifies contour lines of the deflection increment field δw. Repeating these experiments for various initial pressures and a number of incremental steps, one can obtain a comprehensive description of the non-linear deformation at different levels of finite strains.

The physical and geometric parameters of the plate are given in Table 5.

Table 5. Plate parameters.

Length, cm	Thickness, μm	Young Modulus, GPa	Poisson's Ratio
6	184	128	0.35

Experimentally obtained fields of deflection increments were compared with the results of calculations according to the proposed solution. Intermediate theoretical results for initial pressure $p_0 = 3$ kPa are shown in Figure 9. A comparison of theoretical and experimental fields of deflection increments for three different initial pressures, $p_0 = 0$, $p_0 = 3$ kPa, and $p_0 = 6$ kPa, and additional loading with equal increments in 10 Pa, is shown on the chart in Figure 10. On this chart, one can see experimentally and theoretically obtained maximal deflections (in the center of the plate), represented in terms of contour line quantities. This representation of the displacements is convenient for comparison with the holographic experimental method, because in the latter we directly obtain contour lines of the displacement with a step equal to half the measuring laser wavelength.

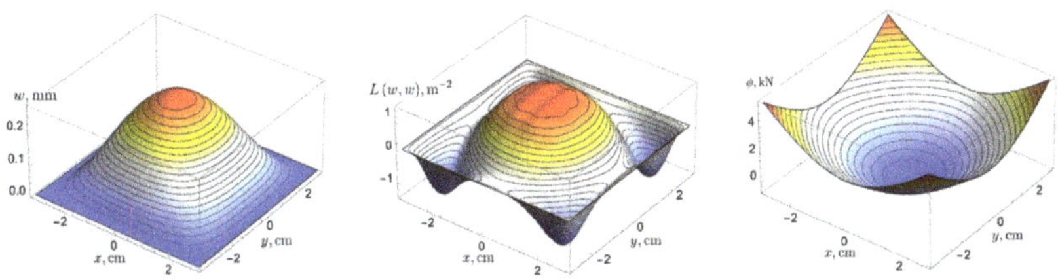

(**a**) Plate deflections, w (**b**) Incompatibility, $L(w_0, w_0)$ (**c**) Airy stress function, ϕ

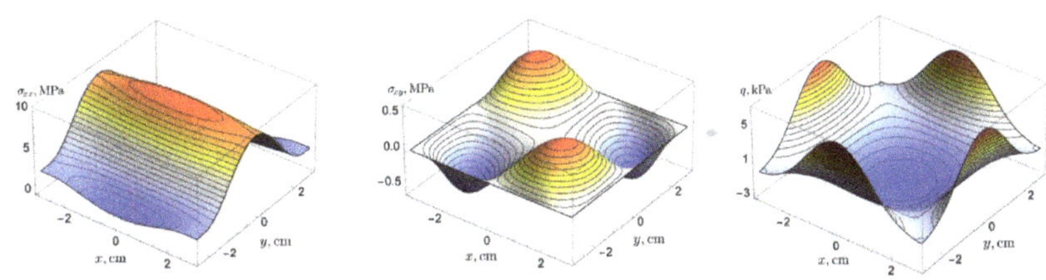

(**d**) Normal stresses, $\frac{\partial^2\phi}{\partial y^2}$ (**e**) Tangent stresses, $-\frac{\partial^2\phi}{\partial x \partial y}$ (**f**) Dummy loading, q

Figure 9. Distributions for the elements of the solution at initial pressure 3 kPa without incremental loading.

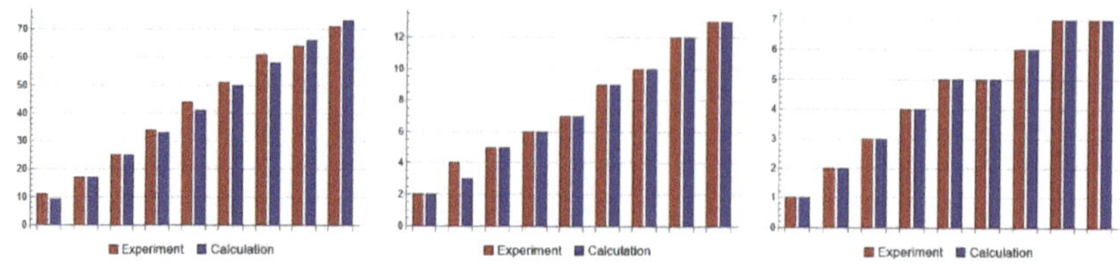

(a) $\delta w,\, p_0 = 0$ (b) $\delta w,\, p_0 = 3,$ kPa (c) $\delta w,\, p_0 = 6\,\text{kPa}$

Figure 10. Experimental and theoretical deflections of the plate central point in terms of contour line order (order increases by 1 when displacements increase by 266 nm).

A more detailed juxtaposition of the theoretical and experimental contours for the 5th and 8th steps of incremental loading is presented in Figure 11. The above results demonstrate a good agreement between theoretical results, obtained with the Föppl-von Kármán model and proposed solution, and the experimental data.

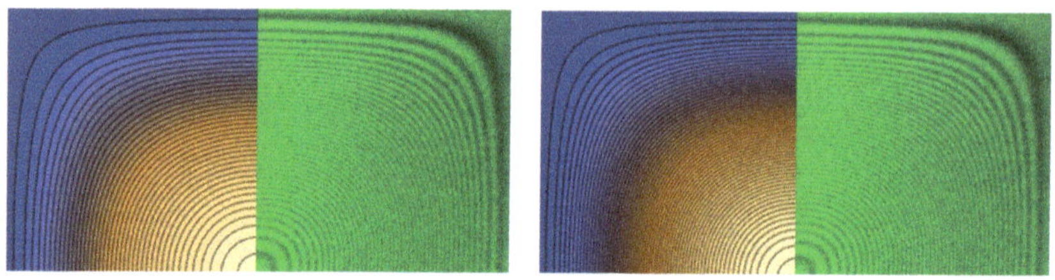

(a) $\delta w,\, p_0 = 0,\, \delta p = 50\,\text{Pa}$ (b) $\delta w,\, p_0 = 0,\, \delta p = 80\,\text{Pa}$

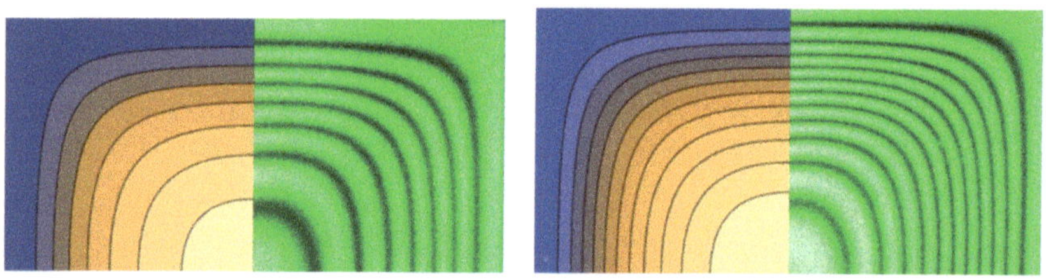

(c) $\delta w,\, p_0 = 3\,\text{kPa},\, \delta p = 50\,\text{Pa}$ (d) $\delta w,\, p_0 = 3\,\text{kPa},\, \delta p = 80\,\text{Pa}$

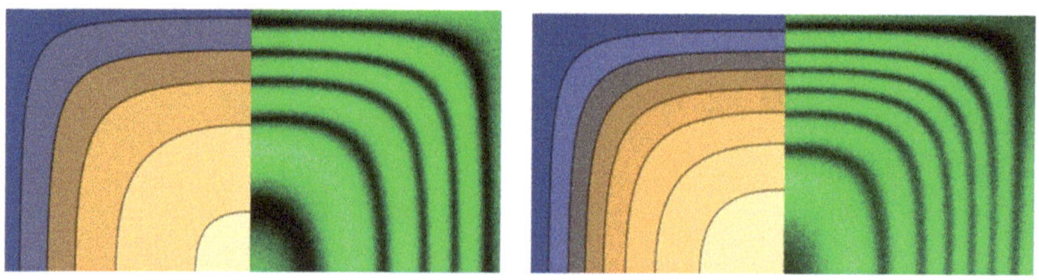

(e) $\delta w,\, p_0 = 6\,\text{kPa},\, \delta p = 50\,\text{Pa}$ (f) $\delta w,\, p_0 = 6\,\text{kPa},\, \delta p = 80\,\text{Pa}$

Figure 11. Juxtaposition of theoretical and experimental deflections under different initial and incremental pressure. Step of contour lines is taken to be equal to half laser wavelength (266 nm).

6. Conclusions

To sum up, we would like to highlight the key advantages of the proposed solution:

- The proposed solution, which reduces the boundary-value problem to the infinite-dimensional system of cubic algebraic equations with coefficients determined analytically via finite formula, can be an effective tool for modeling and design of flexible plates.

- The theoretical model accurately reproduces the tendency of the plate effective stiffness increasing with the increasing of in-plane tensile stresses state. Note that compressive in-plane stresses may decrease the effective stiffness even up to zero.
- The theoretical modeling shows a good agreement with the experiment. The comparison analysis demonstrates that the proposed solution allow us to describe plates bending both quantitatively and qualitatively.

Author Contributions: Conceptualization and methodology, S.L.; validation, S.L.; writing—original draft preparation, S.L., A.D. and N.D. All authors have read and agreed to the published version of the manuscript.

Funding: This research was supported by the Ministry of Science and Higher Education of the Russian Federation under contract No. 075-15-2021-1350, dated 5 October 2021 (internal number 15.SIN.21.0004).

Institutional Review Board Statement: Not applicable.

Informed Consent Statement: Not applicable.

Data Availability Statement: Data are contained within the article.

Conflicts of Interest: The authors declare no conflict of interest.

Abbreviations

The following abbreviations are used in this manuscript:

FEM	finite element method
LHS	left hand side
RHS	right hand side

Appendix A

The list of the auxiliary integrals using in the article is provided below (note that parameter k is integer and ρ, ν, ξ satisfy transcendental equation $\tan(.) + \tanh(.) = 0$):

$$I_0(\rho) = \int_{-1}^{1} (\cos(\rho v)\cosh(\rho) - \cos(\rho)\cosh(\rho v))dv = \frac{4\sin(\rho)\cosh(\rho)}{\rho}$$

$$I_1(\rho,\nu) = \cos(\rho)\cos(\nu)\int_{-1}^{1} \cosh(\rho v)\cosh(\nu v)dv$$

$$= \begin{cases} \dfrac{2\cosh(\rho)\cosh(\nu)(\rho\sin(\rho)\cos(\nu) - \nu\sin(\nu)\cos(\rho))}{\nu^2-\rho^2} & \rho \neq \nu \\ \dfrac{\rho\cos^2(\rho) - \sin(\rho)\cos(\rho)\cosh^2(\rho)}{\rho} & \rho = \nu \end{cases}$$

$$I_2(\rho,\nu) = \cos(\rho)\cosh(\nu)\int_{-1}^{1} \cosh(\rho v)\cos(\nu v)dv$$

$$= \frac{2\cos(\rho)\cosh(\nu)(\rho\sinh(\rho)\cos(\nu) + \nu\sin(\nu)\cosh(\rho))}{\nu^2+\rho^2}$$

$$I_3(\rho, \nu) = \cosh(\rho)\cosh(\nu)\int_{-1}^{1}\cos(\rho v)\cos(\nu v)dv$$
$$= \begin{cases} -\dfrac{2\cosh(\rho)\cosh(\nu)(\rho\sin(\rho)\cos(\nu) - \nu\sin(\nu)\cos(\rho))}{\nu^2-\rho^2} & \rho \neq \nu \\ \dfrac{\rho\cosh^2(\rho) + \sin(\rho)\cos(\rho)\cosh^2(\rho)}{\rho} & \rho = \nu \end{cases}$$

$$I_4(\rho, k) = \cosh(\rho)\int_{-1}^{1}\cos(\rho v)\cos(\pi k v)dv = \frac{2(-1)^k\rho\sin(\rho)\cosh(\rho)}{\rho^2-\pi^2k^2}$$

$$I_5(\rho, k) = \cos(\rho)\int_{-1}^{1}\cosh(\rho v)\cos(\pi k v)dv = -\frac{2(-1)^k\rho\sin(\rho)\cosh(\rho)}{\rho^2+\pi^2k^2}$$

$$I_6(\rho, \nu) = \cos(\rho)\cos(\nu)\int_{-1}^{1}\sinh(\rho v)\sinh(\nu v)dv$$
$$= \begin{cases} \dfrac{2\cosh(\rho)\cosh(\nu)(\rho\cos(\rho)\sin(\nu) - \nu\cos(\nu)\sin(\rho))}{\nu^2-\rho^2} & \rho \neq \nu \\ -\dfrac{\rho\cos^2(\rho) + \sin(\rho)\cos(\rho)\cosh^2(\rho)}{\rho} & \rho = \nu \end{cases}$$

$$I_7(\rho, \nu) = \cos(\rho)\cosh(\nu)\int_{-1}^{1}\sinh(\rho v)\sin(\nu v)dv$$
$$= \frac{2\cosh(\rho)\cosh(\nu)(\rho\cos(\rho)\sin(\nu) + \nu\cos(\nu)\sin(\rho))}{\nu^2+\rho^2}$$

$$I_8(\rho, \nu) = \cosh(\rho)\cosh(\nu)\int_{-1}^{1}\sin(\rho v)\sin(\nu v)dv$$
$$= \begin{cases} \dfrac{2\cosh(\rho)\cosh(\nu)(\rho\cos(\rho)\sin(\nu) - \nu\cos(\nu)\sin(\rho))}{\nu^2-\rho^2} & \rho \neq \nu \\ \dfrac{\rho\cosh^2(\rho) - \sin(\rho)\cos(\rho)\cosh^2(\rho)}{\rho} & \rho = \nu \end{cases}$$

$$I_9(\rho, \nu, \xi) = \cosh(\rho)\cosh(\nu)\cosh(\xi)\int_{-1}^{1}\cos(\rho v)\cos(\nu v)\cos(\xi v)dv$$
$$= \frac{1}{2}\left[\frac{\sin(\rho-\nu-\xi)}{\rho-\nu-\xi} + \frac{\sin(\rho+\nu-\xi)}{\rho+\nu-\xi} + \frac{\sin(\rho-\nu+\xi)}{\rho-\nu+\xi} + \frac{\sin(\rho+\nu+\xi)}{\rho+\nu+\xi}\right]$$
$$\times\cosh(\rho)\cosh(\nu)\cosh(\xi)$$

$$I_{10}(\rho, \nu, \xi) = \cos(\rho)\cosh(\nu)\cosh(\xi)\int_{-1}^{1}\cosh(\rho v)\cos(\nu v)\cos(\xi v)dv$$
$$= -\frac{2\cosh(\rho)\cosh(\nu)\cosh(\xi)}{\left(\rho^2+(\nu-\xi)^2\right)\left(\rho^2+(\nu+\xi)^2\right)}$$
$$\times\Big[2\rho\nu\xi\sin(\rho)\sin(\nu)\sin(\xi) + \rho\sin(\rho)\cos(\nu)\cos(\xi)\left(\rho^2+\nu^2+\xi^2\right)$$
$$-\nu\cos(\rho)\sin(\nu)\cos(\xi)\left(\rho^2+\nu^2-\xi^2\right) - \xi\cos(\rho)\cos(\nu)\sin(\xi)\left(\rho^2-\nu^2+\xi^2\right)\Big]$$

$$I_{11}(\rho, \nu, \xi) = \cos(\rho)\cos(\nu)\cosh(\xi)\int_{-1}^{1}\cosh(\rho v)\cosh(\nu v)\cos(\xi v)dv$$
$$= -\frac{2\cosh(\rho)\cosh(\nu)\cosh(\xi)}{(\rho^2-\nu^2)^2+\xi^2(2\rho^2+2\nu^2+\xi^2)}$$
$$\times \Big[2\rho\nu\xi\sin(\rho)\sin(\nu)\sin(\xi)+\rho\sin(\rho)\cos(\nu)\cos(\xi)\left(\rho^2-\nu^2+\xi^2\right)$$
$$+\nu\cos(\rho)\sin(\nu)\cos(\xi)\left(-\rho^2+\nu^2+\xi^2\right)-\xi\cos(\rho)\cos(\nu)\sin(\xi)\left(\rho^2+\nu^2+\xi^2\right)\Big]$$

$$I_{12}(\rho, \nu, \xi) = \cos(\rho)\cos(\nu)\cos(\xi)\int_{-1}^{1}\cosh(\rho v)\cosh(\nu v)\cosh(\xi v)dv$$
$$= \frac{1}{2}\left[\frac{\sinh(\rho-\nu-\xi)}{\rho-\nu-\xi}+\frac{\sinh(\rho+\nu-\xi)}{\rho+\nu-\xi}+\frac{\sinh(\rho-\nu+\xi)}{\rho-\nu+\xi}+\frac{\sinh(\rho+\nu+\xi)}{\rho+\nu+\xi}\right]$$
$$\times\cos(\rho)\cos(\nu)\cos(\xi)$$

$$I_{13}(\rho, \nu, \xi) = \cosh(\rho)\cosh(\nu)\cosh(\xi)\int_{-1}^{1}\cos(\rho v)\sin(\nu v)\sin(\xi v)dv$$
$$= \frac{1}{2}\left[-\frac{\sin(\rho-\nu-\xi)}{\rho-\nu-\xi}+\frac{\sin(\rho+\nu-\xi)}{\rho+\nu-\xi}+\frac{\sin(\rho-\nu+\xi)}{\rho-\nu+\xi}-\frac{\sin(\rho+\nu+\xi)}{\rho+\nu+\xi}\right]$$
$$\times\cosh(\rho)\cosh(\nu)\cosh(\xi)$$

$$I_{14}(\rho, \nu, \xi) = \cos(\rho)\cosh(\nu)\cosh(\xi)\int_{-1}^{1}\cosh(\rho v)\sin(\nu v)\sin(\xi v)dv$$
$$= -\frac{2\cosh(\rho)\cosh(\nu)\cosh(\xi)}{\left(\rho^2+(\nu-\xi)^2\right)\left(\rho^2+(\nu+\xi)^2\right)}$$
$$\times \Big[2\rho\nu\xi\sin(\rho)\cos(\nu)\cos(\xi)+\rho\sin(\rho)\sin(\nu)\sin(\xi)\left(\rho^2+\nu^2+\xi^2\right)$$
$$+\nu\cos(\rho)\cos(\nu)\sin(\xi)\left(\rho^2+\nu^2-\xi^2\right)+\xi\cos(\rho)\sin(\nu)\cos(\xi)\left(\rho^2-\nu^2+\xi^2\right)\Big]$$

$$I_{15}(\rho, \nu, \xi) = \cosh(\rho)\cos(\nu)\cosh(\xi)\int_{-1}^{1}\cos(\rho v)\sinh(\nu v)\sin(\xi v)dv$$
$$= -\frac{2\cosh(\rho)\cosh(\nu)\cosh(\xi)}{(\rho^2+\nu^2)^2+\xi^2(-2\rho^2+2\nu^2+\xi^2)}$$
$$\times \Big[2\rho\nu\xi\sin(\rho)\cos(\nu)\cos(\xi)+\rho\sin(\rho)\sin(\nu)\sin(\xi)\left(\rho^2+\nu^2-\xi^2\right)$$
$$-\nu\cos(\rho)\cos(\nu)\sin(\xi)\left(\rho^2+\nu^2+\xi^2\right)-\xi\cos(\rho)\sin(\nu)\cos(\xi)\left(-\rho^2+\nu^2+\xi^2\right)\Big]$$

$$I_{16}(\rho, \nu, \xi) = \cos(\rho)\cos(\nu)\cosh(\xi)\int_{-1}^{1}\cosh(\rho v)\sinh(\nu v)\sin(\xi v)dv$$
$$= -\frac{2\cosh(\rho)\cosh(\nu)\cosh(\xi)}{(\rho^2-\nu^2)^2+\xi^2(2\rho^2+2\nu^2+\xi^2)}$$
$$\times \Big[2\rho\nu\xi\sin(\rho)\cos(\nu)\cos(\xi)-\rho\sin(\rho)\sin(\nu)\sin(\xi)\left(\rho^2-\nu^2+\xi^2\right)$$
$$-\nu\cos(\rho)\cos(\nu)\sin(\xi)\left(-\rho^2+\nu^2+\xi^2\right)-\xi\cos(\rho)\sin(\nu)\cos(\xi)\left(\rho^2+\nu^2+\xi^2\right)\Big]$$

$$I_{17}(\rho,\nu,\xi) = \cosh(\rho)\cos(\nu)\cos(\xi)\int_{-1}^{1}\cos(\rho v)\sinh(\nu v)\sinh(\xi v)dv$$
$$= -\frac{2\cosh(\rho)\cosh(\nu)\cosh(\xi)}{\left(\rho^2+(\nu-\xi)^2\right)\left(\rho^2+(\nu+\xi)^2\right)}$$
$$\times\Big[2\rho\nu\xi\sin(\rho)\cos(\nu)\cos(\xi) - \rho\sin(\rho)\sin(\nu)\sin(\xi)\left(\rho^2+\nu^2+\xi^2\right)$$
$$+\nu\cos(\rho)\cos(\nu)\sin(\xi)\left(\rho^2+\nu^2-\xi^2\right)+\xi\cos(\rho)\sin(\nu)\cos(\xi)\left(\rho^2-\nu^2+\xi^2\right)\Big]$$

$$I_{18}(\rho,\nu,\xi) = \cos(\rho)\cos(\nu)\cos(\xi)\int_{-1}^{1}\cosh(\rho v)\sinh(\nu v)\sinh(\xi v)dv$$
$$= \frac{1}{2}\left[\frac{\sinh(\rho-\nu-\xi)}{\rho-\nu-\xi} - \frac{\sinh(\rho+\nu-\xi)}{\rho+\nu-\xi} - \frac{\sinh(\rho-\nu+\xi)}{\rho-\nu+\xi} + \frac{\sinh(\rho+\nu+\xi)}{\rho+\nu+\xi}\right]$$
$$\times\cos(\rho)\cos(\nu)\cos(\xi)$$

$$I_G(\rho,\nu) = I_3(\rho,\nu) - I_2(\rho,\nu) + I_2(\nu,\rho) - I_1(\rho,\nu)$$

$$I_U(\rho,\nu) = I_6(\rho,\nu) + I_7(\rho,\nu) + I_7(\nu,\rho) + I_8(\rho,\nu)$$

$$I_{M1}(\rho,\nu,\xi) = I_9(\rho,\nu,\xi) - I_{10}(\rho,\nu,\xi) + I_{10}(\nu,\rho,\xi) - I_{10}(\xi,\rho,\nu)$$
$$- I_{11}(\rho,\nu,\xi) + I_{11}(\rho,\xi,\nu) - I_{11}(\nu,\xi,\rho) + I_{12}(\rho,\nu,\xi)$$

$$I_{M2}(\rho,\nu,\xi) = -I_{13}(\rho,\nu,\xi) + I_{14}(\rho,\nu,\xi) - I_{15}(\rho,\nu,\xi) - I_{15}(\rho,\xi,\nu)$$
$$+ I_{16}(\rho,\nu,\xi) + I_{16}(\rho,\xi,\nu) - I_{17}(\rho,\nu,\xi) + I_{18}(\rho,\nu,\xi)$$

$$I_{M3}(\rho,\nu,\xi) = I_9(\rho,\nu,\xi) - I_{10}(\rho,\nu,\xi) - I_{10}(\nu,\rho,\xi) + I_{11}(\rho,\nu,\xi)$$

$$I_{M4}(\rho,\nu,\xi) = I_9(\rho,\nu,\xi) - I_{10}(\rho,\nu,\xi) + I_{10}(\nu,\rho,\xi) - I_{11}(\rho,\nu,\xi)$$

$$I_{M5}(\rho,\nu,\xi) = I_{13}(\rho,\nu,\xi) - I_{14}(\rho,\nu,\xi) + I_{15}(\rho,\nu,\xi) - I_{16}(\rho,\nu,\xi)$$

$$I_M(\rho_1,\nu_1,\xi_1,\rho_2,\nu_2,\xi_2) = \nu_1^2\nu_2^2 I_{M1}(\rho_1,\nu_1,\xi_1)I_{M1}(\rho_2,\nu_2,\xi_2)$$
$$- 2\nu_1\nu_2\xi_1\xi_2 I_{M2}(\rho_1,\nu_1,\xi_1)I_{M2}(\rho_2,\nu_2,\xi_2)$$
$$+ \xi_1^2\xi_2^2 I_{M1}(\rho_1,\xi_1,\nu_1)I_{M1}(\rho_2,\xi_2,\nu_2)$$

$$I_B(\rho_1,\nu_1,\rho_2,\nu_2,k) = \frac{1}{2k\cosh(\pi k)\cosh(\pi/2)}$$
$$\times\Big\{4k^2\nu_2^2 I_{M3}(\rho_1,\nu_1,\pi k)I_{M4}\left(\rho_2,\nu_2,\frac{\pi}{2}\right)$$
$$+\nu_1^2 I_{M4}(\rho_1,\nu_1,\pi k)I_{M3}\left(\rho_2,\nu_2,\frac{\pi}{2}\right)$$
$$-(-1)^k\nu_1^2 I_G(\rho_1,\nu_1)I_{M3}\left(\rho_2,\nu_2,\frac{\pi}{2}\right)\cosh(\pi k)$$
$$-4k\nu_1\nu_2 I_{M5}(\rho_1,\nu_1,\pi k)I_{M5}\left(\rho_2,\nu_2,\frac{\pi}{2}\right)\Big\}$$

References

1. Föppl, A. *Vorlesungen über Technische Mechanik*; B. G. Teubner Verlag: Leipzig, Germany, 1907; Volume 5.
2. Ciarlet, P.G. A justification of the von Kármán equations. *Arch. Ration. Mech. Anal.* **1980**, *73*, 349–389. [CrossRef]
3. Kármán, T. *Festigkeitsprobleme im Maschinenbau*; B. G. Teubner Verlag: Leipzig, Germany, 1910.
4. Vol'mir, A.S. *Flexible Plates and Shells*; Wright-Patterson Air Force Base: Dayton, OH, USA, 1967.
5. Lychev, S.; Digilov, A.; Demin, G.; Gusev, E.; Kushnarev, I.; Djuzhev, N.; Bespalov, V. Deformations of Single-Crystal Silicon Circular Plate: Theory and Experiment. *Symmetry* **2024**, *16*, 137. [CrossRef]
6. Lychev, S.; Digilov, A.; Bespalov, V.; Djuzhev, N. Incompatible Deformations in Hyperelastic Plates. *Mathematics* **2024**, *12*, 596. [CrossRef]
7. Nádai, A. *Die Elastischen Platten*; Springer: Berlin/Heidelberg, Germany, 1925.
8. Timoshenko, S. *Vibration Problems in Engineering*; D. Van Nostrand Co.: Ann Arbor, MI, USA, 1928.
9. Way, S. Bending of circular plates with large deflection. *Trans. Am. Soc. Mech. Eng.* **1934**, *56*, 627–633. [CrossRef]
10. Friedrichs, K.O.; Stoker, J.J. The non-linear boundary value problem of the buckled plate. *Am. J. Math.* **1941**, *63*, 839–888. [CrossRef]
11. Friedrichs, K.O.; Stoker, J.J. The non-linear boundary value problem of the buckled plate. *Proc. Natl. Acad. Sci. USA* **1939**, *25*, 535–540. [CrossRef]
12. Zhang, Y. Large deflection of clamped circular plate and accuracy of its approximate analytical solutions. *Sci. China Phys. Mech. Astron.* **2016**, *59*, 1–11. [CrossRef]
13. Kaiser, R. Rechnerische und experimentelle Ermittlung der Durchbiegungen und Spannungen von quadratischen Platten bei freier Auflagerung an den Rändern, gleichmäßig verteilter Last und großen Ausbiegungen. *ZAMM-J. Appl. Math. Mech./Z. Angew. Math. Mech.* **1936**, *16*, 73–98. [CrossRef]
14. Way, S. Uniformly loaded, clamped, rectangular plates with large deflection. In Proceedings of the 5th International Congress for Applied Mechanics, Cambridge, MA, USA, 12–16 September 1938; pp. 123–128.
15. Panov, D.I. Application of acad. B.G. Galerkin's method to certain nonlinear problems of the theory of elasticity. *Prikl. Matem. i Mech* **1939**, *3*, 72–81
16. Levy, S. *Bending of Rectangular Plates with Large Deflections*; Technical Notes No. 846; NACA: Columbia, SC, USA, 1942.
17. Levy, S. *Square Plate with Clamped Edges Under Normal Pressure Producing Large Deflections*; Technical Notes No. 847; NACA: Columbia, SC, USA, 1942.
18. Levy, S. *Bending with Large Deflection of a Clamped Rectangular Plate with Length-Width Ratio of 1.5 Under Normal Pressure*; Technical Notes No. 853; NACA: Columbia, SC, USA, 1942.
19. Woolley, R.M.; Corrick, J.N.; Levy, S. *Clamped Long Rectangular Plate Under Combined Axial Load and Normal Pressure*; Technical Note No. 1047; NACA: Columbia, SC, USA, 1946.
20. Levy, S.; Goldenberg, D.; Zibritosky, G. *Simply Supported Long Rectangular Plate Under Combined Axial Load and Normal Pressure*; Technical Notes No. 949; NACA: Columbia, SC, USA, 1944.
21. Wang, C.T. *Nonlinear Large-Deflection Boundary-Value Problems of Rectangular Plates*; Technical Note No. 1425; NACA: Columbia, SC, USA, 1948.
22. Wang, C.T. *Bending of Rectangular Plates with Large Deflections*; Technical Note No. 1462; NACA: Columbia, SC, USA, 1948.
23. Southwell, R.V.; Green, J.R. Relaxation Methods Applied to Engineering Problems, VIII A. Problems Relating to Large Transverse Displacements of Thin Elastic Plates. *Phil. Tran. Roy. Soc.* **1944**, *1*, 137–176.
24. Timoshenko, S.; Woinowsky-Krieger, S. *Theory of Plates and Shells*; McGraw-Hill: New York, NY, USA, 1959.
25. Bauer, F.; Bauer, L.; Becker, W.; Reiss, E.L. *Bending of Rectangular Plates With Finite Deflections*; Courant Institute of Mathematical Sciences: New York, NY, USA, 1964.
26. Bauer, F.; Bauer, L.; Becker, W.; Reiss, E.L. Bending of Rectangular Plates With Finite Deflections1. *J. Appl. Mech.* **1965**, *32*, 821–825. [CrossRef]
27. Hooke, R. Approximate analysis of the large deflection elastic behaviour of clamped, uniformly loaded, rectangular plates. *J. Mech. Eng. Sci.* **1969**, *11*, 256–268. [CrossRef]
28. Li-zhou, P.; Shu, W. A perturbation-variational solution of the large deflection of rectangular plates under uniform load. *Appl. Math. Mech.* **1986**, *7*, 727–740. [CrossRef]
29. Li-zhou, P.; Wei-zhong, C. The solution of rectangular plates with large deflection by spline functions. *Appl. Math. Mech.* **1990**, *11*, 429–439. [CrossRef]
30. He, X.T.; Cao, L.; Wang, Y.Z.; Sun, J.Y.; Zheng, Z.L. A biparametric perturbation method for the Föppl–von Kármán equations of bimodular thin plates. *J. Math. Anal. Appl.* **2017**, *455*, 1688–1705. [CrossRef]
31. Isanbaeva, F.S.; Kornishin, M.S. Large deflections of plates with freely moved edges. *Issled. Teor. Plast. Oboloc.* **1965**, *3*, 3–17.
32. Rushton, K.R. Simply supported plates with corners free to lift. *J. Strain Anal.* **1969**, *4*, 306–311. [CrossRef]
33. Scholes, A.; Bernstein, E.L. Bending of normally loaded simply supported rectangular plates in the large-deflection range. *J. Strain Anal.* **1969**, *4*, 190–198. [CrossRef]

34. Petit, P.H. A Numerical Solution to the General Large Deflection Plate Equations. *SAE Trans.* **1966**, *74*, 236–243.
35. Brown, J.C.; Harvey, J.M. Large deflections of rectangular plates subjected to uniform lateral pressure and compressive edge loading. *J. Mech. Eng. Sci.* **1969**, *11*, 305–317. [CrossRef]
36. Miyoshi, T. A mixed finite element method for the solution of the von Kármán equations. *Numer. Math.* **1976**, *26*, 255–269. [CrossRef]
37. Brezzi, F. Finite element approximations of the von Kármán equations. *RAIRO Anal. Numérique* **1978**, *12*, 303–312. [CrossRef]
38. Miyoshi, T. Some aspects of a mixed finite element method applied to fourth order partial differential equations. In *Proceedings of the Computing Methods in Applied Sciences: Second International Symposium, Versailles, France, 15–19 December 1975*; Springer: Berlin/Heidelberg, Germany, 2008; pp. 237–256.
39. Bartels, S. Numerical solution of a Föppl–von Kármán model. *SIAM J. Numer. Anal.* **2017**, *55*, 1505–1524. [CrossRef]
40. Kamiya, N.; Sawaki, Y. An integral equation approach to finite deflection of elastic plates. *Int. J.-Non-Linear Mech.* **1982**, *17*, 187–194. [CrossRef]
41. Ye, T.Q.; Liu, Y. Finite deflection analysis of elastic plate by the boundary element method. *Appl. Math. Model.* **1985**, *9*, 183–188. [CrossRef]
42. Tanaka, M.; Matsumoto, T.; Zheng, Z.D. Incremental analysis of finite deflection of elastic plates via boundary-domain-element method. *Eng. Anal. Bound. Elem.* **1996**, *17*, 123–131. [CrossRef]
43. Wang, W.; Ji, X.; Tanaka, M. A dual reciprocity boundary element approach for the problems of large deflection of thin elastic plates. *Comput. Mech.* **2000**, *26*, 58–65. [CrossRef]
44. Waidemam, L.; Venturini, W.S. BEM formulation for von Kármán plates. *Eng. Anal. Bound. Elem.* **2009**, *33*, 1223–1230. [CrossRef]
45. Boresi, A.P.; Turner, J.P. Large deflections of rectangular plates. *Int. J.-Non-Linear Mech.* **1983**, *18*, 125–131. [CrossRef]
46. Al-Shugaa, M.A.; Al-Gahtani, H.J.; Musa, A.E.S. Automated Ritz method for large deflection of plates with mixed boundary conditions. *Arab. J. Sci. Eng.* **2020**, *45*, 8159–8170. [CrossRef]
47. Al-Shugaa, M.A.; Al-Gahtani, H.J.; Musa, A.E.S. Ritz method for large deflection of orthotropic thin plates with mixed boundary conditions. *J. Appl. Math. Comput. Mech.* **2020**, *19*, 5–16. [CrossRef]
48. Al-Shugaa, M.A.; Musa, A.E.S.; Al-Gahtani, H.J. Ritz Method-Based Formulation for Analysis of FGM Thin Plates Undergoing Large Deflection with Mixed Boundary Conditions. *Arab. J. Sci. Eng.* **2024**, *49*, 14225–14239. [CrossRef]
49. Enem, J.I. Nonlinear analysis of isotropic rectangular thin plates using Ritz method. *Saudi. J. Eng. Technol.* **2022**, *7*, 502–512.
50. Gorder, R.A.V. Analytical method for the construction of solutions to the Föppl–von Kármán equations governing deflections of a thin flat plate. *Int. J.-Non-Linear Mech.* **2012**, *47*, 1–6. [CrossRef]
51. Zhong, X.X.; Liao, S.J. Analytic Solutions of Von Kármán Plate under Arbitrary Uniform Pressure—Equations in Differential Form. *Stud. Appl. Math.* **2017**, *138*, 371–400. [CrossRef]
52. Zhong, X.X.; Liao, S.J. Analytic approximations of Von Kármán plate under arbitrary uniform pressure—equations in integral form. *Sci. China Phys. Mech. Astron.* **2018**, *61*, 1–11. [CrossRef]
53. Yu, Q.; Xu, H.; Liao, S. Coiflets solutions for Föppl-von Kármán equations governing large deflection of a thin flat plate by a novel wavelet-homotopy approach. *Numer. Algorithms* **2018**, *79*, 993–1020. [CrossRef]
54. Nekhotyaev, V.V.; Sachenkov, A.V. Large deflections of thin elastic plates. *Issled. Teor. Plast. Oboloc.* **1973**, *10*, 27–40.
55. Sathyamoorthy, M. *Nonlinear Analysis of Structure*; CRC Press: Boca Raton, FL, USA, 2017.
56. Khoa, D.N. Calculation of Flexible Rectangular Plates Using the Method of Successive Approximation. Ph.D. Thesis, Moscow State University of Civil Engineering, Moscow, Russia, 2023.
57. Berger, H.M. A New Approach to the Analysis of Large Deflections of Plates. Ph.D. Thesis, California Institute of Technology, Pasadena, CA, USA, 1954.
58. Nowinski, J. Note on an analysis of large deflections of rectangular plates. *Appl. Sci. Res. Sect. A* **1963**, *11*, 85–96. [CrossRef]
59. Datta, S. Large deflection of a circular plate on elastic foundation under a concentrated load at the center. *J. Appl. Mech.* **1975**, *42*, 503–505. [CrossRef]
60. Datta, S. Large deflection of a semi-circular plate on elastic foundation under a uniform load. In *Proceedings of the Indian Academy of Sciences-Section A*; Springer: Berlin/Heidelberg, Germany, 1976; Volume 83, pp. 21–32.
61. Karmakar, B.M. Non-linear dynamic behaviour of plates on elastic foundations. *J. Sound Vib.* **1977**, *54*, 265–271. [CrossRef]
62. Bychkov, P.S.; Lychev, S.A.; Bout, D.K. Experimental technique for determining the evolution of the bending shape of thin substrate by the copper electrocrystallization in areas of complex shapes. *Vestn. Samar. Univ.-Estestv.-Nauchnaya Seriya* **2019**, *25*, 48–73. [CrossRef]
63. Bout, D.K.; Bychkov, P.S.; Lychev, S.A. The theoretical and experimental study of the bending of a thin substrate during electrolytic deposition. *PNRPU Mech. Bull.* **2020**, *1*, 17–31. [CrossRef]
64. Lychev, S.A.; Digilov, A.V.; Pivovaroff, N.A. Bending of a circular disk: From cylinder to ultrathin membrane. *Vestn. Samar. Univ.-Estestv.-Nauchnaya Seriya* **2023**, *29*, 77–105. [CrossRef]
65. Banerjee, B.; Datta, S. A new approach to an analysis of large deflections of thin elastic plates. *Int. J.-Non-Linear Mech.* **1981**, *16*, 47–52. [CrossRef]

66. Yamaki, N. Influence of large amplitudes on flexural vibrations of elastic plates. *ZAMM-J. Appl. Math. Mech./Z. Angew. Math. Mech.* **1961**, *41*, 501–510. [CrossRef]
67. Iyengar, K.T.S.R.; Naqvi, M.M. Large deflections of rectangular plates. *Int. J.-Non-Linear Mech.* **1966**, *1*, 109–122. [CrossRef]
68. Dai, H.H.; Paik, J.K.; Atluri, S.N. The global nonlinear galerkin method for the analysis of elastic large deflections of plates under combined loads: A scalar homotopy method for the direct solution of nonlinear algebraic equations. *Comput. Mater. Contin.* **2011**, *23*, 69.
69. Dai, H.H.; Paik, J.K.; Atluri, S.N. The Global Nonlinear Galerkin Method for the Solution of von Karman Nonlinear Plate Equations: An Optimal & Faster Iterative Method for the Direct Solution of Nonlinear Algebraic Equations F (x) = 0, using x = λ [αF + (1 − α) B^TF]. *Comput. Mater. Contin.* **2011**, *23*, 155.
70. Dai, H.H.; Yue, X.; Atluri, S.N. Solutions of the von Kármán plate equations by a Galerkin method, without inverting the tangent stiffness matrix. *J. Mech. Mater. Struct.* **2014**, *9*, 195–226. [CrossRef]
71. Wang, X.; Liu, X.; Wang, J.; Zhoue, Y. A wavelet method for bending of circular plate with large deflection. *Acta Mech. Solida Sin.* **2015**, *28*, 83–90. [CrossRef]
72. Zhang, L.; Wang, J.; Zhou, Y.H. Wavelet solution for large deflection bending problems of thin rectangular plates. *Arch. Appl. Mech.* **2015**, *85*, 355–365. [CrossRef]
73. Ciarlet, P.G. *Mathematical Elasticity: Theory of Plates*; Elsevier: Amsterdam, The Netherlands, 1997.
74. Lychev, S.A. Incompatible Deformations of Elastic Plates. *Uchenye Zap. Kazan. Univ. Seriya-Fiz.-Mat. Nauk.* **2023**, *165*, 361–388. [CrossRef]
75. Morozov, N.F. On the non-linear theory of thin plates. *Dokl. Akad. Nauk. SSSR* **1957**, *5*, 968–971.
76. Leray, J.; Schauder, J. Topologie et équations fonctionnelles. *Ann. Sci. l'École Norm. Supérieure* **1934**, *13*, 45–78. [CrossRef]
77. Berger, M.S.; Fife, P.C. On von Kármán's equations and the buckling of a thin elastic plate. *Bull. Am. Math. Soc.* **1966**, *72*, 1006–1011. [CrossRef]
78. Knightly, G.H. An existence theorem for the von Kármán equations. *Arch. Ration. Mech. Anal.* **1967**, *27*, 233–242. [CrossRef]
79. Knightly, G.H.; Sather, D. On nonuniqueness of solutions of the von Kármán equations. *Arch. Ration. Mech. Anal.* **1970**, *36*, 65–78. [CrossRef]
80. Hlaváček, I.; Naumann, J. Inhomogeneous boundary value problems for the von Kármán equations. I. *Appl. Math.* **1974**, *19*, 253–269. [CrossRef]
81. Hlaváček, I.; Naumann, J. Inhomogeneous boundary value problems for the von Kármán equations. II. *Appl. Math.* **1975**, *20*, 280–297. [CrossRef]
82. Lurie, A.I. *Theory of Elasticity*; Springer Science & Business Media: Berlin/Heidelberg, Germany, 2010.
83. Ciarlet, P.G. *Mathematical Elasticity: Three-Dimensional Elasticity*; Elsevier: Amsterdam, The Netherlands, 1988.
84. Shamina, V.A. Determination of the Displacement vector from the Components of the Deformation Tensor in Nonlinear Continuum Mechanics. *Mech. Solids* **1974**, *9*, 9–16.
85. Pietraszkiewicz, W.; Badur, J. Finite rotations in the description of continuum deformation. *Int. J. Eng. Sci.* **1983**, *21*, 1097–1115. [CrossRef]
86. Georgievskii, D.V. Constructing generalized Cesàro formulas for finite plane strains. *J. Appl. Mech. Tech. Phys.* **2014**, *55*, 494–498. [CrossRef]
87. Yavari, A. Compatibility equations of nonlinear elasticity for non-simply-connected bodies. *Arch. Ration. Mech. Anal.* **2013**, *209*, 237–253. [CrossRef]
88. Muskhelishvili, N.I. *Some Basic Problems of the Mathematical Theory of Elasticity*; Noordhoff Groningen: Groningen, The Netherlands, 1953.
89. Kalandiya, A.I. *Mathematical Methods of Two-Dimensional Elasticity*; Mir Publishers: Moscow, Russia, 1975.
90. Kolmogorov, A.N.; Fomin, S.V. *Introductory Real Analysis*; Courier Corporation: Chelmsford, MA, USA, 1975.
91. Cantor, G. Ein beitrag zur mannigfaltigkeitslehre. *J. Die Reine Angew. Math. (Crelles J.)* **1878**, *1878*, 242–258.
92. Koyalovich, B.M. A study of infinite systems of linear algebraic equations. *Izv. Fiz.-Mat. Inst.* **1930**, *3*, 41–167.
93. Kantorovich, L.V.; Krylov, V.I. *Approximate Methods of Higher Analysis*; Dover Publications: Mineola, NY, USA, 2018.
94. Vaĭnberg, M.M.; Trenogin, V.A. *Theory of Branching of Solutions of Non-Linear Equations*; Noordhoff International Pub.: Groningen, The Netherlands, 1974.
95. Vest, C.M. *Holographic Interferometry*; Wiley: Hoboken, NJ, USA, 1979.

symmetry

MDPI

Article

Bending Analysis of Symmetrical Porous Functionally Graded Sandwich Panels

Zhicheng Huang [1,*], Yingjie Chen [1], Xingguo Wang [1] and Fulei Chu [2]

[1] School of Mechanical and Electronic Engineering, Jingdezhen Ceramic University, Jingdezhen 333403, China
[2] School of Mechanical Engineering, Tsinghua University, Beijing 100084, China
* Correspondence: huangzhicheng@jci.edu.cn

Abstract: The study of the mechanical behavior of functionally graded material (FGM) sandwich plates under thermo-mechanical loading is of great significance for advanced structural design. This study systematically verifies the applicability of the shear strain functions proposed by Reddy and Touratier in the nonlinear bending analysis of porous FGM sandwich plates. Using the existing four-variable shear deformation theory framework, the governing equations are derived through the principle of minimum potential energy, and the Navier method is applied for a numerical solution. For the first time, the study systematically compared the effects of three different porosity distribution patterns on dimensionless deflection, and verified the reliability of the model by comparing it with literature data. The results demonstrate that the adopted shear strain functions can accurately predict the influence of key parameters, including layer thickness ratio, aspect ratio, side-to-thickness ratio, volume fraction index, and porosity, on the deflection performance of sandwich plates. This research provides an important verification basis for the theoretical analysis and engineering application of FGM sandwich plates, particularly offering quantitative evidence for assessing the influence of porosity effects on theoretical prediction accuracy.

Keywords: functionally graded materials; sandwich panels; thermal–mechanical loading; porosity; Navier method

1. Introduction

Sandwich structures [1] have been widely applied in numerous fields, such as aerospace, automotive industry, construction and civil engineering, shipbuilding and marine engineering, and sustainable energy, due to their high strength, lightweight, and excellent compressive and flexural properties. In recent years, with the increasing demand for extreme-environment engineering, porous functionally graded material (FGM) sandwich panels have demonstrated unique advantages in the following areas: Aerospace: As a thermal protection system for hypersonic vehicles, their gradient heat resistance can withstand transient thermal shocks exceeding 1500 °C during atmospheric re-entry [2]. In aircraft, FGM sandwich panels significantly reduce thermal stress concentration through gradient pore design, improving the reliability of thermal protection systems. New energy vehicles: Used in lightweight designs for battery pack brackets, achieving a balance between structural strength and heat dissipation efficiency through gradient porosity [3]. FGM sandwich panels with a Pore Model (uniform distribution) exhibit excellent mechanical performance under collision and thermal runaway scenarios. Biomedical: In bionic bone implants, controllable porous structures can match the mechanical compatibility of human bones, reducing stress-shielding effects [4]. Gradient porosity designs promote bone cell growth and nutrient transport, significantly improving the biocompatibility of implants. However, as the complexity of industrial scenarios increases, traditional sandwich structures struggle to meet reliability requirements under thermo-mechanical coupled loads.

Traditional sandwich structures are typically composed of two rigid and high-strength outer panels with a relatively lightweight, low-density core layer in between, and are bonded together using specific adhesives. This design has notable limitations: Due to the significant difference in material properties between the panels and the core layer, large interlayer stresses can easily develop during use, leading to delamination or even complete separation of the layers. This issue significantly impacts the structural performance and service life of the sandwich structure.

To address the issues associated with traditional sandwich panels, functionally graded materials (FGM) [5] have been introduced into the design of sandwich structures. FGM is a type of composite material with a non-uniform microstructure, composed of materials with different properties (typically ceramics and metals), allowing for a continuous gradient in material performance. Ceramic materials offer excellent heat resistance, while metal materials provide good mechanical strength. As a result, FGM can simultaneously meet the requirements for toughness and strength in complex environments (such as under large temperature differentials). This characteristic significantly compensates for the shortcomings of traditional sandwich panels, greatly enhancing their overall performance. In recent years, research on the mechanical properties of FGM sandwich structures has developed in three main directions: 1. Vibration and Buckling Characteristics Analysis: For example, Swaminathan K. et al. [6] used the finite element method to reveal the sensitivity of porosity to the vibration characteristics of FGM sandwich panels. Tran T. T. et al. [7] explored the influence of porous layers on thermo-mechanical buckling based on higher-order shear deformation theory. 2. Thermo-Mechanical Coupled Field Response Studies: Nguyen et al. [8] developed a higher-order finite element model to analyze the nonlinear thermo-mechanical behavior of bidirectional FGM sandwich panels. Zenkour et al. [9] and Mantari et al. [7,10] solved thermoelastic bending problems using improved plate theories. 3. Manufacturing Defects and Performance Correlation: It is noteworthy that most existing research focuses on ideal material assumptions (such as uniform porosity distribution). However, the impact of porosity randomness in actual manufacturing processes (such as differences in laser sintering [11] and powder metallurgy [12]) on performance has not yet been fully quantified. Although Mechab et al. [13] and Wang et al. [14] have attempted to introduce porosity variables, their research remains limited to single physical fields (pure mechanical or pure thermal loads). There is a lack of literature systematically exploring the synergistic effects of pore configuration and material gradients in thermo-mechanical coupled fields. This gap significantly hinders the reliability assessment of porous FGM sandwich panels in extreme environments, such as thermal shocks during spacecraft re-entry.

It is particularly important to note that the generation mechanism of pores is closely related to the manufacturing process. For example, powder metallurgy [12] can adjust porosity (0.1–15%) by controlling the packing density of metal/ceramic particles and sintering parameters, but it tends to result in closed-pore structures. On the other hand, 3D printing technology [15] can achieve directional open-pore configurations through layer-by-layer deposition, but the pore distribution is significantly influenced by the printing path and interlayer bonding strength. This process dependency means that porosity not only exists as a material defect but can also become a new dimension for optimizing thermo-mechanical performance through active design, such as gradient porosity [16]. However, most current area, aerospace, as a core component of thermo-mechanical models [17,18], still simplifies porosity as a uniformly distributed parameter, failing to reflect the spatial gradient characteristics and process constraints of porosity in actual manufacturing.

Based on this, this study focuses on the analysis of the nonlinear bending behavior of porous functionally graded material (FGM) sandwich panels under thermo-mechanical coupled loads. By incorporating refined shear deformation plate theory, a novel porous FGM sandwich panel model is proposed to systematically investigate the effects of four pore configurations on deflection characteristics. In this paper, the governing equations are derived using the principle of minimum potential energy, and the Navier method is employed to obtain the analytical solution of these equations under simply supported boundary conditions. Additionally, a parametric study is conducted on the porous FGM sandwich panels, thoroughly analyzing the effects of layer thickness ratio, aspect ratio, edge thickness ratio, volume fraction index, and porosity on the deflection performance of FGM sandwich panels under thermo-mechanical loads.

2. Model Establishment and Formula

This paper studies the porous functional gradient sandwich plate, which is a symmetrical structure composed of upper and lower surface layers of functional gradient materials and a middle layer of ceramic material (as shown in Figure 1).

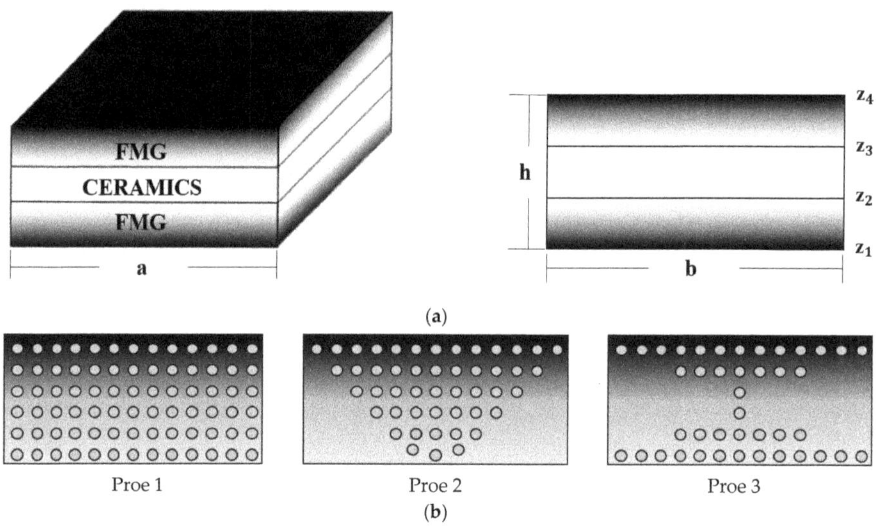

Figure 1. (**a**) Functionally graded sandwich panels, where a, b, and h are the length, width, and height, respectively. (**b**) Porosity distribution model.

2.1. FMG Sandwich Panel

The lower surface layer transitions from metal ($z = z_1$) to ceramic ($z = z_2$), the upper surface layer transitions from metal ($z = z_4$) to ceramic ($z = z_3$), and the middle layer (z_2–z_3) is a ceramic layer. The volume fraction of ceramic in the functionally gradient sandwich plate is [19]

$$\begin{cases} V_c^{(1)}(z) = \left(\frac{z-z_1}{z_2-z_1}\right)^s z \in [z_1, z_2] \\ V_c^{(2)}(z) = 1 \quad z \in [z_2, z_3] \\ V_c^{(3)}(z) = \left(\frac{z-z_4}{z_3-z_4}\right)^s z \in [z_3, z_4] \end{cases} \tag{1}$$

where $V_c^{(1)}, V_c^{(2)}$, and $V_c^{(3)}$ are the volume contents of the first, second, and third layers of the sandwich panel, respectively. s is the volume fraction index, also known as the gradient index.

2.2. Material Properties

Many porosity distribution models have been proposed by researchers to calculate the material properties of porous functionally gradient sandwich panels. In this paper, the following three porosity models are used for calculation, and the specific forms are as follows [9,20–22].

Pore model 1:

$$p(z) = p_c\left(V_c(z) - \frac{xi}{2}\right) + p_m(V_m(z) - \frac{xi}{2}) \tag{2}$$

Pore model 2:

$$\begin{cases} p^{(1)} = p_c {V_c}^{(1)}(z) + p_m {V_m}^{(1)}(z) - \log\left(1 + \frac{xi}{2}\right)(p_c + p_m)\left(1 - \frac{|2z - z_2 - z_1|}{z_2 - z_1}\right) \\ p^{(2)} = p_c {V_c}^{(2)}(z) + p_m {V_m}^{(2)}(z) \\ p^{(3)} = p_c {V_c}^{(3)}(z) + p_m {V_m}^{(3)}(z) - \log\left(1 + \frac{xi}{2}\right)(p_c + p_m)\left(1 - \frac{|2z - z_4 - z_3|}{z_4 - z_3}\right) \end{cases} \tag{3}$$

Pore model 3:

$$\begin{cases} p^{(1)} = p_c {V_c}^{(1)}(z) + p_m {V_m}^{(1)}(z) - \frac{xi}{2}(p_c + p_m)\left(1 - \frac{z - z_2}{z_1 - z_2}\right) \\ p^{(2)} = p_c {V_c}^{(2)}(z) + p_m {V_m}^{(2)}(z) \\ p^{(3)} = p_c {V_c}^{(3)}(z) + p_m {V_m}^{(3)}(z) - \frac{xi}{2}(p_c + p_m)\left(1 - \frac{z - z_4}{z_3 - z_4}\right) \end{cases} \tag{4}$$

where p_c and p_m represent the material properties of ceramic and metal (such as Young's modulus, Poisson's ratio, thermal expansion coefficient, etc.). V_c and V_m represent the total volume fraction of ceramic and metal, and their relationship always satisfies $V_c + V_m = 1$. xi is the porosity.

2.3. Displacement and Shape Function

Based on the four-variable plate theory, with further assumptions. The lateral displacement w is decomposed into a bending component w_a and a shear component w_b $(w = w_a + w_b)$

The following displacement field is obtained [23]

$$\begin{cases} u(x,y,z) = u_1(x,y) - z\frac{\partial w_a}{\partial x} - f(z)\frac{\partial w_b}{\partial x} \\ v(x,y,z) = v_1(x,y) - z\frac{\partial w_a}{\partial y} - f(z)\frac{\partial w_b}{\partial y} \\ w(x,y,z) = w_a(x,y) + w_b(x,y) \end{cases} \tag{5}$$

Among them u, v, w are the displacements in the x, y, z directions, respectively; u_1 and v_1 are the tensile components in the x and y directions, respectively; w_a and w_b are the bending and shear components, respectively; and $f(z)$ is the shape function about z, where $f(z) = z - \Psi(z)$. In high-order shear deformation theory, the commonly used shape functions $\Psi(z)$ are:

Reissner [24]

$$\Psi(z) = \frac{5z}{4}\left(1 - \frac{4z^2}{3H^2}\right) \tag{6}$$

Reddy [25]

$$\Psi(z) = z\left(1 - \frac{4z^2}{3H^2}\right) \tag{7}$$

Touratier [26]

$$\Psi(z) = \frac{h}{\pi} sin\left(\frac{\pi z}{h}\right) \tag{8}$$

2.4. Constitutive Equation

The relationship between strain and displacement:

depends on

$$\begin{pmatrix} \varepsilon_{xx} \\ \varepsilon_{yy} \\ \varepsilon_{zz} \end{pmatrix} = \begin{pmatrix} \frac{\partial u}{\partial x} \\ \frac{\partial v}{\partial y} \\ \frac{\partial w}{\partial z} \end{pmatrix}, \begin{pmatrix} \gamma_{xy} \\ \gamma_{yz} \\ \gamma_{xz} \end{pmatrix} = \begin{pmatrix} \frac{\partial v}{\partial x} + \frac{\partial u}{\partial y} \\ \frac{\partial w}{\partial y} + \frac{\partial v}{\partial z} \\ \frac{\partial w}{\partial x} + \frac{\partial u}{\partial z} \end{pmatrix} \tag{9}$$

obtains

$$\begin{gathered} \begin{pmatrix} \varepsilon_{xx} \\ \varepsilon_{yy} \\ \gamma_{xy} \end{pmatrix} = \begin{pmatrix} \varepsilon_{xx}^{(0)} \\ \varepsilon_{yy}^{(0)} \\ \gamma_{xy}^{(0)} \end{pmatrix} + z \begin{pmatrix} \kappa_{xx}^{a} \\ \kappa_{yy}^{a} \\ \kappa_{xy}^{a} \end{pmatrix} + f(z) \begin{pmatrix} \kappa_{xx}^{b} \\ \kappa_{yy}^{b} \\ \kappa_{xy}^{b} \end{pmatrix}, \ \varepsilon_{zz} = 0 \\ \begin{pmatrix} \gamma_{xz} \\ \gamma_{yz} \end{pmatrix} = [1 - f'(z)] \begin{pmatrix} \gamma_{xz}^{(2)} \\ \gamma_{yz}^{(2)} \end{pmatrix} \end{gathered} \tag{10}$$

while

$$\begin{gathered} \begin{pmatrix} \varepsilon_{xx}^{1} \\ \varepsilon_{yy}^{1} \\ \gamma_{xy}^{1} \end{pmatrix} = \begin{pmatrix} \frac{\partial u_1}{\partial x} \\ \frac{\partial v_1}{\partial y} \\ \frac{\partial u_1}{\partial y} + \frac{\partial v_1}{\partial x} \end{pmatrix}, \begin{pmatrix} \kappa_{xx}^{1} \\ \kappa_{yy}^{1} \\ \kappa_{xy}^{1} \end{pmatrix} = - \begin{pmatrix} \frac{\partial^2 w_a}{\partial x^2} \\ \frac{\partial^2 w_a}{\partial y^2} \\ 2\frac{\partial^2 w_a}{\partial x \partial y} \end{pmatrix} \\ \begin{pmatrix} \kappa_{xx}^{2} \\ \kappa_{yy}^{2} \\ \kappa_{xy}^{2} \end{pmatrix} = - \begin{pmatrix} \frac{\partial^2 w_b}{\partial x^2} \\ \frac{\partial^2 w_b}{\partial y^2} \\ 2\frac{\partial^2 w_b}{\partial x \partial y} \end{pmatrix}, \begin{pmatrix} \gamma_{xz}^{1} \\ \gamma_{yz}^{1} \end{pmatrix} = \begin{pmatrix} \frac{\partial w_b}{\partial x} \\ \frac{\partial w_b}{\partial y} \end{pmatrix} \end{gathered} \tag{11}$$

Adding the thermal effect, the relationship between the stress component and the deformation component is

$$\begin{gathered} \begin{pmatrix} \sigma_{xx} \\ \sigma_{yy} \\ \tau_{yz} \\ \tau_{xz} \\ \tau_{xy} \end{pmatrix}^{(s)} = \begin{bmatrix} a_{11} & a_{12} & 0 & 0 & 0 \\ a_{12} & a_{22} & 0 & 0 & 0 \\ 0 & 0 & a_{44} & 0 & 0 \\ 0 & 0 & 0 & a_{55} & 0 \\ 0 & 0 & 0 & 0 & a_{66} \end{bmatrix}^{(s)} \begin{pmatrix} \varepsilon_{xx} - \alpha T \\ \varepsilon_{yy} - \alpha T \\ \gamma_{yz} \\ \gamma_{xz} \\ \gamma_{xy} \end{pmatrix}^{(s)} \\ (s = 1, 2, 3) \end{gathered} \tag{12}$$

where

$$a_{11}^{(s)} = a_{22}^{(s)} = \frac{E^{(s)}(z)}{1 - \left(\ell^{(s)}\right)^2}, a_{12}^{(s)} = \vartheta^{(s)} a_{11}^{(s)}, a_{44}^{(s)} = a_{55}^{(s)} = a_{66}^{(s)} = \frac{E^{(s)}(z)}{2\left(1 + \ell^{(s)}\right)} \tag{13}$$

where $E, \vartheta,$ and α are Young's modulus, Poisson's ratio, and thermal expansion coefficient of the porous functionally graded sandwich plate, respectively.

2.5. Governing Equations

The deformation potential energy of the sandwich plate is

$$\begin{gathered} U = \frac{1}{2} \int_V \left[\sigma_{xx}^{(s)} (\varepsilon_{xx} - \alpha T) + \sigma_{yy}^{(s)} (\varepsilon_{yy} - \alpha T) + \tau_{xy}^{(s)} \gamma_{xy}^{(s)} + \tau_{xz}^{(s)} \gamma_{xz}^{(s)} + \tau_{yz}^{(s)} \gamma_{yz}^{(s)} \right] dV \\ (s = 1, 2, 3) \end{gathered} \tag{14}$$

The work performed by the external force is

$$W = \int_{\Omega} qwd\Omega \tag{15}$$

According to the principle of minimum potential energy

$$\delta U - \delta W = 0 \tag{16}$$

Substituting Equations (11), (15), and (16) into (17) and integrating z, we obtain:

$$\begin{aligned} &\int_{\Omega}\Big[N_{xx}\delta\varepsilon_{xx}^{1} + N_{yy}\delta\varepsilon_{yy}^{1} + N_{xy}\delta\gamma_{xy}^{1} + M_{xx}^{a}\delta\kappa_{xx}^{1} + M_{yy}^{a}\delta\kappa_{yy}^{1} + M_{xy}^{a}\delta\kappa_{xy}^{1} \\ &\qquad + M_{xx}^{b}\delta\kappa_{xx}^{2} + M_{yy}^{b}\delta\kappa_{yy}^{2} + M_{xy}^{b}\delta\kappa_{xy}^{2} + Q_{xz}^{b}\delta\kappa_{xz}^{2} + Q_{yz}^{b}\delta\kappa_{yz}^{2}\Big]d\Omega \\ &\qquad - \int_{\Omega} q\delta w(\delta w_a + \delta w_b)d\Omega = 0 \end{aligned} \tag{17}$$

In the formula

$$\begin{bmatrix} N_{xx} & N_{yy} & N_{xy} \\ M_{xx}^{a} & M_{yy}^{a} & M_{xy}^{a} \\ M_{xx}^{b} & M_{yy}^{b} & M_{xy}^{b} \end{bmatrix} = \sum_{n=1}^{3}\int_{z_n}^{z_{n+1}} \begin{pmatrix} 1 \\ z \\ f(z) \end{pmatrix} \begin{pmatrix} \sigma_{xx} & \sigma_{yy} & \sigma_{xy} \end{pmatrix}^{(s)} dz \tag{18}$$

$$\begin{bmatrix} Q_{xz}^{b} \\ Q_{yz}^{b} \end{bmatrix} = \sum_{n=1}^{3}\int_{z_n}^{z_{n+1}} [1 - f'(z)] \begin{pmatrix} \tau_{xz} \\ \tau_{yz} \end{pmatrix}^{(s)} dz \tag{19}$$

Substituting (12) into (18) and integrating by parts, and setting the coefficients before $\delta u_1, \delta v_1, \delta w_1, \delta w_2$ to zero, the equilibrium equation can be derived:

$$\begin{aligned} &\delta u_1 : \frac{\partial N_{xx}}{\partial x} + \frac{\partial N_{xy}}{\partial y} = 0 \\ &\delta v_1 : \frac{\partial N_{xy}}{\partial x} + \frac{\partial N_{yy}}{\partial y} = 0 \\ &\delta w_a : \frac{\partial^2 M_{xx}^{a}}{\partial x^2} + 2\frac{\partial^2 M_{xy}^{a}}{\partial x \partial y} + \frac{\partial^2 M_{yy}^{a}}{\partial y^2} + q = 0 \\ &\delta w_b : \frac{\partial^2 M_{xx}^{b}}{\partial x^2} + 2\frac{\partial^2 M_{xy}^{b}}{\partial x \partial y} + \frac{\partial^2 M_{yy}^{b}}{\partial y^2} + \frac{\partial Q_{xz}^{b}}{\partial x} + \frac{\partial Q_{yz}^{b}}{\partial y} + q = 0 \end{aligned} \tag{20}$$

Substituting (11) and (13) into (19) and (20), we obtain:

$$\begin{gathered} \begin{pmatrix} N \\ M^{a} \\ M^{b} \end{pmatrix} = \begin{bmatrix} A & B & C \\ B & D & G \\ C & G & H \end{bmatrix} \begin{pmatrix} \varepsilon^{1} \\ \kappa^{a} \\ \kappa^{b} \end{pmatrix} - \begin{pmatrix} N^{T} \\ M^{aT} \\ M^{bT} \end{pmatrix} \\ \begin{bmatrix} Q_{yz}^{2} \\ Q_{xz}^{2} \end{bmatrix} = \begin{bmatrix} E_{44} & 0 \\ 0 & E_{55} \end{bmatrix} \begin{pmatrix} \gamma_{yz}^{2} \\ \gamma_{xz}^{2} \end{pmatrix} \end{gathered} \tag{21}$$

where

$$N = \begin{pmatrix} N_{xx} \\ N_{yy} \\ N_{xy} \end{pmatrix} M^{a} = \begin{pmatrix} M_{xx}^{a} \\ M_{yy}^{a} \\ M_{xy}^{a} \end{pmatrix} M^{b} = \begin{pmatrix} M_{xx}^{b} \\ M_{yy}^{b} \\ M_{xy}^{b} \end{pmatrix} \tag{22}$$

$$N^{T} = \begin{pmatrix} N_{x}^{T} \\ N_{y}^{T} \\ 0 \end{pmatrix}^{T} M^{aT} = \begin{pmatrix} M_{x}^{aT} \\ M_{y}^{aT} \\ 0 \end{pmatrix}^{T} M^{bT} = \begin{pmatrix} M_{x}^{bT} \\ M_{y}^{bT} \\ 0 \end{pmatrix}^{T} \tag{23}$$

$$\varepsilon^{1} = \begin{pmatrix} \varepsilon_{xx}^{1} \\ \varepsilon_{yy}^{1} \\ \gamma_{xy}^{1} \end{pmatrix} \kappa^{a} = \begin{pmatrix} \kappa_{xx}^{a} \\ \kappa_{yy}^{a} \\ \kappa_{xy}^{a} \end{pmatrix} \kappa^{b} = \begin{pmatrix} \kappa_{xx}^{b} \\ \kappa_{yy}^{b} \\ \kappa_{xy}^{b} \end{pmatrix} \tag{24}$$

and

$$A=\begin{pmatrix}A_{11}&A_{12}&0\\A_{12}&A_{22}&0\\0&0&A_{66}\end{pmatrix}B=\begin{pmatrix}B_{11}&B_{12}&0\\B_{12}&B_{22}&0\\0&0&B_{66}\end{pmatrix}A^1=\begin{pmatrix}A^1_{11}&A^1_{12}&0\\A^1_{12}&A^1_{22}&0\\0&0&A^1_{66}\end{pmatrix}$$
$$C=\begin{pmatrix}C_{11}&C_{12}&0\\C_{12}&C_{22}&0\\0&0&C_{66}\end{pmatrix}D=\begin{pmatrix}D_{11}&D_{12}&0\\D_{21}&D_{22}&0\\0&0&D_{66}\end{pmatrix}A^2=\begin{pmatrix}B^1_{11}&B^1_{12}&0\\B^1_{12}&B^1_{22}&0\\0&0&B^1_{66}\end{pmatrix} \tag{25}$$

The coefficients and $E_{44},\ E_{55}$ in the above matrix are expressed as:

$$\begin{pmatrix}A_{11}&A_{12}&A_{66}\\B_{11}&B_{12}&B_{66}\\A^1_{11}&A^1_{12}&A^1_{66}\\C_{11}&C_{12}&C_{66}\\A^2_{11}&A^2_{12}&A^2_{66}\\D_{11}&D_{12}&D_{66}\end{pmatrix}=\sum_{s=1}^{3}\int_{z_s}^{z_{s+1}}\begin{pmatrix}1\\z\\z^2\\f(z)\\zf(z)\\f^2(z)\end{pmatrix}\begin{pmatrix}a_{11}^{(s)}&a_{12}^{(s)}&a_{66}^{(s)}\end{pmatrix}dz$$
$$\binom{E_{44}}{E_{55}}=\sum_{s=1}^{3}\int_{z_s}^{z^z_{s+1}}[1-f'(z)]\binom{a_{44}^{(s)}}{a_{55}^{(s)}}dz \tag{26}$$

The part of Equation (22) related to the heat load can be written as:

$$\begin{pmatrix}N_x^T&M_x^{aT}&M_x^{bT}\\N_Y^T&M_Y^{aT}&M_y^{bT}\end{pmatrix}=\sum_{s=1}^{3}\int_{z_{s-1}}^{z_s}\begin{pmatrix}(a_{11}+a_{12})\alpha T\\(a_{12}+a_{22})\alpha T\end{pmatrix}^{(s)}\begin{pmatrix}1&z&f(z)\end{pmatrix}dz \tag{27}$$

The temperature field used in this paper is as follows [17,27,28]:

$$T(x,y,z)=T_1(x,y)+\frac{z}{h}T_2(x,y)+\frac{\Psi(z)}{h}T_3(x,y) \tag{28}$$

2.6. Closed-Form Solution of Functionally Graded Sandwich Panels

This paper considers a simply supported rectangular functionally gradient sandwich plate with length, width, and height of a, b, and h, respectively.

The boundary conditions for the four simply supported sides are as follows:

$$x=0,a:\ v_1=w_a=w_b=0,\frac{\partial w_a}{\partial y}=\frac{\partial w_b}{\partial y}=0,N_{xx}=M^a_{xx}=M^b_{xx}=0$$
$$y=0,b:\ u_1=w_a=w_b=0,\frac{\partial w_a}{\partial x}=\frac{\partial w_b}{\partial x}=0,N_{yy}=M^a_{yy}=M^b_{yy}=0$$

The force load and temperature field borne by the plate are both double-sinusoidal distributions:

$$q=q_0 sin(\lambda x)sin(\ell y) \tag{29}$$

$$\begin{pmatrix}T_1\\T_2\\T_3\end{pmatrix}=\begin{pmatrix}t_1\\t_2\\t_3\end{pmatrix}sin(\lambda x)sin(\ell y) \tag{30}$$

Among them q_0,t_1,t_2,t_3 are constants and $\lambda=\frac{\pi}{a},\ell=\frac{\pi}{b}$.

Using the Navier solution, the displacement expression is taken as follows:

$$\begin{pmatrix}u_1\\v_1\\w_a\\w_b\end{pmatrix}=\begin{pmatrix}U\cos(\lambda x)\sin(\ell y)\\V\sin(\lambda x)\cos(\ell y)\\W_a\sin(\lambda x)\sin(\ell y)\\W_b\sin(\lambda x)\sin(\ell y)\end{pmatrix} \tag{31}$$

In the formula U, V, W_a, and W_b are all unknown constants.

Substituting (31) and (32) into the operator equation, we can obtain:

$$[\Theta][\Delta] = [\coprod] \quad (32)$$

where $\{\Delta\} = \left\{ U \quad V \quad W_a \quad W_b \right\}^T$

The elements of the symmetric stiffness matrix $[\Theta]$ are

$$\begin{aligned}
\Theta_{11} &= A_{11}\lambda^2 + A_{66}\ell^2 \\
\Theta_{12} &= \lambda\ell(A_{12} + A_{66}) \\
\Theta_{13} &= -\lambda\left[B^1_{11}\lambda^2 + (B^1_{12} + 2B^1_{66})\ell^2\right] \\
\Theta_{14} &= -\lambda\left[C^1_{11}\lambda^2 + (C^1_{12} + 2C^1_{66})\ell^2\right] \\
\Theta_{22} &= A_{66}\lambda^2 + A_{66}\ell^2 \\
\Theta_{23} &= -\ell\left[B^1_{22}\ell^2 + (B^1_{12} + 2B^1_{66})\lambda^2\right] \\
\Theta_{24} &= -\ell\left[C_{11}\ell^2 + (C_{12} + 2C_{66})\lambda^2\right] \\
\Theta_{33} &= A^1_{11}\lambda^4 + 2(A^1_{12} + 2A^1_{66})\lambda^2\ell^2 + A^1_{22}\ell^4 \\
\Theta_{34} &= D_{11}\lambda^4 + 2(D_{12} + 2D_{66})\lambda^2\ell^2 + D_{22}\ell^4 \\
\Theta_{44} &= A^2_{11}\lambda^4 + 2(A^2_{12} + 2A^2_{66})\lambda^2\ell^2 + A^2_{22}\ell^4 + E_{44}\ell^4 + E_{55}\lambda^4
\end{aligned}$$

The elements of the generalized force vector $\{\coprod\}$ are

$$\begin{aligned}
\coprod_1 &= -\lambda(A^T t_1 + B^T t_2 + {}^aB_T t_3) \\
\coprod_2 &= -\ell(A^T t_1 + B^T t_2 + {}^aB_T t_3) \\
\coprod_3 &= q_0 + h(\lambda^2 + \ell^2)(B^T t_1 + D^T t_2 + {}^aD_T t_3) \\
\coprod_4 &= q_0 + h(\lambda^2 + \ell^2)(C^T t_1 + F^T t_2 + {}^aF_T t_3)
\end{aligned}$$

and

$$\begin{aligned}
\{A^T, B^T, D^T\} &= \sum_{s=1}^{3} \int_{z_s}^{z_{s+1}} \frac{E^{(s)}(z)}{1-\left(\ell^{(s)}\right)^{(2)}} \left(1 + \ell^{(s)}\right)\alpha^{(s)}\{1, \bar{z}, \bar{z}^2\}dz \\
\left\{{}^aB^T, {}^aD^T\right\} &= \sum_{s=1}^{3} \int_{z_s}^{z_{s+1}} \frac{E^{(s)}(z)}{1-\left(\ell^{(s)}\right)^{(2)}} \left(1 + \ell^{(s)}\right)\alpha^{(s)}\overline{\Psi}(z)\{1, \bar{z}\}dz \\
\left\{C^T, F^T, {}^aF^T\right\} &= \sum_{s=1}^{3} \int_{z_s}^{z_{s+1}} \frac{E^{(s)}(z)}{1-\left(\ell^{(s)}\right)^{(2)}} \left(1 + \ell^{(s)}\right)\alpha^{(s)}\overline{f}(z)\{1, \bar{z}, \overline{\Psi}(z)\}dz
\end{aligned} \quad (33)$$

In the equation $(\bar{z} = \frac{z}{h}, \overline{f}(z) = \frac{f(z)}{h}, \overline{\Psi}(z) = \frac{\Psi(z)}{h})$.

Finally, the displacement field can be obtained using Equation (33).

3. Model Validation and Numerical Results

In this summary, specific material parameter values are cited to verify the accuracy of the proposed method, and the effects of volume content, size, temperature variation, thermal–mechanical loading and porosity on the bending problem of functionally graded sandwich panels are analyzed.

The functionally graded sandwich panel materials used in this Section are shown in Table 1.

Table 1. Material properties of ceramics and metals in functionally graded materials.

Material	E	ϑ	α (10^{-6}/K)
Al	70	0.3	23
ZrO_2	117	1/3	7.11

Unless otherwise specified, the following parameters are used in this paper: $a = b$, $a/h = 10$, $q_0 = 100$, $t_1 = 0$, $t_2 = 100$, $t_3 = 100$. The shape function adopts Reissner, and the pore structure adopts pore one.

In this paper, the dimensionless deflection is [28]

$$\overline{w} = \frac{10^3}{q_0 a^4/(E_0 h^3) + 10^3 \alpha_0 t_2 a^2/h} w\left(\frac{a}{2}, \frac{b}{2}\right) \tag{34}$$

where $E_0 = 1GPa, \alpha_0 = 10^{-6}/K$.

The structure of the functionally graded sandwich panel can be represented by the thickness of each layer. Several thicknesses are used in this paper as shown in Table 2.

Table 2. Layer thickness ratio.

Type	1-0-1	1-1-1	1-2-1	2-1-2
z_1	$-\frac{h}{2}$	$-\frac{h}{2}$	$-\frac{h}{2}$	$-\frac{h}{2}$
z_2	0	$-\frac{h}{6}$	$-\frac{h}{4}$	$-\frac{h}{10}$
z_3	0	$\frac{h}{6}$	$\frac{h}{4}$	$\frac{h}{10}$
z_4	$\frac{h}{2}$	$\frac{h}{2}$	$\frac{h}{2}$	$\frac{h}{2}$

In this Section, three different Reissner Reddy Touratier type functions are used to calculate the dimensionless deflection of functionally gradient sandwich plates (without pores/with pores) under thermal–mechanical loads with different layer thickness ratios and different gradient indices. The dimensionless deflection of functionally graded sandwich panels with different volume fractions and different layer thickness ratios under thermal–mechanical loads. And the porosity value is between [0, 0.3], because the porosity of metal–ceramic functionally gradient materials (FGMs) is usually determined by the preparation process (such as powder metallurgy, and gradient sintering). A large number of studies have shown that due to the change in composition gradient and incomplete densification during sintering, the porosity of actual FGM materials is generally controlled at 0–30%, so the porosity xi is taken as 0, 0.1, 0.15, 0.2, and 0.25, to cover typical working conditions and systematically analyze the influence of porosity. In order to verify the accuracy of the model, this paper degenerates the functionally graded sandwich panels with pores into functionally graded sandwich panels without pores. The results corresponding to the three types of functions are compared with the FSDPT and SSDPT in the literature [29]. The solutions obtained in this paper are all close to the results of FSDPT and SSDPT. The calculation results are shown in Table 3.

It can be seen from Table 3 that, under the condition of a given layer thickness ratio, the dimensionless deflection of the functionally gradient sandwich plate increases with the increase in the gradient index s. This is because as the gradient index s increases, the ceramic content in the functionally gradient material layer gradually decreases, resulting in a decrease in the overall stiffness of the material, which in turn causes an increase in the dimensionless deflection. In addition, the results obtained by solving the Reissner-type function are within 0.3% of the results in the literature [29] (FSDPT), the results obtained by solving the Reddy-type function are within 1.5% of the results in the literature [29] (SSDPT), and the results of the Touratier-type function are between FSDPT and SSPPT, thus verifying the accuracy and reliability of the proposed method.

It can be seen from Table 4 that (1) When the porosity is zero, the deflection values of the three different pore structures are the same. This is because when the porosity is zero, the four pore structure models all degenerate into a pore-free model, and their material parameters are completely consistent, so the corresponding dimensionless deflections are also the same. (2) Under the conditions of a given volume fraction and the same layer thickness ratio, the deflection variation trends of different pore structures are different as the porosity increases. Specifically, the dimensionless deflection of pore 1 decreases with the increase in porosity, while the dimensionless deflections of pores 2 and 3 increase with the increase in porosity. This phenomenon can be explained by the following mechanical mechanisms: (1) Stress Redistribution Effect: Pore distribution patterns alter the stress transfer paths. Pore Structure 1 (uniform distribution) promotes uniform stress dispersion, whereas Pore Structures 2 (gradient distribution) and 3 (interface clustering) induce localized stress concentrations (e.g., in gradient transition zones or interfaces), thereby reducing overall structural stiffness. (2) Effective Stiffness Variation: The presence of pores reduces the effective stiffness of the material, but the extent of this reduction depends on the distribution pattern. Pore Structure 1 uniformly weakens global stiffness, leading to decreased deflection with porosity (possibly due to a "pre-relaxation" effect from uniform pores). In contrast, Pore Structures 2 and 3 exhibit abrupt local stiffness changes caused by gradient or clustered distributions, exacerbating bending deformation. (3) Neutral Axis Shift: Pore distribution modifies the neutral axis position of the sandwich plate. Pore Structure 2 shifts the neutral axis toward the denser side, causing an imbalance in stiffness between tensile and compressive zones. Pore Structure 3 induces a sudden neutral axis shift at interfaces, generating additional bending moments and increasing deflection. (4) Local Buckling Modes: Gradient or clustered distributions in Pore Structures 2 and 3 may trigger local buckling (e.g., gradient buckling or interface buckling), while uniformly distributed Pore Structure 1 primarily exhibits global buckling. The differing sensitivity of these buckling modes to deflection accounts for the observed divergence.

Table 3. Dimensionless deflection of functionally graded sandwich panels at different volume fraction indices ($a/h = 10$, $xi = 0$).

		$\bar{w}$			
s	theory	1-0-1	1-1-1	2-1-2	3-1-3
0	paper [29] (SSDPT)	0.79678	0.79678	0.79678	0.79678
	paper [29] (FSDPT)	0.89573	0.89573	0.89573	0.89573
	Reissner	0.89543	0.89543	0.89543	0.89543
	Reddy	0.80817	0.80817	0.80817	0.80817
	Touratier	0.79678	0.79678	0.79678	0.79678
1	paper [29] (SSDPT)	1.06284	1.01126	1.03621	1.04502
	paper [29] (FSDPT)	1.19072	1.13244	1.16056	1.17053
	Reissner	1.19411	1.13626	1.16425	1.17414
	Reddy	1.07769	1.02537	1.05067	1.05961
	Touratier	1.06284	1.01126	1.03621	1.04503
2	paper [29] (SSDPT)	1.12160	1.06809	1.09609	1.10517
	paper [29] (FSDPT)	1.25730	1.19570	1.22776	1.23823
	Reissner	1.25988	1.19991	1.23130	1.24148
	Reddy	1.13730	1.08291	1.11135	1.12058
	Touratier	1.12161	1.06809	1.09609	1.10517
3	paper [29] (SSDPT)	1.14165	1.09231	1.11979	1.12808
	paper [29] (FSDPT)	1.28074	1.22323	1.25504	1.26472
	Reissner	1.28228	1.22701	1.25781	1.26709
	Reddy	1.15769	1.10748	1.13542	1.14386
	Touratier	1.14166	1.09231	1.11979	1.12808
5	paper [29] (SSDPT)	1.15441	1.11266	1.13799	1.14485
	paper [29] (FSDPT)	1.29610	1.24683	1.27649	1.28462
	Reissner	1.29653	1.24976	1.27816	1.28584
	Reddy	1.17072	1.12815	1.15395	1.16095
	Touratier	1.15441	1.11266	1.13799	1.14485

Table 4. Dimensionless deflection of porous symmetric functionally graded sandwich panels with different porosities (s = 2).

Volume Fraction	Type Function	Pore 1				Pore 2			Pore 3		
		xi = 0	xi = 0.1	xi = 0.15	xi = 0.2	xi = 0.1	xi = 0.15	xi = 0.2	xi = 0.1	xi = 0.15	xi = 0.2
1-0-1	Reissner	1.25988	1.13080	1.06891	1.00894	1.34635	1.74373	2.10809	1.69370	1.93663	2.19699
	Reddy	1.13730	1.02173	0.96640	0.91285	1.23730	1.59340	1.92076	1.52665	1.74501	1.97919
	Touratier	1.12161	1.00781	0.95335	0.90065	1.21230	1.56145	1.88216	1.50507	1.72012	1.95071
1-1-1	Reissner	1.19991	1.06955	1.00671	0.94549	1.96094	2.53031	3.14298	1.95532	2.41580	2.93388
	Reddy	1.08291	0.96610	0.90985	0.85512	1.77430	2.28917	2.84568	1.76210	2.17677	2.64404
	Touratier	1.06809	0.95305	0.89766	0.84377	1.74727	2.25191	2.79569	1.73655	2.14463	2.60409
1-2-1	Reissner	1.14440	1.01349	0.95018	0.88834	2.27972	3.14814	4.21453	2.22296	2.96961	4.31152
	Reddy	1.03269	0.91533	0.85863	0.80328	2.05833	2.84445	3.83492	2.00441	2.67823	3.95517
	Touratier	1.01858	0.90298	0.84713	0.79263	2.02737	2.79849	3.75552	1.97388	2.63628	3.85594
2-1-2	Reissner	1.23130	1.10156	1.03918	0.97856	1.74183	2.19808	2.66548	1.82628	2.17166	2.54937
	Reddy	1.11135	0.99514	0.93934	0.88517	1.58252	1.99484	2.41851	1.64575	1.95646	2.29660
	Touratier	1.09609	0.98166	0.92671	0.87340	1.55618	1.95974	2.37354	1.62233	1.92825	2.26299

These findings demonstrate that the influence of pore distribution patterns on the mechanical behavior of sandwich plates extends beyond porosity magnitude. The spatial configuration of pores decisively governs structural performance by modifying stress fields, stiffness distributions, and buckling modes. This conclusion provides a theoretical foundation for the active design of gradient porosity in engineering applications.

Table 5 lists the dimensionless deflection of FMG sandwich panels under thermal–mechanical loads with different layer thickness ratios when the edge thickness ratio a/h is 5, 10, 15, 20, and 25. It can be seen that (1) When the layer thickness ratio of the sandwich panel is constant, as the edge thickness ratio a/h increases, the calculated dimensionless deflection w decreases. (2) When the edge thickness ratio (a/h) of the sandwich panel is constant, the deflection of different layer thickness ratios increases with the increase in ceramic content, and the corresponding dimensionless deflection w decreases. And the thickness ratio of each layer to the ceramic content is (1-0-1 > 2-1-2 > 1-1-1 > 1-2-1).

Table 5. Dimensionless deflection of porous FMG sandwich panels at different edge thickness ratios and layer thickness ratios (s = 2, xi = 0.25).

a/*h*	1-0-1	1-1-1	1-2-1	2-1-2
5	0.97698	0.91263	0.85449	0.94628
10	0.95104	0.88597	0.82799	0.91982
15	0.91489	0.84979	0.79257	0.88346
20	0.87513	0.81010	0.75378	0.84352
25	0.83659	0.77166	0.71623	0.80482

Table 6 lists the dimensionless deflection of FMG sandwich panels under thermal–mechanical loads with different layer thickness ratios when the aspect ratio a/b = 1, 2, 3, 4, and 5. It can be seen that when the layer thickness ratio of the sandwich panel is constant, as the aspect ratio (a/b) increases, the calculated dimensionless deflection w gradually decreases, especially when $a/b > 1$, the decrease in dimensionless deflection increases significantly.

Table 6. Dimensionless deflection of porous FMG sandwich panels with different aspect ratios and layer thickness ratios (s = 2, xi = 0.25).

a/*b*	1-0-1	1-1-1	1-2-1	2-1-2
1	0.95100	0.88600	0.82800	0.91980
2	0.36770	0.34340	0.32150	0.35610
3	0.18190	0.17030	0.15960	0.17640
4	0.10670	0.10010	0.09401	0.10360
5	0.06973	0.06564	0.06177	0.06780

Parameter Analysis

In this subsection, the deflection distribution of sandwich plates along the x and y axes, the influence of the edge thickness ratio a/h, the aspect ratio a/b, the volume fraction index s, and the porosity xi on the dimensionless deflection of different types of sandwich plates, and the influence of temperature on the dimensionless deflection are studied.

Figure 2 shows the distribution of dimensionless deflection along the x and y axes when the volume fraction index $s = 2$. Figure 3 shows the dimensionless deflection of the porous FMG sandwich plate as the edge thickness ratio a/h changes under the action of different volume fraction exponents s when the porosity $xi = 0.25$. Figure 4 shows the dimensionless deflection of the porous FMG sandwich plate as the aspect ratio a/b changes under the action of different volume fraction exponents s when the porosity $xi = 0.25$. Figure 5 shows the dimensionless deflection of the porous FMG sandwich plate as the edge thickness ratio a/h changes under the action of different porosities xi when the volume fraction index $s = 2$. Figure 6 shows that when the volume fraction index $s = 2$, the dimensionless deflection of the porous FMG sandwich plate varies with the aspect ratio a/b under different porosities xi. Figure 7 shows the effect of thermal load and mechanical load on dimensionless deflection when the volume fraction index $s = 2$ and porosity $xi =$ 0.15.

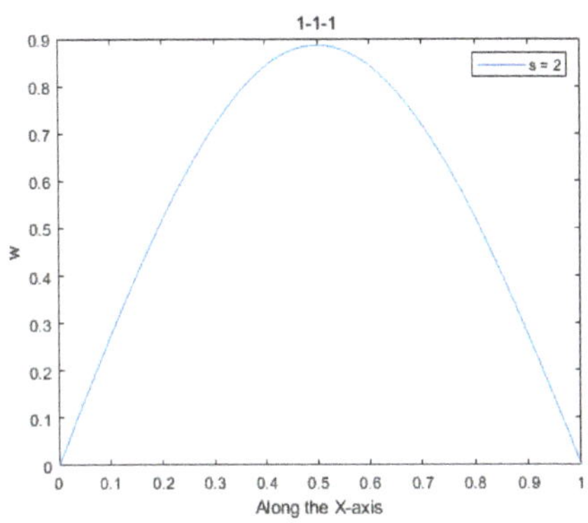

(a)

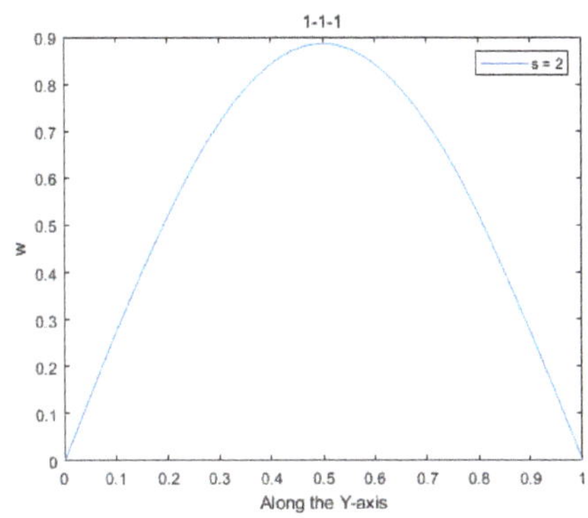

(b)

Figure 2. (**a**) Dimensionless deflection distribution along the X axis. (**b**) Dimensionless deflection distribution along the Y axis.

From Figure 2, it can be observed that (1) The dimensionless deflection is symmetrically distributed along both the x axis and the y axis. This symmetry reflects the uniform load-bearing characteristics of the sandwich plate under simply supported boundary conditions and also validates the consistency between the geometric symmetry of the model and the boundary conditions. (2) By combining Figure 2a,b, it is found that under simply supported conditions, the maximum deflection of the sandwich plate is located at (2/a, 2/b) of the plate. This result indicates that the sandwich plate experiences the highest bending stress in the central region, which aligns with the deflection distribution patterns in classical plate theory under simply supported boundary conditions. Furthermore, the location of the maximum deflection provides valuable insights for stress concentration analysis in critical areas of engineering design, such as optimizing the layout of support points in aerospace thermal protection systems.

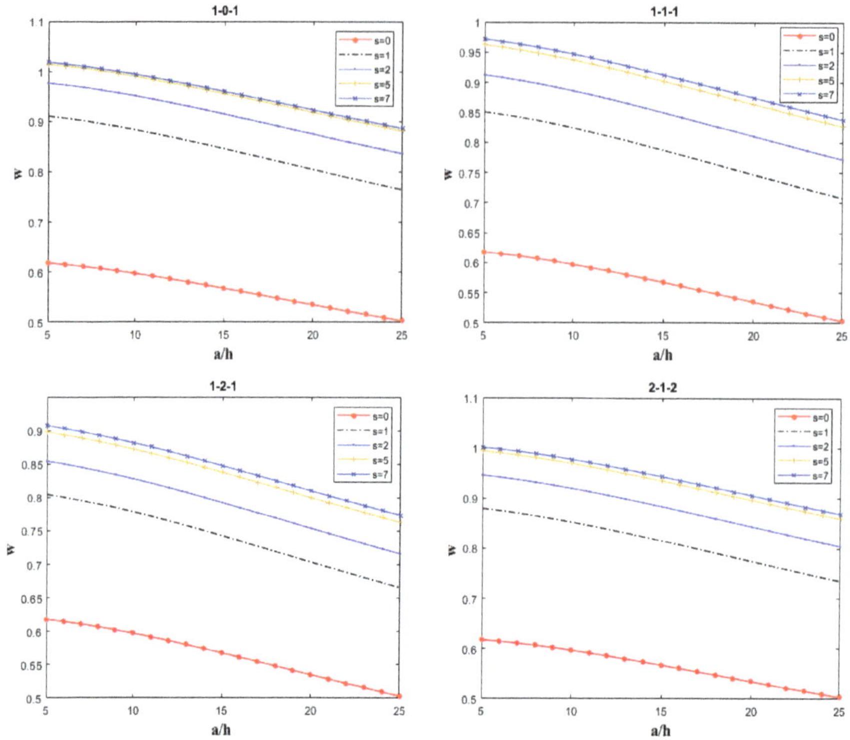

Figure 3. Variations in the dimensionless deflection of a porous FMG sandwich plate with edge thickness ratio under different volume fraction index *k* (*xi* = 0.25).

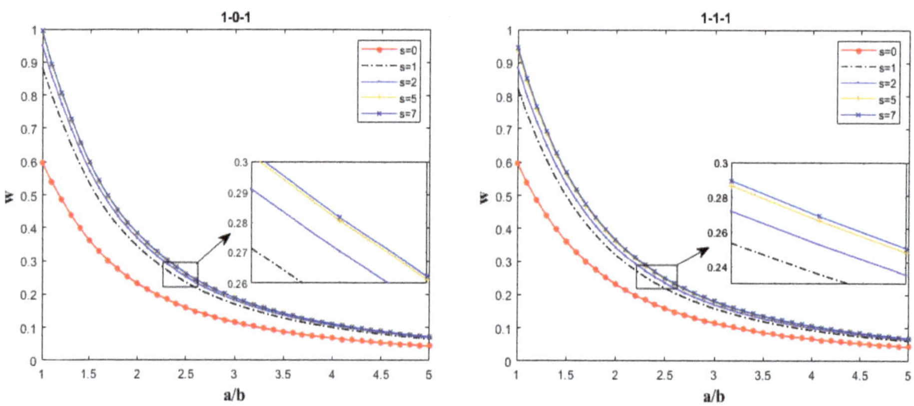

Figure 4. *Cont.*

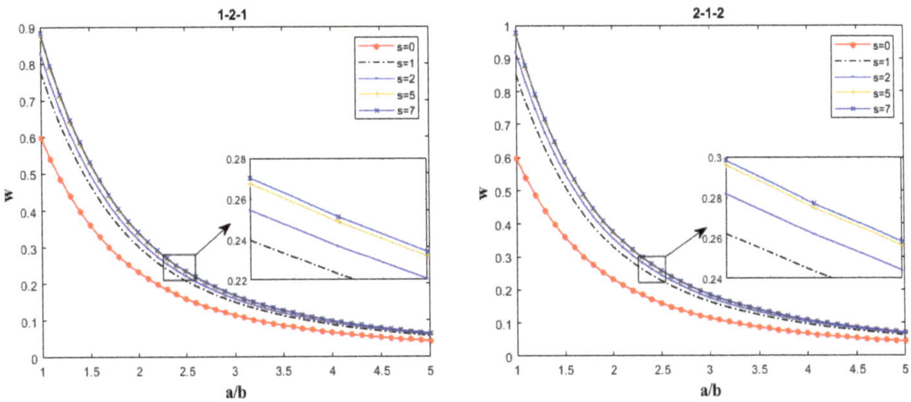

Figure 4. Variations in the dimensionless deflection of a porous FMG sandwich plate with aspect ratio under different gradient index s (xi = 0.25).

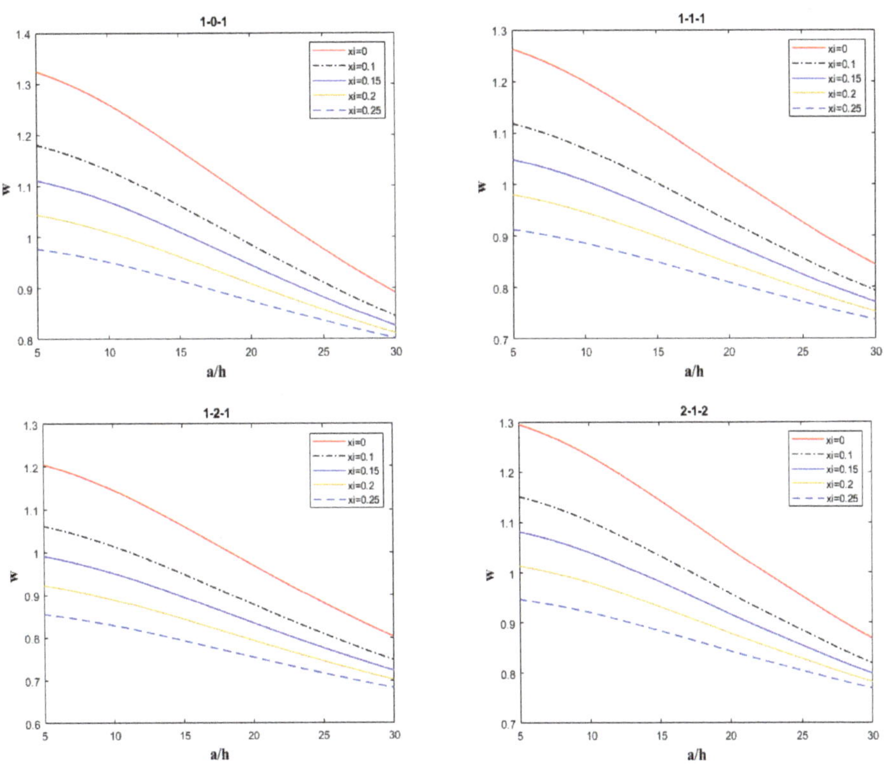

Figure 5. Variations in the dimensionless deflection of a porous FMG sandwich plate with edge thickness ratio under different porosity xi (s = 2).

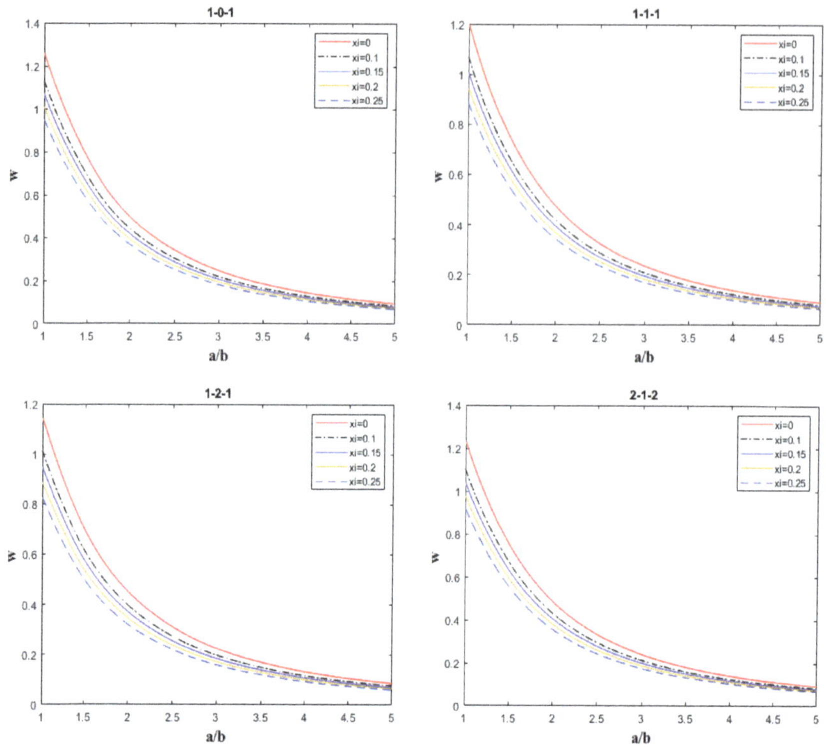

Figure 6. Variations in the dimensionless deflection of a porous FMG sandwich plate with aspect ratio under different porosity *xi* ($s = 2$).

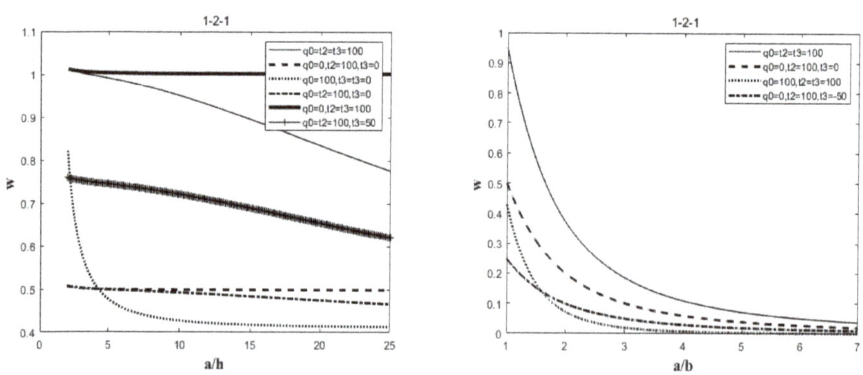

(**a**) (**b**)

Figure 7. (**a**) Effect of different edge thickness ratios on the dimensionless deflection of FMG sandwich panels under mechanical and thermal loading. (**b**) Effect of different lengths and widths on the dimensionless deflection of FMG sandwich panels under mechanical and thermal loads.

From Figure 3, it can be observed that (1) When the porosity $xi = 0.25$, the dimensionless deflection w of the porous functionally graded sandwich plate under different volume fraction exponents s shows that, regardless of the layer thickness ratio structure, the deflection reaches its maximum value at $s = 7$. This phenomenon may be related to the reduced stiffness caused by intensified material gradient variations at high volume fraction exponents, indicating a significant decline in the bending performance of the sandwich plate under extreme gradient conditions. This finding provides important guidance for the selection of volume fraction exponents in engineering design. (2) When the porosity is constant, regardless of the specific value of the volume fraction exponent k, the dimensionless deflection w of the sandwich plate shows a decreasing trend with an increasing edge thickness ratio. This result suggests that increasing the edge thickness ratio can effectively enhance the overall stiffness of the sandwich plate, thereby reducing deflection. This law provides a theoretical basis for structural optimization in engineering practice.

As can be seen from Figure 4: (1) When the porosity xi = 0.2, the dimensionless deflection w of the porous functional gradient sandwich plate under the action of different volume fraction index s shows that, regardless of the aspect ratio structure, the deflection reaches the maximum value when the volume fraction index $s = 7$. This phenomenon may be related to the reduction in stiffness caused by the intensified material gradient change under a high-volume fraction index, indicating that the bending resistance of the sandwich plate is significantly reduced under extreme gradient conditions. (2) When the porosity is constant, regardless of the volume fraction index, the dimensionless deflection w of the sandwich plate shows a downward trend with the increase in the aspect ratio. This result shows that increasing the aspect ratio within a certain range can effectively improve the overall stiffness of the sandwich plate, thereby reducing the deflection.

It can be seen from Figure 5 that (1) The deflection is maximum when the porosity $xi = 0$, and the deflection is minimum when the porosity $xi = 0.25$. (2) It can be seen that when the porosity is between 0 and 0.25, the deflection of the functionally gradient sandwich plate decreases with the increase in the edge thickness ratio (a/h). This rule shows that increasing the edge thickness ratio can effectively improve the bending resistance of the sandwich panel and thus reduce the deflection.

Figure 6 shows that (1) It is found that the deflection is maximum when the porosity $xi = 0$, and minimum when the porosity $xi = 0.25$. This result further validates the significant impact of porosity on the stiffness of the sandwich plate, providing theoretical support for porosity optimization in material design. (2) It is also found that when the porosity is between 0 and 0.25, the deflection of the functionally gradient sandwich plate decreases with the increase in the aspect ratio (a/b), and the deflection values tend to be similar. This phenomenon shows that the increase in the aspect ratio within a certain range can significantly improve the overall stiffness of the sandwich panel and thus reduce the deflection.

It can be seen from Figure 7 that Figure 7a shows the effect of the edge thickness ratio (a/h) of the functionally gradient sandwich plate ($s = 2$, $xi = 0.15$) containing pores (1-2-1) on the dimensionless deflection. (1) The dimensionless deflection of the sandwich plate that only bears a thermal load is larger, and the dimensionless deflection of the sandwich plate that only bears a mechanical load is smaller. (2) When thermal–mechanical loads are included ($q0 = t2 = t3 = 100$ and $q0 = t2 = 100$, $t3 = 50$), the deflections are between those when only mechanical loads are considered and those when only thermal loads are considered. Moreover, the deflection of thermal–mechanical loads decreases with the increase in the edge thickness ratio (a/h). (3) For sandwich plates subjected only to mechanical loads, the deflection value does not change much when $a/h \geq 10$. (4) By comparing (a) and (b), it is found that the aspect ratio (a/b) has a more significant effect on the dimensionless deflection of sandwich plates than the edge thickness ratio (a/h).

4. Conclusions

In this paper, a more accurate higher-order shear deformation theory incorporating porosity is applied to investigate the deflection of functionally graded sandwich plates under thermo-mechanical loading. The governing equations are derived based on the principle of minimum potential energy and solved using Navier's method, yielding an analytical solution for simply supported functionally graded sandwich plates with porosity. In the numerical examples Section, the porous model is reduced to zero porosity and compared with results from the literature. Subsequently, a detailed investigation was conducted on the influence of layer thickness ratio, aspect ratio, edge thickness ratio, volume fraction index, and porosity on the deflection performance of functionally graded sandwich plates under thermo-mechanical loading. The proposed theory involves only four unknown variables, which significantly reduces the computational effort compared to other theories requiring five or six unknown variables, thereby substantially improving computational efficiency.

The main findings are as follows:

1. Under a given layer thickness ratio, the dimensionless deflection of the functionally graded sandwich plate increases with the rise in the gradient index s.
2. For a given volume fraction and identical layer thickness ratio, the deflection trends of different pore structures vary with increasing porosity. Specifically, the dimensionless deflection of Pore Type 1 decreases as porosity increases, whereas the dimensionless deflections of Pore Type 2 and Pore Type 3 exhibit an increasing trend with higher porosity.
3. At a given porosity level, regardless of the specific value of the volume fraction index s, the dimensionless deflection w of the sandwich plate decreases with increasing edge thickness ratio or aspect ratio.
4. For a fixed gradient index, irrespective of the porosity xi within the range [0, 0.3], the dimensionless deflection w of the sandwich plate decreases as the edge thickness ratio or aspect ratio increases.
5. Compared to mechanical loading and thermal loading, the dimensionless deflection under thermal loading alone is greater than that under mechanical loading alone.
6. Under thermo-mechanical loading, the deflection of the functionally graded sandwich plate is significantly more sensitive to changes in the aspect ratio than to variations in the edge thickness ratio.
7. By adjusting the gradient index and porosity of the sandwich panel, the bending stiffness of the structure can be improved without significantly increasing the weight. For example, in aerospace applications, the optimized design of the gradient index can effectively alleviate thermal stress concentration while reducing the amount of material used; in new energy vehicle battery brackets, the gradient porosity design can balance structural strength and heat dissipation efficiency, thereby extending battery life.

Author Contributions: Z.H., Writing—Review and editing, Funding; Y.C., Writing—original draft, Visualization; X.W., Data curation, F.C., Validation. All authors have read and agreed to the published version of the manuscript.

Funding: This research was funded by the Natural Science Foundation of China (11862007, 52265020).

Data Availability Statement: Data are contained within the article.

Conflicts of Interest: The authors declare no conflicts of interest.

References

1. Guo, H.; Yuan, H.; Zhang, J.; Ruan, D. Review of sandwich structures under impact loadings: Experimental, numerical and theoretical analysis. *Thin Walled Struct.* **2024**, *196*, 111541. [CrossRef]
2. Alkunte, S.; Fidan, I.; Naikwadi, V.; Gudavasov, S.; Ali, M.A.; Mahmudov, M.; Hasanov, S.; Cheepu, M. Advancements and Challenges in Additively Manufactured Functionally Graded Materials: A Comprehensive Review. *J. Manuf. Mater. Process.* **2024**, *8*, 23. [CrossRef]
3. Nguyen-Van, V.; Peng, C.; Liu, J.; Tran, P.; Nguyen-Xuan, H. Performance evaluations of functionally graded porous structures. In *Machine Learning Aided Analysis Design and Additive Manufacturing of Functionally Graded Porous Composite Structures*; Elsevier: Amsterdam, The Netherlands, 2024; pp. 315–346.
4. Mechab, I.; Mechab, B.; Benaissa, S.; Serier, B.; Bouiadjra, B.B. Free vibration analysis of FGM nanoplate with porosities resting on Winkler Pasternak elastic foundations based on two-variable refined plate theories. *J. Braz. Soc. Mech. Sci. Eng.* **2016**, *38*, 2193–2211. [CrossRef]
5. Zenkour, A.M. Hygrothermal effects on the bending of angle-ply composite plates using a sinusoidal theory. *Compos. Struct.* **2012**, *94*, 3685–3696. [CrossRef]
6. Thai, H.T.; Vo, T.P. A new sinusoidal shear deformation theory for bending, buckling, and vibration of functionally graded plates. *Appl. Math. Model.* **2013**, *37*, 3269–3281. [CrossRef]
7. Mahi, A.; Tounsi, A. A new hyperbolic shear deformation theory for bending and free vibration analysis of isotropic, functionally graded, sandwich and laminated composite plates. *Appl. Math. Model.* **2015**, *39*, 2489–2508. [CrossRef]
8. Mantari, J.L.; Granados, E.V. A refined FSDT for the static analysis of functionally graded sandwich plates. *Thin Walled Struct.* **2015**, *90*, 150–158. [CrossRef]
9. Thai, C.H.; Kulasegaram, S.; Tran, L.V.; Nguyen-Xuan, H. Generalized shear deformation theory for functionally graded isotropic and sandwich plates based on isogeometric approach. *Comput. Struct.* **2014**, *141*, 94–112. [CrossRef]
10. Mantari, J.L.; Monge, J.C. Buckling, free vibration and bending analysis of functionally graded sandwich plates based on an optimized hyperbolic unified formulation. *Int. J. Mech. Sci.* **2016**, *119*, 170–186. [CrossRef]
11. Nguyen, T.N.; Thai, C.H.; Nguyen-Xuan, H. A novel computational approach for functionally graded isotropic and sandwich plate structures based on a rotation-free meshfree method. *Thin Walled Struct.* **2016**, *107*, 473–488. [CrossRef]
12. Swaminathan, K.; Hirannaiah, S.; Rajanna, T. Vibration and stability characteristics of functionally graded sandwich plates with/without porosity subjected to localized edge loadings. *Mech. Based Des. Struct. Mach.* **2023**, *51*, 6254–6292. [CrossRef]
13. Ertenli, M.F.; Esen, İ. The effect of the various porous layers on thermomechanical buckling of FGM sandwich plates. *Mech. Adv. Mater. Struct.* **2024**, *31*, 10935–10961. [CrossRef]
14. Nguyen, V.C.; Tran, H.Q.; Pham, V.V. Nonlinear static analysis of bi-directional functionally graded sandwich plates in thermal environments by a higher-order finite element model. *Thin Walled Struct.* **2023**, *188*, 110819. [CrossRef]
15. Zenkour, A.M.; Alghamdi, N.A. Bending analysis of functionally graded sandwich plates under the effect of mechanical and thermal loads. *Mech. Adv. Mater. Struct.* **2010**, *17*, 419–432. [CrossRef]
16. Mantari, J.L.; Granados, E.V. Thermoelastic behavior of advanced composite sandwich plates by using a new 6 unknown quasi-3D hybrid type HSDT. *Compos. Struct.* **2015**, *126*, 132–144. [CrossRef]
17. Mantari, J.L.; Granados, E.V. Thermoelastic analysis of advanced sandwich plates based on a new quasi-3D hybrid type HSDT with 5 unknowns. *Compos. Part B Eng.* **2015**, *69*, 317–334. [CrossRef]
18. Wang, Z.X.; Shen, H.S. Nonlinear analysis of sandwich plates with FGM face sheets resting on elastic foundations. *Compos. Struct.* **2011**, *93*, 2521–2532. [CrossRef]
19. Yang, J.; Liew, K.M.; Kitipornchai, S. Dynamic stability of laminated FGM plates based on higher-order shear deformation theory. *Comput. Mech.* **2004**, *33*, 305–315. [CrossRef]
20. Hadji, L.; Tounsi, A. Static deflections and stress distribution of functionally graded sandwich plates with porosity. *Smart Struct. Syst. Int. J.* **2021**, *28*, 343–354.
21. Sah, S.K.; Ghosh, A. Effect of porosity on the thermal buckling analysis of power and sigmoid law functionally graded material sandwich plates based on sinusoidal shear deformation theory. *Int. J. Struct. Stab. Dyn.* **2022**, *22*, 2250063. [CrossRef]
22. Gupta, A.; Talha, M. Influence of porosity on the flexural and free vibration responses of functionally graded plates in thermal environment. *Int. J. Struct. Stab. Dyn.* **2018**, *18*, 1850013. [CrossRef]
23. Shahsavari, D.; Shahsavari, M.; Li, L.; Karami, B. A novel quasi-3D hyperbolic theory for free vibration of FG plates with porosities resting on Winkler/Pasternak/Kerr foundation. *Aerosp. Sci. Technol.* **2018**, *72*, 134–149. [CrossRef]
24. Wattanasakulpong, N.; Ungbhakorn, V. Linear and nonlinear vibration analysis of elastically restrained ends FGM beams with porosities. *Aerosp. Sci. Technol.* **2014**, *32*, 111–120. [CrossRef]
25. Reddy, J.N. A simple higher-order theory for laminated composite plates. *J. Appl. Mech.* **1984**, *51*, 745–752. [CrossRef]
26. Reissner, E. On tranverse bending of plates, including the effect of transverse shear deformation. *Int. J. Solids Struct.* **1975**, *11*, 569–573. [CrossRef]

27. Touratier, M. An efficient standard plate theory. *Int. J. Eng. Sci.* **1991**, *29*, 901–916. [CrossRef]
28. Tounsi, A.; Houari MS, A.; Benyoucef, S. A refined trigonometric shear deformation theory for thermoelastic bending of functionally graded sandwich plates. *Aerosp. Sci. Technol.* **2013**, *24*, 209–220. [CrossRef]
29. Houari MS, A.; Tounsi, A.; Bég, O.A. Thermoelastic bending analysis of functionally graded sandwich plates using a new higher order shear and normal deformation theory. *Int. J. Mech. Sci.* **2013**, *76*, 102–111. [CrossRef]

symmetry

MDPI

Article

Effect of Laying Angle on the Stress Distribution and Stiffness Degradation of Symmetrically Cracked Laminates

Zhicheng Huang [1,*], Biao Xiong [1], Xingguo Wang [1] and Fulei Chu [2]

[1] School of Mechanical and Electronic Engineering, Jingdezhen Ceramic University, Jingdezhen 333403, China
[2] School of Mechanical Engineering, Tsinghua University, Beijing 100084, China
* Correspondence: huangzhicheng@jci.edu.cn

Abstract: In this paper, the effects of laying angle on the stress distribution and stiffness degradation of glass epoxy laminates with symmetrical cracks are systematically studied by variational analysis and numerical simulation. Based on the principle of minimum complementary energy, the control equation was established and solved using MATLAB programming. The stress field and stiffness attenuation characteristics of 90°-layer crack density (CD = 0.5 cr/mm) and laminates with different angles $[\theta_m/90_n]_s$ were analyzed. The results show that the crack significantly aggravates the stress concentration effect, and when the laying angle exceeds 45°, the crack growth rate and stiffness degradation rate are significantly improved. Specifically, when the laying angle is 30° and $L_1/t_1 = 50$, the stiffness degradation rate (E_x/E_{x0}) of the laminates only decreases to 0.987, while when the laying angle increases to 60° and $L_1/t_1 = 3$, the stiffness degradation rate suddenly decreases to 0.754 under the same conditions, indicating that small-angle laying (<45°) can effectively alleviate local stress concentration and delay stiffness degradation. By comparing the experimental data of existing references, the model error is verified to be less than 1%, and the reliability of the method is confirmed. It is further proposed that in the design of symmetrical laminated plates, the crack propagation resistance can be significantly optimized by controlling the laying angle of the main bearing layer in the range of 30°~45°, which provides a quantitative basis for crack suppression and structural life improvement in engineering.

Keywords: symmetrically cracked laminates; laying angle; variational analysis; stress distribution; stiffness degradation

1. Introduction

Composite laminates are widely used in aerospace, automotive manufacturing, marine engineering, and other high-performance structures due to their light weight, high strength, flexible design, and excellent corrosion resistance. As a typical composite material [1], glass epoxy is composed of glass fiber and epoxy matrix. Glass fiber has the characteristics of high strength and high modulus, which endows the composite with good mechanical properties and enables it to withstand large loads. The epoxy has excellent adhesion and chemical resistance, which can effectively bond the glass fibers together, ensuring the integrity of the composite material and protecting the fibers from the external environment. The material structure of this paper is shown in Figure 1. However, complex environments and external load conditions will lead to cracks, delamination, and other damage to the laminated plate, which will not only change the stress distribution of the composite material, but also significantly reduce its structural stiffness and bearing capacity, thus threatening the

https://doi.org/10.3390/sym17040495

safety and reliability of engineering structures. In order to ensure the long-term stability of laminates [2] under actual working conditions, it is particularly important to study the influence of cracks on the mechanical properties of laminates.

Figure 1. Glass epoxy laminate structure.

In recent years, many researchers have carried out research on the mechanical behavior and stiffness degradation of laminates with cracks. Through variational stress analysis and experimental verification, Fikry et al. revealed [3] the influence of matrix cracks on the stiffness degradation of fiber-reinforced polymer (FRP) laminates with different off-axis angles. Ramezani et al. [4] demonstrated that the addition of carbon nanofibers (CNFs) can effectively inhibit crack propagation in the matrix of fiber polymer composites and significantly reduce the crack density and stiffness degradation. Ghadami [5] proposed a new anisotropic nonlinear multi-scale damage model based on the TALREJA method. Experimental verification showed that the model is significantly better than the existing theories for stress–strain prediction and reference state optimization. Chaupa [6] used five different convolutional neural network (CNN) models with transfer learning techniques to discriminate whether randomly oriented chopped glass fiber composite laminates were damaged or not. Fakoor [7] predicted the critical angle between fiber and crack directions in orthotropic materials through stress series expansion and numerical methods, combined with experimental validation, thereby establishing a theoretical foundation for preventing stiffness degradation and catastrophic failure. Farrokhabadi [8] calculated the changes in the mechanical properties of symmetric laminates in the presence of matrix cracks by considering the equivalent damage layer and using the relationships of classical laminate theory (CLT). Mohammadi [9] used the variational method to simulate the damage state and verified the propagation of matrix crack with experiments. Pakdel et al. [10] established an energy-based criterion for predicting the distribution pattern of matrix cracks in laminates under uniaxial tensile loading with competitive damage modes. Qi, Wenxuan et al.'s research [11] is based on the continuum damage mechanics (CDM) theory; the proposed model is effective and reliable in predicting the occurrence and evolution of matrix cracks and the corresponding stiffness degradation in composite laminates under fatigue loading. Rezaei Jafari [12] quantitatively predicted the dynamic association between stiffness degradation and strain energy release rate during crack propagation in a sub-matrix by developing an energy-based analytical model with a crack-closure technique. Kassapoglou et al. [13] proposed a model to accurately calculate the stress of the matrix crack layer under a combined strain state. Based on the energy density criterion, the predicted crack spacing is in good agreement with the test results. Yuan, Mingqing et al. [14] proposed a data-driven method to predict the stiffness degradation of cross-ply laminates with matrix cracking, which provides an efficient method for the optimization of damage and the mechanical properties of composites. Zhang et al. [15] proposed that the shear stress of laminates varies linearly within a 90° layer. Berthelot [16] argues that the relationship between the longitudinal displacement of the laminate and the transverse coordinate is quadratic within the 90° layer and that the scaling factor of this relationship is an exponential function of the transverse crack density. M. Kashtalyan et al. [17] proposed an improved two-dimensional shear lag analysis method. Relevant scholars have discussed the stiffness degradation for calculating the internal stress field of cracked laminates using the shear lag model and its modified model and the energy principle [18], and described the stiffness degradation [19–21] process. In subsequent studies, scholars have continuously explored and summarized the variational method [22,23], which is used to determine the stress field of cross-laminated plates with cracks [24,25] present in 90° layers under tensile and shear loads [26].

In this paper, the general expression of the stress component is obtained through the boundary conditions of the damage analysis element and the equilibrium differential equation, and the governing equation and supplementary equation are established using the variation in the minimum complementary energy principle. Programming calculations are undertaken with MATLAB2021. We calculate and analyze cracks under ideal conditions, calculating the normal stress distribution and the change in the stiffness degradation [27] of 90° laminated plates. By citing the models of other researchers and comparing the experimental results, the accuracy of this model is verified.

2. Mathematical Modeling and Theoretical Formulation

2.1. Analysis Unit Model

In symmetrical laminates $[\theta_m/90_n]_s$, the upper and lower θ layers have the same mechanical properties, while the middle 90° layer has different mechanical properties due to having different ply angles. Under the action of load, the transverse cracks in the 90° layer will be evenly distributed along the main direction until they penetrate the entire 90° layer of the laminate. Therefore, the cracks in the angle-ply symmetric laminate have periodicity in the main direction. Select the volume element RVE in the damaged area, as shown in Figure 2. $t_1 = nt_0$ and $t_2 = mt_0$ are the thickness of the 90° layer and the θ layer. Subscripts 1 and 2 represent the 90° layer and the θ layer, respectively. The distance between the two cracks is $2L_1$. $h = t_1 + t_2$. The laminate is laid in sequence from top to bottom at θ, 90°, and θ, and has upper and lower symmetry in the structure. Therefore, only the upper half of the volume element needs to be studied $(z \geq 0)$.

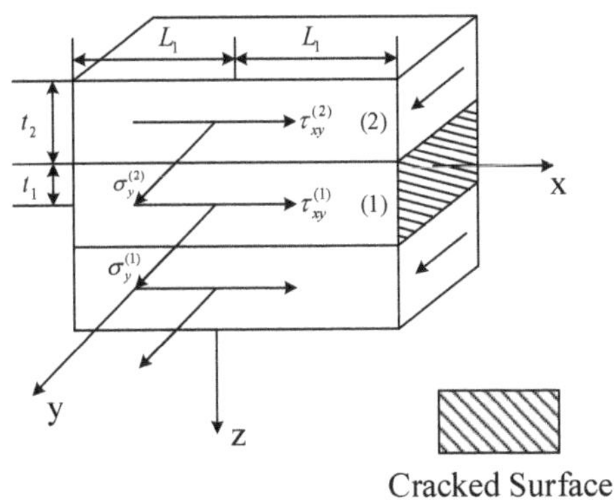

Figure 2. Model of cracked element.

2.2. Boundary Conditions and Representations of the Stress Components

At the boundary $x = \pm L_1$, the tensile stress and shear stress of layers (1) and (2) meet the resultant force of

$$t_1 \sigma_x^{(1)}\Big|_{x=\pm L_1} + t_2 \sigma_x^{(2)}\Big|_{x=\pm L_1} = h\overline{\sigma_x} \tag{1}$$

$$t_1 \tau_{xy}^{(1)}\Big|_{x=\pm L_1} + t_2 \tau_{xy}^{(2)}\Big|_{x=\pm L_1} = 0 \tag{2}$$

The superscripts (1) and (2) in the formula represent the 90° layers and θ layers, respectively. $\overline{\sigma_x}$ is the average tensile stress.

For an open crack, the axial tensile stress $\overline{\sigma_x}$ at the boundary of $x = \pm L_1$, and layers (1) and (2) under tensile load are

$$\sigma_x^{(1)}\Big|_{x=\pm L_1} = 0 \tag{3}$$

$$\tau_{xy}^{(1)}\Big|_{x=\pm L_1} = 0 \tag{4}$$

From Equations (1) and (2), we obtain

$$\sigma_x^{(2)}\Big|_{x=\pm L_1} = \frac{h}{t_2}\overline{\sigma_x} \tag{5}$$

$$\tau_{xy}^{(2)}\Big|_{x=\pm L_1} = 0 \tag{6}$$

According to the requirements of symmetry, the out-of-plane shear stress on the middle plane $(z = 0)$ of the laminated plate is 0; that is,

$$\tau_{xz}^{(1)}\Big|_{Z=0} = \tau_{yz}^{(1)}\Big|_{Z=0} = 0 \tag{7}$$

On the surface $(z = h)$ of the laminate, the normal tensile stress and out-of-plane shear stress are 0; that is,

$$\sigma_z^{(2)}\Big|_{Z=h} = \tau_{xz}^{(2)}\Big|_{Z=h} = \tau_{yz}^{(2)}\Big|_{Z=h} = 0 \tag{8}$$

The element is subjected to tensile load. At the boundary, the out-of-plane shear stress is 0; that is,

$$\tau_{xz}^{(1)}\Big|_{x=\pm L_1} = \tau_{xz}^{(2)}\Big|_{x=\pm L_1} = 0 \tag{9}$$

Since the RVE material and load are uniformly distributed in the y direction, the stress component is assumed to be a binary function that depends only on x and z. Combined with Equations (1)–(9) and their interlaminate stress continuity conditions, the general expressions of the layer (1) and (2) stress components in the damage element can be deduced as follows:

$$\begin{cases} \sigma_x^{(1)} = -\frac{1}{t_1} L_1 \overline{\sigma_x} \varphi(s) \\ \sigma_y^{(1)} = \frac{1}{t_1} L_1 \overline{\sigma_x} \phi(s) \\ \sigma_z^{(1)} = \frac{1}{2}\left(h - \frac{z^2}{t_1}\right) \frac{1}{L_1} \overline{\sigma_x} \varphi''(s) \\ \tau_{zx}^{(1)} = \frac{z}{t_1} \overline{\sigma_x} \varphi'(s) \\ \tau_{zy}^{(1)} = \frac{z}{t_1} \overline{\sigma_x} \psi'(s) \\ \tau_{xy}^{(1)} = -\frac{1}{t_1} L_1 \overline{\sigma_x} \psi(s) \end{cases} \tag{10}$$

$$\begin{cases} \sigma_x^{(2)} = \frac{1}{t_2}L_1\overline{\sigma_x}\varphi(s) \\ \sigma_y^{(2)} = -\frac{1}{t_2}L_1\overline{\sigma_x}\phi(s) \\ \sigma_z^{(2)} = \frac{(h-z)^2}{2t_2}\frac{1}{L_1}\overline{\sigma_x}\varphi''(s) \\ \tau_{zx}^{(2)} = \frac{h-z}{t_2}\overline{\sigma_x}\varphi'(s) \\ \tau_{zy}^{(2)} = \frac{h-z}{t_2}\overline{\sigma_x}\psi'(s) \\ \tau_{xy}^{(2)} = \frac{1}{t_2}L_1\overline{\sigma_x}\psi(s) \end{cases} \tag{11}$$

where $\varphi(s)$, $\psi(s)$, $\phi(s)$ is an independent pending function. Substituting Equation (10) and Equation (11) into the boundary conditions of Equations (1)–(9), respectively, we can obtain

$$\varphi(s)|_{s=\pm1} = \varphi'(s)|_{s=\pm1} = \psi(s)|_{s=\pm1} = 0 \tag{12}$$

2.3. *Governing and Supplementary Equations and Their Solutions*

For the hyperstatic boundary, the variational method with governing equations and supplementary equations is equivalent to the variational analysis under natural boundary conditions. According to the principle of minimum complementary energy,

$$\delta U = \delta \sum_{i=1}^{2} U^{(i)} = 0 \tag{13}$$

where U is the total residual energy of the laminated plate, and the superscript (i) is the (i) layer of the single-layer plate.

$$U = \int_{-L_1}^{L_1}\int_0^{t_1} \mu^{(1)} dzdx + \int_{-L_1}^{L_1}\int_{t_1}^{h} \mu^{(2)} dzdx \tag{14}$$

in which

$$\mu^{(i)} = \frac{1}{2}[\sigma_x^{(i)}\varepsilon_x^{(i)} + \sigma_y^{(i)}\varepsilon_y^{(i)} + \sigma_z^{(i)}\varepsilon_z^{(i)} + \tau_{zx}^{(i)}\gamma_{zx}^{(i)} + \tau_{zy}^{(i)}\gamma_{zy}^{(i)} + \tau_{xy}^{(i)}\gamma_{xy}^{(i)}](i = 1, 2) \tag{15}$$

where ε_x, ε_y, ε_z, γ_{zx}, γ_{zy}, and γ_{xy} are normal and shear strain components.

Substituting Equations (10) and (11) into (15) and applying Hooke's law and Equation (13) to give non-zero $\delta\varphi$, $\delta\psi$, $\delta\phi$, the governing equation can be obtained.

$$A_{11}\varphi''''(s) + B_{11}\varphi''(s) + C_{11}\varphi(s) + B_{12}\psi''(s) + C_{12}\psi(s) + B_{13}\phi''(s) + C_{13}\phi(s) + \lambda_1\lambda = D_1 \tag{16}$$

$$B_{21}\varphi''(s) + C_{21}\varphi(s) + B_{22}\psi''(s) + C_{22}\psi(s) + C_{23}\phi(s) + \lambda_2\lambda = D_2 \tag{17}$$

$$B_{31}\varphi''(s) + C_{31}\varphi(s) + C_{32}\psi(s) + C_{33}\phi(s) + \lambda_3\lambda = D_3 \tag{18}$$

According to Equation (18), we can obtain an equation that uses the sum of functions to express functions

$$\phi(s) = \frac{D_3}{C_{33}} - \frac{1}{C_{33}}[B_{31}\varphi''(s) + C_{31}\varphi(s) + C_{32}\psi(s)] \tag{19}$$

Based on the basic theory of linear differential equations, the solution of non-homogeneous linear differential equations is composed of the general solution of its corresponding homogeneous equations and the special solution of non-homogeneous equations, and the general solution of general exponential function (Ce^{rs}) is assumed to be the following form:

$$\begin{vmatrix} a_{11}r^4 + b_{11}r^2 + c_{11} & b_{12}r^2 + c_{12} \\ b_{21}r^2 + c_{21} & b_{22}r^2 + c_{22} \end{vmatrix} = 0 \tag{20}$$

According to Equation (20), the characteristic equation has three pairs of positive and negative characteristic roots, which are recorded as $\pm r_j, j = 1,2,3$.

The eigenvector $\vec{D_{(j)}} = [D_{(j)1} \quad 1]^T$ can be obtained from the following equation:

$$(a_{11}r_j^4 + b_{11}r_j^2 + c_{11})D_{(j)} = -(b_{12}r_j^2 + c_{12}) \tag{21}$$

Combined with the general solution of the secondary equations and the special solution and eigenvector of the non-homogeneous equations, the following results can be obtained:

$$\begin{aligned} \varphi(s) &= \sum_{j=1}^{3} (K_j e^{r_j s} + K_{-j} e^{-r_j s}) D_{(j)1} + \varphi^* \\ \psi(s) &= \sum_{j=1}^{3} (K_j e^{r_j s} + K_{-j} e^{-r_j s}) + \psi^* \end{aligned} \tag{22}$$

For different roots, where K_j and K_{-j} are unknown constants, $j = 1,2,3$, and φ^* and ψ^* are undetermined constants, respectively.

According to the boundary conditions of $K_j = K_{-j}$, Equation (26) can be rewritten as

$$\begin{aligned} \varphi(s) &= \sum_{j=1}^{3} A_j D_{(j)1} \cosh(r_j s) + \varphi^* \\ \psi(s) &= \sum_{j=1}^{3} A_j \cosh(r_j s) + \psi^* \end{aligned} \tag{23}$$

where $A_j = 2K_j$.

According to Equation (12),

$$\varphi(1) = \sum_{j=1}^{3} A_j D_{(j)1} \cosh(r_j) + \varphi^* = 0 \tag{24}$$

$$\psi(1) = \sum_{j=1}^{3} A_j \cosh(r_j) + \psi^* = 0 \tag{25}$$

$$\varphi'(1) = \sum_{j=1}^{3} A_j D_{(j)1} \sinh(r_j) = 0 \tag{26}$$

2.4. *Effective Mechanical Properties*

Generally, the relationship between the stress and strain of laminates is obtained by measuring the surface deformation of specimens through load tests. In order to estimate the performance of the material, the average strain on the surface of the representative volume element (RVE) can be calculated, and the crack-free corner ply laminates follow Hooke's law. The average strain component of the RVE surface can be expressed as

$$\bar{\varepsilon}_J^{(2)} = \frac{1}{2L_1} \int_{-L_1}^{L_1} \varepsilon_J^{(2)} \Big|_{z=h} dx = \frac{1}{2L_1} \int_{-L_1}^{L_1} [S_{J1}^{(2)} \sigma_x^{(2)} + S_{J2}^{(2)} \sigma_y^{(2)} + S_{J6}^{(2)} \tau_{xy}^{(2)}] dx, J = 1,2,6 \tag{27}$$

where $S_{J1}^{(2)}$, $S_{J2}^{(2)}$, and $S_{J6}^{(2)}$ are the elastic parameter of RVE.

Substituting Equation (23) into Equation (27) and combining this with Equation (19), we can obtain the effective plane flexibility elements of cracked laminated plates:

$$\bar{S}_{J1} = \frac{\bar{\varepsilon}_J^{(2)}}{\bar{\sigma}_x} = \frac{L_1}{t_2} (\sum_{i=1}^{3} P_{Ji} A_i \frac{\sinh(r_i)}{r_i} + P_{J4}\varphi^* + P_{J5}\psi^* + P_{J6}), J = 1,2,6 \tag{28}$$

where

$$P_{Ji} = (S_{J1}^{(2)} + S_{J2}^{(2)}\frac{C_{31}}{C_{33}})D_{(i)1} + S_{J6}^{(2)} + S_{J2}^{(2)}\frac{C_{32}}{C_{33}}, i = 1,2,3,$$

$$P_{J4} = S_{J1}^{(2)} + S_{J2}^{(2)}\frac{C_{31}}{C_{33}},$$

$$P_{J5} = S_{J6}^{(2)} + S_{J2}^{(2)}\frac{C_{32}}{C_{33}},$$

$$P_{J6} = S_{J1}^{(2)}\frac{h}{L_1} + S_{J2}^{(2)}\frac{D_3}{C_{33}}.$$

Then, the effective Young's modulus in the X direction of the cracked laminated plate is

$$\overline{E}_{xy} = \frac{1}{\overline{S}_{11}}. \tag{29}$$

3. Numerical Results and Simulation

In order to calculate the effective mechanical properties of the 90° layer in composite laminates with cracks and compare the data with other studies, this paper selects glass epoxy resin as the material. The material parameters are from the articles of Hashin [28], Huang et al. [29,30], and P.A. Carraro [31]. The following table lists the material constants of the two kinds of glass/epoxy for the orthogonal laminates. The effective mechanical properties were calculated and compared with the experimental data from researchers Hashin [28] and Huang et al. [29,30] to verify the correctness of the model. The material constant of laminates $[0/90_3]_s$ and $[0/90_2]_s$ were calculated using glass/epoxy2 in Table 1. The results are shown in Tables 2 and 3.

Table 1. Material constant.

	E_A/GPa	E_T/GPa	G_A/GPa	G_T/GPa	v_A	v_T	t_0/mm
Glass/epoxy1	42.0	13.0	3.60	4.50	0.25	0.38	0.330
Glass/epoxy2	41.7	13.0	3.40	4.58	0.30	0.42	0.100

Table 2. Young's modulus attenuation of laminates: $[0/90_3]_s$.

L_1/t_1	Present	Huang et al. [30]	Huang et al. [29]	Hashin [28]
	E_x/E_{x0}	E_x/E_{x0}	E_x/E_{x0}	E_x/E_{x0}
50	0.987	0.978	0.980	0.980
20	0.958	0.948	0.952	0.951
10	0.914	0.902	0.908	0.907
5	0.837	0.821	0.832	0.830
3	0.754	0.735	0.748	0.745
2	0.669	0.653	0.665	0.661
1	0.556	0.551	0.553	0.548
0.5	0.532	0.529	0.529	0.524

Table 3. Young's modulus attenuation of laminates: $[0/90_2]_s$.

L_1/t_1	Present	Huang et al. [30]	Huang et al. [29]	Hashin [28]
	E_x/E_{x0}	E_x/E_{x0}	E_x/E_{x0}	E_x/E_{x0}
50	0.998	0.989	0.990	0.990
20	0.984	0.974	0.976	0.975
10	0.961	0.950	0.954	0.953
5	0.918	0.905	0.911	0.910
3	0.868	0.854	0.861	0.859
2	0.822	0.812	0.816	0.813
1	0.784	0.778	0.778	0.775
0.5	0.779	0.773	0.773	0.770

It can be seen from Tables 2 and 3 that the calculated Young's modulus is highly consistent with the data of Hashin [28], Huang [29,30], and others. Although there is a small error, the error range is less than 1%, which is within the normal allowable range. This result can be attributed to the fact that the principal axis of the orthogonal laminate material is aligned with the coordinate axis. Moreover, the tension is carried out in the direction of the principal axis, and there is no shear stress between the laminates.

For cases $\theta = 20°, 35°, 45°$, and $50°$, when the crack density is 0.5 (cr/mm) in the 90° plies, the material constant of the laminate is glass/epoxy 1, and the stress distribution results are shown in Figures 2–13.

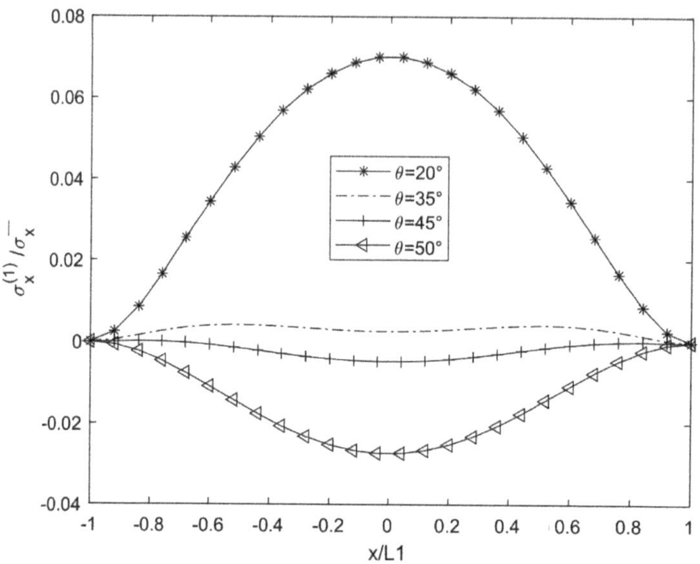

Figure 3. Graph illustrating σ_x in 90° ply for $[\theta_1/90_2]_s$ laminate.

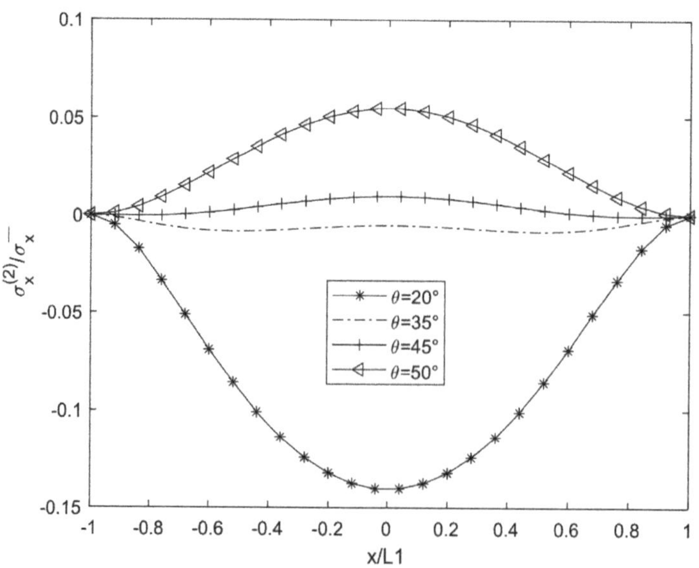

Figure 4. Graph illustrating σ_x in θ ply for $[\theta_1/90_2]_s$ laminate.

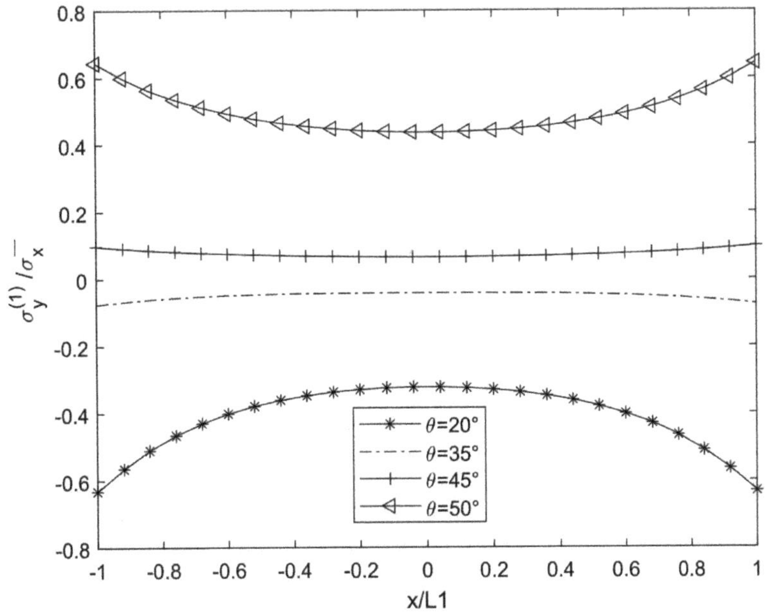

Figure 5. Graph illustrating σ_y in 90° ply for $[\theta_1/90_2]_s$ laminate.

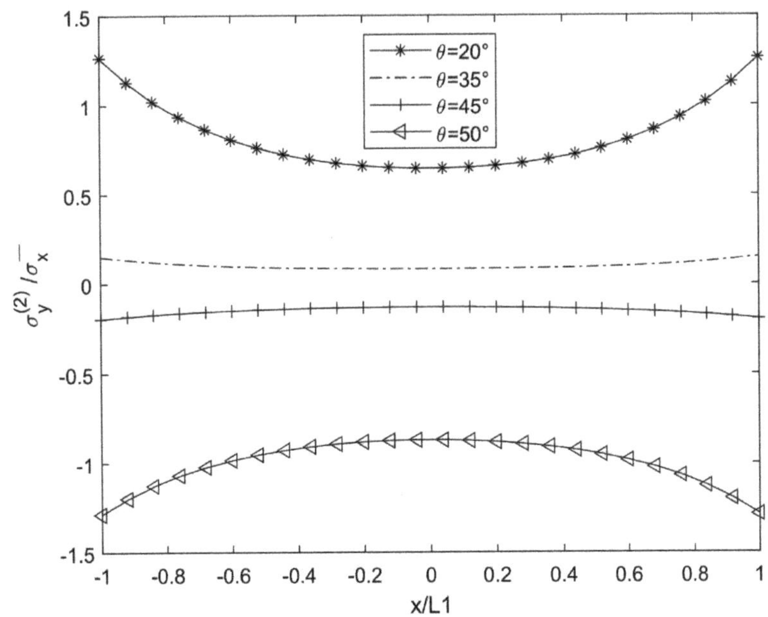

Figure 6. Graph illustrating σ_y in θ ply for $[\theta_1/90_2]_s$ laminate.

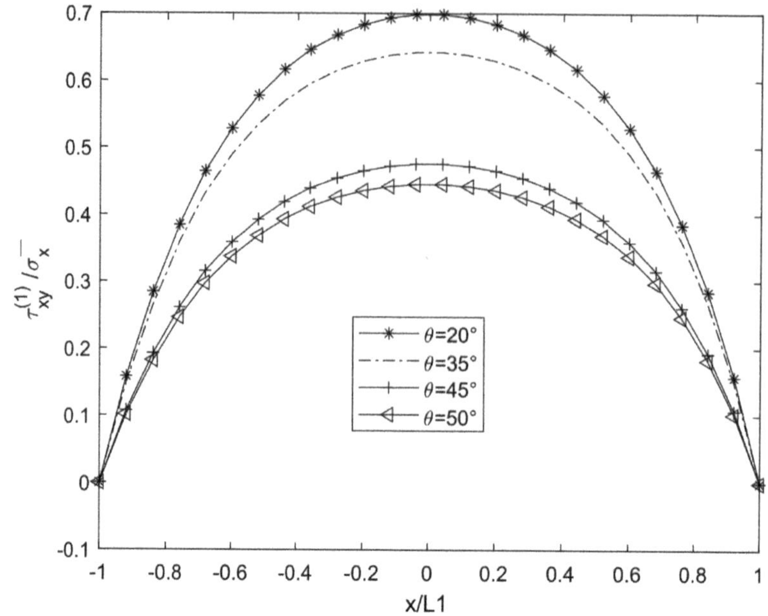

Figure 7. Graph illustrating τ_{xy} in 90° ply for $[\theta_1/90_2]_s$ laminate.

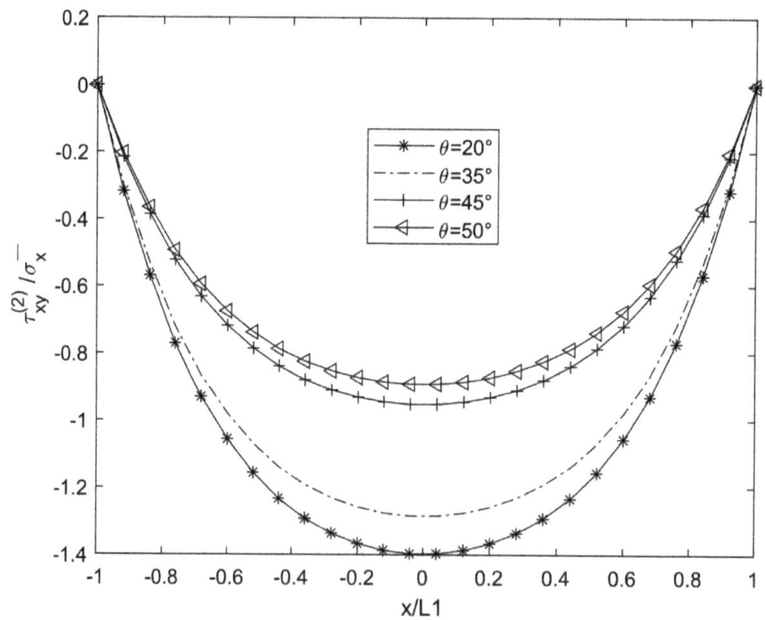

Figure 8. Graph illustrating τ_{xy} in θ ply for $[\theta_1/90_2]_s$ laminate.

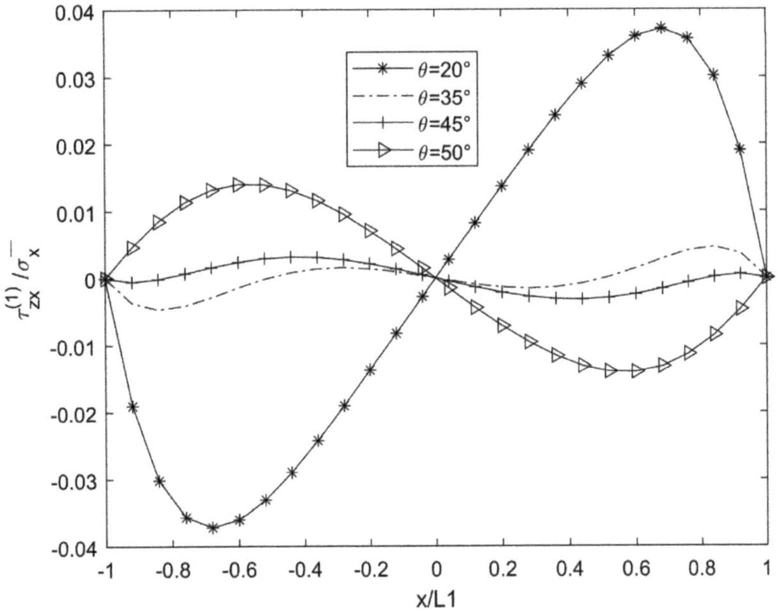

Figure 9. Graph illustrating τ_{zx} in 90° ply for $[\theta_1/90_2]_s$ laminate.

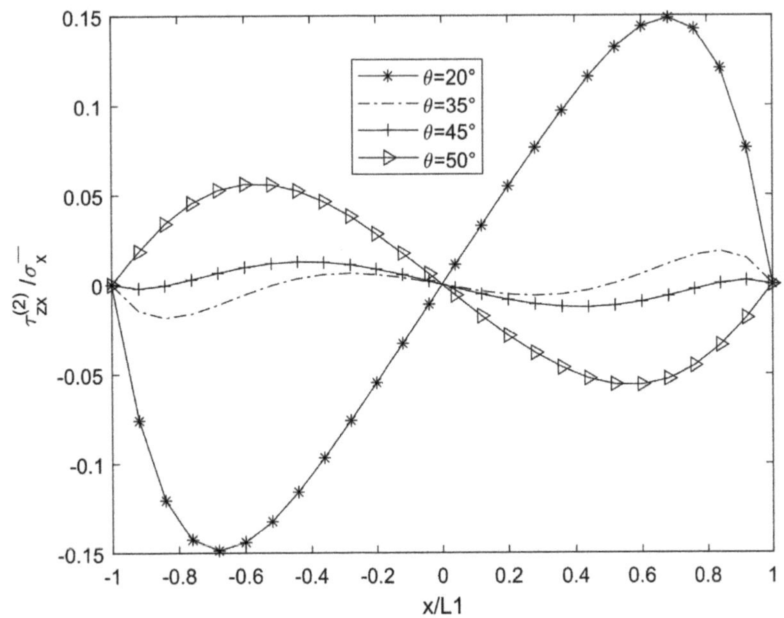

Figure 10. Graph illustrating τ_{zx} in θ ply for $[\theta_1/90_2]_s$ laminate.

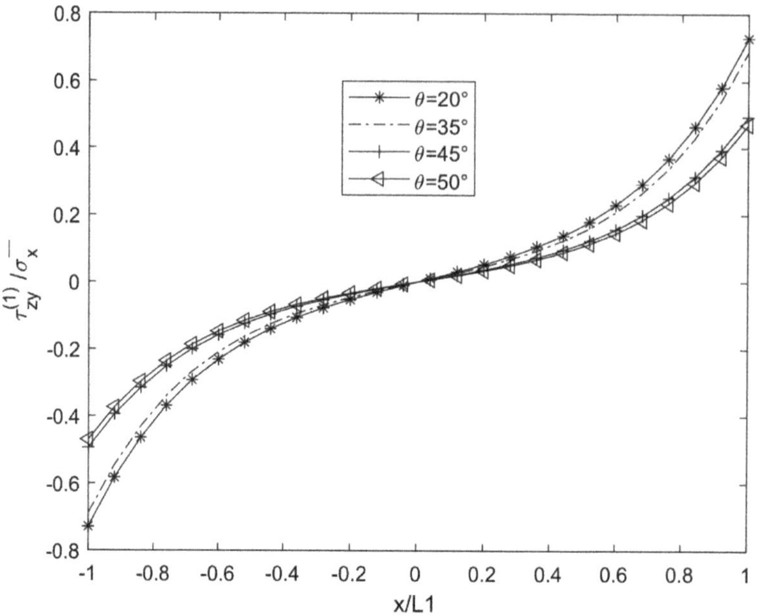

Figure 11. Graph illustrating τ_{zy} in 90° ply for $[\theta_1/90_2]_s$ laminate.

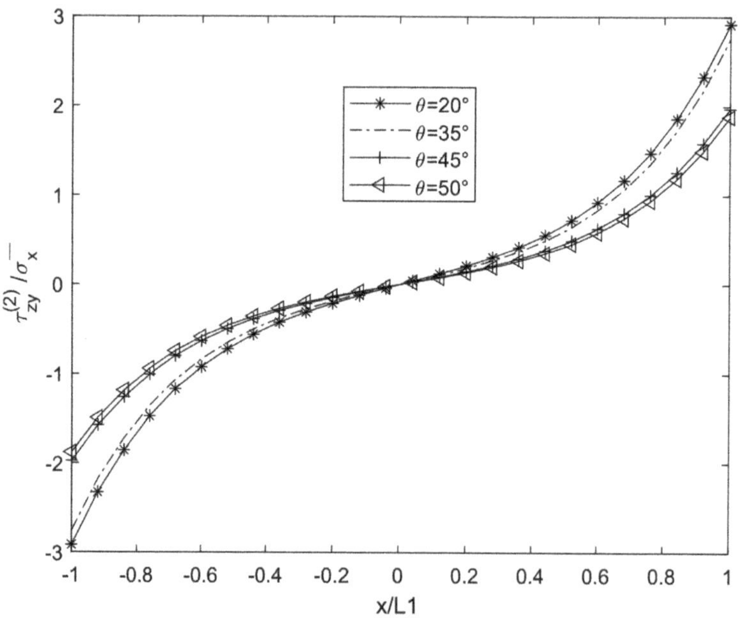

Figure 12. Graph illustrating τ_{zy} in θ ply for $[\theta_1/90_2]_s$ laminate.

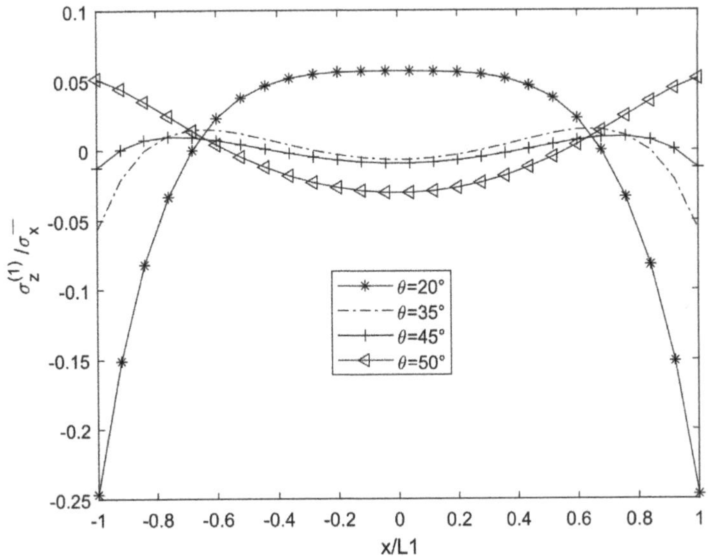

Figure 13. Graph illustrating σ_z in 90° ply for $[\theta_1/90_2]_s$ laminate.

In Figures 3–6, compared with the stress attenuation model of Hashin [28], the stress concentration effect at the crack center of the two models is significant, but the influence of the laying angle on local stress is further quantified. For example, Hashin [28] observed a similar normal stress distribution in orthogonal laminates, but did not consider the quantitative analysis of multi-angle laying. However, the experimental data of Huang [29] show that when the paving angle exceeds 45°, the increase in interlaminate normal stress can reach 15~20%, which is consistent with the stress jump phenomenon when $\theta = 60°$ in Figure 3 of this paper, verifying the sensitivity of the angle to stress concentration.

In Figures 7 and 8, compared with the shear lag theory of Kashtalyan and Soutis [17], the predicted shear stress fluctuates nonlinearly when $\theta > 50°$, which is consistent with the fluctuation amplitude of shear stress when $\theta = 50°$ in Figure 7 of this paper. However, this paper further reveals the modulation effect of interlayer constraint on the fluctuation amplitude using a variational method. However, according to the quadratic displacement hypothesis of Berthelot [16], the shear stress variation trend of the theoretical curve at a low θ value is consistent with that of the results in this paper ($\theta = 30°$), and the difference appears in the high-θ-value region, which may be due to the finer interlayer coordination conditions of this model.

In Figures 9–14, compared with the multi-layer crack model of Carraro [31], this research shows that when $L_1/t_1 < 5$, the stiffness degradation rate is significantly increased, which is highly consistent with $E_x/E_{x0} = 0.754$ (predicted as 0.748 in the literature [31]) when $L_1/t_1 = 3$ in Figure 10, and the error is only 0.8%. Combined with the variational energy method of Vinogradov and Hashin [23], the lower limit theory of stiffness (error < 2%) and the results of this paper (Table 2, error < 1%) jointly verify the reliability of the minimum complementary energy principle in the analysis of cracked laminated plates.

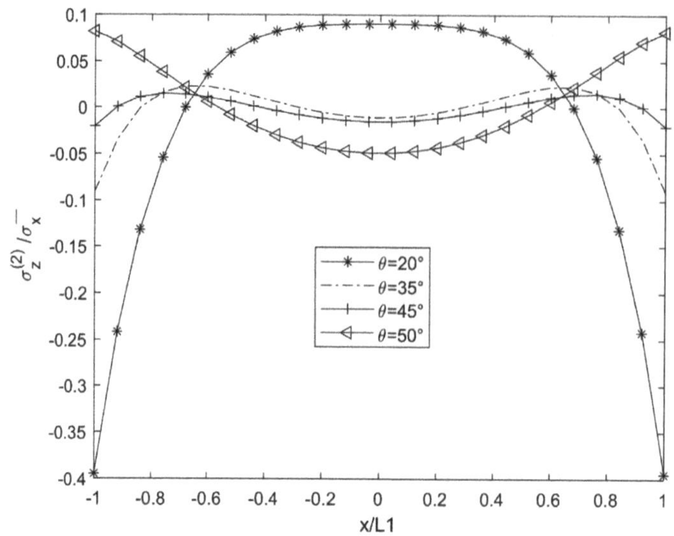

Figure 14. Graph illustrating σ_z in θ ply for $[\theta_1/90_2]_s$ laminate.

4. Conclusions

In this paper, a series of key results are obtained in the field of composite laminates using a variational analysis method. Firstly, in the aspect of predicting stiffness reduction, variational analysis successfully realizes the effective prediction of stiffness reduction in $[0_m/90_n]_s$ laminates with different degrees, providing a reliable means for the prediction of material properties. Secondly, the stress distribution of laminates caused by matrix cracks in 90° laminates is discussed in detail and in depth. The interlaminate stress and in-plane stress distribution of each ply are accurately calculated, which is in good agreement with the existing experimental data and verifies the accuracy and reliability of the analysis method. It is worth emphasizing that the analysis method used in this study has a wide range of applicability, not only to specific ply structures, but also to multi-angle-ply laminates, which greatly expands the application scope of this method in composite research. Through the in-depth analysis of interlaminate stress and in-plane stress distribution, the formation of matrix cracks and local interlaminate delamination can be predicted in advance and effectively, which is of great significance for improving the structural integrity and safety assessment of composite laminates.

Author Contributions: Methodology, X.W.; Writing—original draft, B.X.; Writing—review & editing, Z.H.; Supervision, F.C. All authors have read and agreed to the published version of the manuscript.

Funding: This research was funded by Natural Science Foundation of China, grant number 11862007.

Data Availability Statement: The original contributions presented in this study are included in the article. Further inquiries can be directed to the corresponding author.

Conflicts of Interest: The authors declare no conflict of interest.

References

1. Shen, G.; Hu, G. *Mechanics of Composite Materials*; Tsinghua University Press: Beijing, China, 2013; Volume 9, pp. 30–44.
2. De Luca, A.; Greco, A.; Rezazadeh, N.; Perfetto, D.; Aversano, A. Guided Waves Propagation in Additively Manufactured GF30-PA6 Panel. *Macromol. Symp.* **2024**, *413*, 2400042.
3. Fikry, M.J.M.; Ogihara, S.; Vinogradov, V. The effect of matrix cracking on mechanical properties in FRP laminates. *Mech. Adv. Mater. Mod. Process.* **2018**, *4*, 3.

4. Ramezani, H.; Kazemirad, S.; Shokrieh, M.; Mardanshahi, A. Effects of adding carbon nanofibers on the reduction of matrix cracking in laminated composites: Experimental and analytical approaches. *Polym. Test.* **2021**, *94*, 106988.
5. Ghadami, M. An anisotropic nonlinear multiscale damage model for mechanical behavior of a composite laminate with matrix cracks. *Meccanica* **2023**, *58*, 1809–1822.
6. Chaupal, P.; Rohit, S.; Rajendran, P. Matrix cracking and delamination detection in GFRP laminates using pre-trained CNN models. *J. Braz. Soc. Mech. Sci. Eng.* **2023**, *45*, 136.
7. Fakoor, M.; Rafiee, R.; Sheikhansari, M. The influence of fiber-crack angle on the crack tip parameters in orthotropic materials. *Proc. Inst. Mech. Eng. Part C J. Mech. Eng. Sci.* **2017**, *231*, 418–431.
8. Farrokhabadi, A.; Bahrami, M.; Babaei, R. Predicting the matrix cracking formation in symmetric composite laminates subjected to bending loads. *Compos. Struct.* **2019**, *223*, 110945.
9. Mohammadi, B.; Pakdel, H. Fatigue driven matrix crack propagation in laminated composites. *Mater. Des.* **2018**, *146*, 108–115.
10. Pakdel, H.; Mohammadi, B. Experimental observation and energy based analytical investigation of matrix cracking distribution pattern in angle-ply laminates. *Theor. Appl. Fract. Mech.* **2017**, *92*, 146–154.
11. Qi, W.; Huang, J.; Yao, W.; Shen, H. A damage model for matrix cracks in composite laminates under fatigue loading based on continuum damage mechanics. *Int. J. Fatigue* **2024**, *184*, 108315.
12. Rezaei Jafari, S.; Farrokhabadi, A.; Montesano, J. The effect of staggered matrix crack induced delamination growth on the mechanical properties of cross-ply laminates. *Compos. Struct.* **2021**, *272*, 114196.
13. Socci, C.A.; Kassapoglou, C. Prediction of matrix crack initiation and evolution and their effect on the stiffness of laminates with off-axis plies under in-plane loading. *Compos. Sci. Technol.* **2020**, *200*, 108427.
14. Yuan, M.; Zhao, H.; Liu, S.; Ren, H.; Zhang, B.; Chen, J. Prediction of matrix-cracking-induced stiffness degradation of cross-ply laminates based on data-driven method. *Compos. Sci. Technol.* **2022**, *230*, 109716.
15. Zhang, J.; Fan, J.; Soutis, C. Analysis of multiple matrix cracking in composite laminates. Part 1, In-plane stiffness properties. *Composites* **1992**, *23*, 291–298.
16. Berthelot, J.M. Analysis of the transverse cracking of cross-ply laminates: A generalized approach. *J. Compos. Mater.* **1997**, *31*, 1780–1805.
17. Kashtalyan, M.; Soutis, C. Analysis of composite laminates with intra-and interlaminate damage. *Prog. Aerosp. Sci.* **2005**, *41*, 152–173.
18. Hajikazemi, M.; Sadr, M.; Hosseini-Toudeshky, H.; Mohammadi, B. Thermo-elastic constants of cracked symmetric laminates: A refined variational approach. *Int. J. Mech. Sci.* **2014**, *89*, 47–57.
19. Carraro, P.A.; Maragoni, L.; Quaresimin, M. Stiffness degradation of symmetric laminates with off-axis cracks and delamination: An analytical model. *Int. J. Solids Struct.* **2021**, *213*, 50–62.
20. Wang, F.; Wang, B.; Kong, F.; Ouyang, J.; Ma, T.; Chen, Y. Assessment of degraded stiffness matrices for composite laminates under low-velocity impact based on modified characteristic length model. *Compos. Struct.* **2021**, *272*, 114145. [CrossRef]
21. Herrmann, L.; Mikkelsen, L.P.; Legarth, B.N.; Duddeck, F.; Niordson, C.F. An efficient stiffness degradation model for layered composites with arbitrarily oriented tunneling and delamination cracks. *Compos. Sci. Technol.* **2022**, *230*, 109729.
22. Wu, T.; Wang, G.; Wang, J.; Wang, Y.; Xu, R. Variational principles of laminated plates considering interlayer slip. *Compos. Struct.* **2022**, *281*, 114937. [CrossRef]
23. Vinogradov, V.; Hashin, Z. Variational analysis of cracked angle-ply laminates. *Compos. Sci. Technol.* **2010**, *70*, 638–646.
24. Behera, A.; Thawre, M.M.; Ballal, A. Effect of matrix crack generation on Fatigue life of CFRP multidirectional laminates. *Mater. Today Proc.* **2018**, *9 Pt 3*, 20078–20084. [CrossRef]
25. Li, S.; Hafeez, F. Variation-based cracked laminate analysis revisited and fundamentally extended. *Int. J. Solids Struct.* **2009**, *46*, 3505–3515. [CrossRef]
26. Katerelos, D.G.; Mccartney, L.N.; Galiotis, C. Local strain re-distribution and stiffness degradation in cross-ply polymer composites under tension. *Acta Mater.* **2005**, *53*, 3335–3343. [CrossRef]
27. Mohammad-Zaheri, F.; Mohammadi, B.; Taheri-Behrooz, F. Prediction of stress distribution and stiffness degradation in fiber metal laminates containing matrix cracking. *Compos. Struct.* **2023**, *311*, 116820.
28. Hashin, Z. Analysis of orthogonally cracked laminates under tension. *J. Appl. Mech.* **1987**, *54*, 872–879.
29. Huang, Z.; Yi, S.; Chen, H.; He, X. Parameter analysis of damaged region for laminates with matrix defects. *J. Sandw. Struct. Mater.* **2021**, *23*, 580–620.
30. Huang, Z.Q.; Nie, G.H.; Chan, C.K. An exact solution for stresses in cracked composite laminates and evaluation of the characteristic damage state. *Compos. Part B Eng.* **2011**, *42*, 1008–1014.
31. Carraro, P.A.; Quaresimin, M. A stiffness degradation model for cracked multidirectional laminates with cracks in multiple layers. *Int. J. Solids Struct.* **2015**, *58*, 34–51.

symmetry

MDPI

Article

Bound Optimization by Quadratic Approximation for Heat-Dissipation-Oriented Design of an Air-Cooled Lithium Battery Energy Storage Cabinet

Liqun Wang [1,2,3,4], Yunqing Tang [1,2,3,4,5,*], Jianbin Yu [6], Wei Qin [6], Yangyang Zhang [6], Guoyan Wang [6], Dongjing Liu [5], Yukui Cai [1,2,3,4] and Zhanqiang Liu [1,3,7]

1 School of Mechanical Engineering, Shandong University, Jinan 250061, China
2 State Key Laboratory of Advanced Equipment and Technology for Metal Forming, Shandong University, Jinan 250061, China
3 Key Laboratory of High Efficiency and Clean Mechanical Manufacture of Ministry of Education, Jinan 250061, China
4 Key National Demonstration Center for Experimental Mechanical Engineering Education, Jinan 250061, China
5 Guangxi Key Laboratory of Manufacturing System & Advanced Manufacturing Technology, Guilin University of Electronic Technology, Guilin 541004, China
6 SDEE Contemporary Energy Technology Co., Ltd., Jinan 250061, China
7 School of Airspace Science and Engineering, Shandong University, Weihai 264209, China
* Correspondence: tangyunqing@sdu.edu.cn

Abstract

With the increasing energy density of lithium-ion batteries, the heat dissipation performance of air-cooled battery energy storage cabinets has become a critical determinant of both system performance and service life. This performance depends strongly on the geometry of the airflow channels and their influence on the internal flow distribution. In this study, the internal flow field of a battery energy storage cabinet was analyzed, and the airflow-channel geometry was optimized using the BOBYQA algorithm. The results indicate that the risk of thermal runaway is largely associated with inadequate airflow design, which leads to localized heat accumulation. Geometric optimization of the airflow channels reduced the maximum hotspot temperature from 72.9 °C to 57.6 °C. The hotspots were concentrated at the tops of the battery modules. Modifications to the channel geometry increased the airflow velocity and improved its directionality in these regions, thereby reducing both the hotspot temperature and the extent of the affected area. Moreover, slightly increasing the inlet pressure while reducing the outlet pressure produced a more uniform temperature distribution across the tops of the battery modules.

Keywords: energy storage cabinet; heat dissipation; airflow channel; BOBYQA algorithm; hotspot temperature; geometry optimization

1. Introduction

Lithium batteries have become the predominant choice for energy storage systems due to their exceptional characteristics, including high energy density, long cycle life, low self-discharge rate, and fast charging capabilities [1]. However, they are highly sensitive to ambient temperature fluctuations during charging and discharging cycles, and generate heat as part of these processes. Particularly during high-power charging and discharging, significant amounts of heat are produced within the energy storage system. If this heat is not effectively managed, the battery temperature can rise rapidly, potentially leading to thermal runaway or even fire, posing substantial safety risks to the system [2]. Additionally, uneven temperature distribution within the battery pack can cause variations in electrochemical behavior and electrical imbalances, ultimately affecting the battery's performance. Considerable efforts have been made, both in academia and industry, to understand the impact of temperature on battery performance and lifespan [3–6]. To optimize battery efficiency and extend its service life, Na et al. [7] concluded that the temperature of all batteries should be maintained within the range of 20 °C to 45 °C, with the maximum temperature difference between batteries not exceeding 5 °C. Motloch et al. [8] found that lithium-ion batteries operating within the temperature range of 30 °C to 40 °C experience a reduction in lifespan of approximately 60 days for every 1 °C increase in temperature. Therefore, thermal management is a critical consideration in the design and operation of energy storage systems. The effectiveness of thermal management is directly linked to the performance, lifespan, and safety of the battery compartment. To address this issue, a cooling system is used to transfer heat from the battery compartment to the external environment. Four common thermal management methods for lithium-ion batteries include air-cooled dissipation, liquid-cooled dissipation, heat pipe heat transfer, and phase change material heat transfer [9,10]. Among these, air-cooled heat dissipation has emerged as the most widely adopted thermal management solution for energy storage systems due to its simplicity, ease of maintenance, and cost-effectiveness.

A significant number of studies have been conducted on forced air cooling for battery packs [11–14]. Wang et al. [12] investigated a forced air-cooling strategy for battery thermal management, emphasizing the inconsistency between battery cells and highlighting the importance of optimizing cooling efficiency to address temperature discrepancies across cells. Zhang et al. [13] proposed an optimization method for forced air-cooled lithium-ion battery modules, utilizing multi-vent designs to enhance thermal management performance and improve heat dissipation. Xin et al. [14] investigated a combined air and liquid cooling thermal management scheme, optimizing it to enhance thermal control and cooling efficiency in cylindrical lithium-ion battery packs. Xin et al. [15] studied the influence of three factors, air inlet angle, air outlet angle, and air flow channel width between cells, on the heat dissipation of lithium-ion battery packs, and they found that the layout of the air flow channel had a great impact on the maximum temperature and temperature difference. Xu et al. [16] found that changing the longitudinal battery pack to the horizontal battery pack could shorten the airflow path and improve the heat dissipation performance. For the battery pack with air ducts arranged at the bottom, the heat dissipation performance can be improved by using double "U" type air ducts instead of double "1" type air ducts. Du et al. [17] studied the influence of four kinds of battery arrangements, namely square arrangement, staggered arrangement, and two-trapeze arrangement, on the heat dissipation performance of the battery pack, and they found that the square arrangement was the structure with the best air cooling effect. When the cold air inlet was located at the top of the battery pack, the cooling effect was the best. Sahin [18] studied the effects of various baffles on the cooling performance and pressure drop of air-cooled battery modules and found that Delta winglets provided the best solution, reducing the maximum temperature and temperature difference by 2% (0.6 °C) and 15% (0.7 °C), respectively.

The thermal performance of air-cooled battery thermal management systems (BTMS) is strongly governed by airflow-channel geometry and flow-field organization. Li et al. [19] demonstrated that introducing vortex-generating structures such as Delta winglets can enhance convective heat transfer and improve temperature uniformity, although at the cost of increased pressure drop. Lin et al. [20] showed that combining multiple secondary outlets

https://doi.org/10.3390/sym18010107

with baffles can effectively suppress peak temperatures and reduce temperature differences within battery modules, but the strategy remains highly geometry-dependent and lacks general applicability. To address the limitations of single-parameter adjustments, Feng et al. [21] proposed a unified optimization framework capable of handling both continuous and discrete structural variables, providing a more systematic approach to airflow-channel design. Meanwhile, Argade and De [22] and Dai and Wang [23] highlighted the significant influence of inlet height, angle, and manifold configuration on flow distribution and thermal uniformity, emphasizing the need for holistic rather than localized design strategies. Further enhancement efforts, such as incorporating vortex generators or guide vanes [24,25], have improved local heat transfer but often introduce additional aerodynamic losses and rely heavily on empirical design choices.

In addition to advancements in module-level cooling, recent research has increasingly focused on the design of battery energy storage cabinets, where effective thermal management is crucial. Dong et al. [26] explored rack-level air-cooling strategies within battery storage cabinets and demonstrated that optimizing airflow channel layouts at the container level can significantly enhance heat dissipation efficiency. Similarly, Z Lv et al. [27] developed a multi-level thermal model for battery energy storage systems, including container-level thermal design, and validated its application in commercial energy storage units. These studies highlight the importance of not only designing individual modules but also considering the entire cabinet or enclosure to ensure comprehensive thermal management. Permana et al. [28] integrated CFD simulations with experimental results to analyze thermal performance within battery energy storage cabinets, focusing on airflow distribution and thermal uniformity. They concluded that optimizing the overall cabinet design could prevent hot spots and extend the longevity of the storage system. Future work should build on these insights to develop a more integrated approach to the design of battery energy storage cabinets, addressing both airflow and thermal management at the enclosure or rack level.

This study investigates a commercial air-cooled lithium battery energy storage cabinet to address key limitations of air-cooling systems, including insufficient heat dissipation, nonuniform temperature distribution, and low airflow efficiency, which have not been resolved in a systematic or generalizable way. By parameterizing the cabinet's main geometric features and analyzing the baseline flow and thermal fields, four objectives are pursued: improving heat-flux uniformity, reducing average temperature, enhancing airflow efficiency, and achieving more uniform regional temperature distribution. Unlike previous work that mainly relies on empirical geometric tuning and case-specific designs, this study formulates these objectives as optimization functions and solves them iteratively using the derivative-free BOBYQA (Bound Optimization BY Quadratic Approximation) algorithm. The novelty lies in applying a fully parameterized, data-driven structural optimization framework to an engineering-scale commercial cabinet, thereby filling a clear gap in the literature on integrated thermal–aerodynamic optimization of air-cooled BTMS for real energy storage applications. The optimized geometry significantly improved internal flow characteristics, reducing the maximum hotspot temperature from 72.9 °C to 57.6 °C and decreasing the high-temperature area (T > 30 °C) on the battery module top from 0.55 m^2 to 0.04 m^2.

2. Models and Methods

2.1. Modeling Details

A commercial air-cooled lithium-ion battery energy storage container manufactured by SDEE Contemporary Energy Technology Co., Ltd, located in Jinan, China. was used

 https://doi.org/10.3390/sym18010107

as the baseline configuration for modeling and optimization. The container measures 12,184 mm × 2432 mm× 2880 mm (length × width × height). Air is supplied through two rectangular sidewall inlets (767 mm × 424.6 mm; one on each sidewall) and discharged through two vertically stacked outlets, as shown in Figure 1a.

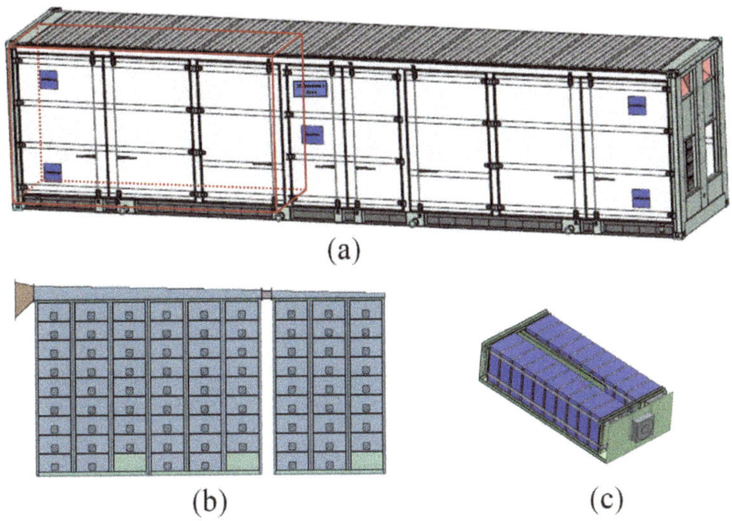

Figure 1. Geometry model of the battery cabinet. (**a**) Battery energy storage cabinet container; (**b**) Single battery energy storage cabinet; (**c**) Battery pack.

The internal layout of the container is shown in Figure 1b, which presents one quarter of the container (one cabinet). The same pack arrangement is mirrored on the opposite side, yielding 78 packs per cabinet and 312 packs in the full container. The inlet plenum is positioned above the battery modules. Cold air enters the duct, flows downward, passes through the battery modules, and exits through the outlets, thereby sustaining forced-air cooling. The structure of each battery pack is shown in Figure 1c. Each pack contains 20 cells, giving a total of 6240 cells in the container. With a heat generation rate of 29 W per cell, the resulting total internal heat load is 180.96 kW. Given the high symmetry of the system, the original geometry can be partitioned into symmetric units based on its inherent symmetry, significantly reducing the computational load. By leveraging this symmetry, both modeling and analysis can be simplified, resulting in more efficient simulations. Firstly, as shown in Figure 1a, the air-cooled battery energy storage container consists of four substructures, each of which exhibits symmetric distribution. Consequently, the airflow and temperature distribution within the ducts are also symmetric. Therefore, it is only necessary to analyze and optimize one of the substructures, with the other substructures adjusted accordingly using symmetry.

Figure 2 summarizes the boundary conditions applied in the finite element simulations. Figure 2a shows the inlet and outlet locations of the battery energy storage cabinet; the airflow ducts are represented by the interior walls of the computational domain. Figure 2b presents the imposed heat-source boundary condition, which accounts for battery heat generation in the thermal solution. The convective heat transfer coefficient between the cold air and the outer walls of the battery modules is set to 30 W/(m^2·K). The simulation is conducted under steady-state heat transfer conditions, focusing on the battery module. To facilitate thermal conduction analysis in the COMSOL 6.3, the battery modules

https://doi.org/10.3390/sym18010107

are simplified into hexahedral features. The average heat generation rate of the battery is 20.993 kW/m^3.

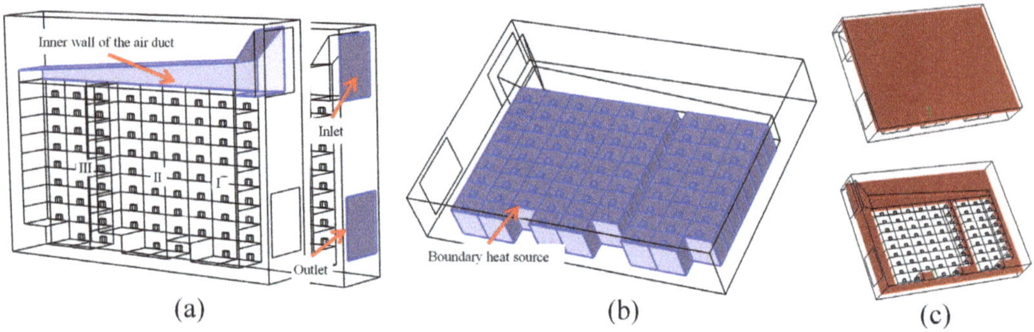

Figure 2. Schematic of model boundary conditions. (**a**) Flow field boundary conditions; (**b**) Boundary heat source. (**c**) Cross-section at the top/middle of the battery pack.

Figure 2 also identifies two characteristic cross-sections for flow-field and temperature-field analysis, as illustrated in Figure 2c. The battery packs are sequentially labeled Battery Packs I, II, and III, beginning from the region near the airflow inlet. Details of the model setup and material parameters are provided in Section S1 of the Supplementary Materials. Table S1 presents the specific parameters of the turbulence equations used. Additionally, to better simulate the impact of temperature on air flow, the dynamic viscosity, specific heat capacity, thermal conductivity, and density of air are defined as functions of temperature, as shown in Figure S1.

2.2. Calculation Methods

When analyzing the flow field characteristics of a battery energy storage cabinet, the Finite Element Method (FEM) is commonly used to solve the governing equations related to fluid dynamics. The primary governing equation involved in flow field analysis is the Navier-Stokes equation, which describes the motion of fluids. The standard form of the Navier-Stokes equation is [29]:

$$\rho(u \cdot \nabla)u = -\nabla p + \mu \nabla^2 u + \rho \tag{1}$$

where ρ is the fluid density, v is the velocity field, p is pressure, μ is dynamic viscosity, and f represents body forces.

The temperature field is governed by the heat conduction equation, which has the following standard form:

$$\rho c_p \frac{DT}{Dt} = k \nabla^2 T + Q \tag{2}$$

where T is the temperature, k is the thermal conductivity, c_p is the specific heat capacity and Q is the heat source term per unit volume.

Based on the operating environment of the battery energy storage cabinet, the objective of the optimization is to solve for the optimal combination of airflow duct geometric parameters that provides the best cooling effect for the battery packs, thereby improving their cycle life. To accurately represent the optimization objectives and prevent scenarios where the highest temperature decreases but the heat accumulation area expands due to unreasonable optimization goals, the following four optimization objectives are set [30]:

(1) Heat Flux Homogenization Objective

https://doi.org/10.3390/sym18010107

$$F_1 = \frac{1}{V}\int_{\Omega}\left(|\nabla T| - |\nabla T_{avg}|\right)^2 dV \tag{3}$$

(2) Average Temperature Minimization Objective

$$F_2 = \frac{1}{V}\int_{\Omega} T dV \tag{4}$$

(3) Airflow Duct Efficiency Optimization Objective

$$F_3 = \int_{\Omega} \Delta p dV + \lambda \int_{\Omega} \epsilon \rho dV \tag{5}$$

(4) Temperature Distribution Uniformity Objective

$$F_4 = \sigma_T = \sqrt{\frac{1}{V}\int_{\Omega}\left(T - \overline{T}\right)^2 dV} \tag{6}$$

Before performing the optimization, the four optimization objectives are discretized as follows:

$$F_1 = \frac{1}{V}\sum_e \left(|\nabla T_e| - |\nabla T_{avg}|\right) V_e \tag{7}$$

$$F_2 = \frac{1}{V}\sum_e T_e V_e \tag{8}$$

$$F_3 = \frac{1}{V}\sum_e (\Delta p_e V_e + \lambda \epsilon_e \rho_e V_e) \tag{9}$$

$$F_4 = \sqrt{\frac{1}{V}\sum_e (T_e - \overline{T})^2 V_e} \tag{10}$$

The BOBYQA optimization algorithm is used, which is an unconstrained optimization method suitable for solving nonlinear problems. During the optimization process, the algorithm gradually searches for the optimal solution by constructing quadratic approximation models. The specific steps are as follows. The BOBYQA optimization algorithm, an unconstrained optimization method suitable for solving nonlinear problems, is employed to search for the optimal solution by gradually constructing quadratic approximation models. In the initialization phase, we assume an initial set of design variables $x_0 = [x_1, x_2, \ldots, x_n]$, typically representing the geometric parameters of the battery energy storage cabinet, flow boundary conditions, and other relevant factors. The optimization objective function and any constraints (if applicable) are defined. At the design point x_0, the values of the four optimization objective functions are calculated, and the gradient for each objective function is then computed to guide the search for the optimal solution.

The core of the BOBYQA algorithm is to construct a quadratic approximation model of the objective function with respect to the design variables. Let the current design point be x_0, where the objective function value $F_0 = F(x_0)$ and its gradient ∇F_0 are computed. The quadratic approximation model is given by:

$$\hat{F}(x) = F_0 + \nabla F_0^T (x - x_0) + \frac{1}{2}(x - x_0)^T H (x - x_0) \tag{11}$$

where F_0 is the value of the objective function at the initial design point x_0. ∇F_0 is the gradient of the objective function at x_0. H is the Hessian matrix, which describes the second

https://doi.org/10.3390/sym18010107

derivatives of the objective function and is used to capture the curvature of the function. The Hessian matrix H can be updated using least squares methods.

The BOBYQA method updates the design variables by minimizing the quadratic approximation model. The goal is to find the point that minimizes the approximation model $F(x)$. Based on the quadratic approximation model of the objective function, we update the design variables as follows:

$$x_1 = x_0 - \alpha H^{-1} \nabla F_0 \tag{12}$$

where H^{-1} is the inverse of the Hessian matrix, α is the step size factor used to adjust the update step and ensure the calue $F_1(x_1)$, the new point is accepted.

At the new design point x_1, the objective function values $F_1(x_1)$, $F_2(x_1)$, $F_3(x_1)$ and $F_4(x_1)$ are calculated. By evaluating $F_1(x_1)$, $F_2(x_1)$, $F_3(x_1)$ and $F_4(x_1)$, the performance of the new design point is assessed. If the objective function values are smaller, the new design point x_1 is accepted, and the current design variables are updated. If the objective function values increase, the new point is rejected, and the step size factor α is reduced or the Hessian matrix is adjusted.

Each time a new design point is accepted, the BOBYQA algorithm updates the quadratic approximation model. The new design point and its corresponding objective function values are used to update the Hessian matrix H, ensuring that the revised quadratic model accurately captures the objective function's behavior near this point. With each iteration, a new evaluation point x_1 is added, and the gradient and objective function values are updated. The Hessian matrix is updated by using the new point and gradient information.

The optimization algorithm has three convergence criteria:

(1) Objective function convergence: If the change in the objective function is smaller than a predefined threshold, the optimization process stops;

(2) Design variable convergence: If the change in the design variables is smaller than a certain threshold, the optimization process stops;

(3) Maximum number of iterations: The optimization process terminates when the maximum number of iterations is reached.

3. Results and Discussion

3.1. Heat Dissipation Performance of the Original Cabinet

Details of the mesh-independence verification are provided in Section S2 of the Supplementary Materials. As shown in Figure S2, to verify that the mesh size used in the model does not affect the results, four different mesh sizes were tested. The flow field results are shown in Figure S3. It can be seen that the airflow characteristics at the outlet of the duct are not fully captured in Figure S2a. The flow field characteristics in Figure S2b–d are generally consistent, but in Figure S2b, the airflow at the top exhibits a distinct sawtooth pattern, indicating that the resolution of the mesh is too low at this size. Figure S2c,b are nearly identical, suggesting that the results converge when the average mesh size is 0.074 m. Figure 3 shows the temperature distribution of the battery energy storage cabinet before optimization. The average temperature in the region is 29.7 °C, and the maximum temperature reaches 72.9 °C. A pronounced hotspot appears at the top of the middle section of Battery Pack 2, indicating that this portion of the pack may be at the highest risk of thermal runaway. In addition, varying degrees of heat accumulation are observed at the tops of Battery Packs 1 and 3, suggesting that the current airflow-duct geometry results in poor heat dissipation in the upper regions of the battery packs. The temperature field is excessively high and non-uniform, highlighting the need for further optimization.

https://doi.org/10.3390/sym18010107

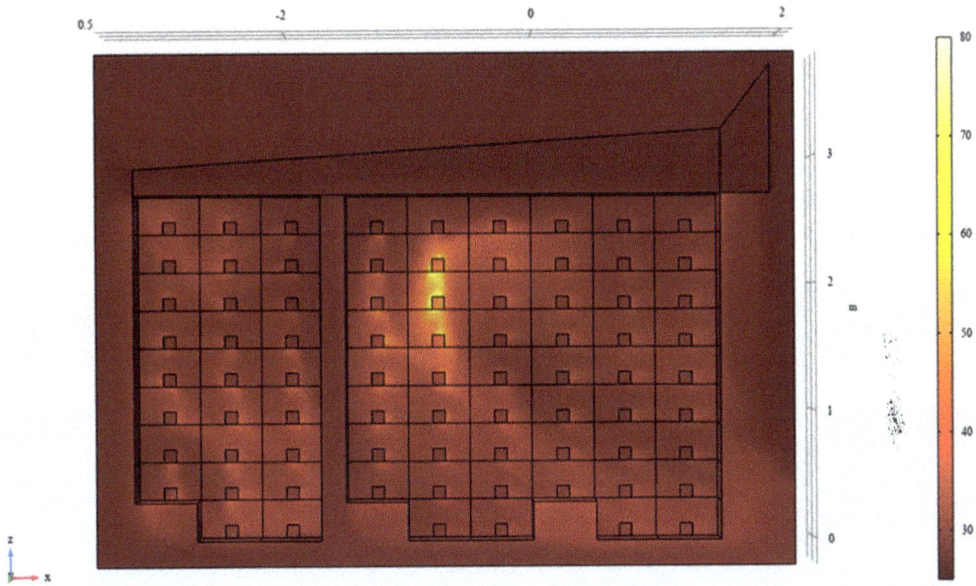

Figure 3. Temperature distribution of the battery cabinet before optimization.

During the forced air cooling process, the heat dissipation effectiveness is directly proportional to the air velocity at the surface. As shown in Figure 4a, the flow velocity contour plot of the structure before optimization reveals that the air velocity is higher at the inlet and outlet of the airflow duct. The highest air velocity, 12.5 m/s, occurs between the intake and the duct entrance. Inside the duct, as the cross-sectional area of the duct gradually decreases, the air velocity increases. The air flows out through the channels between the battery packs and the rear outlet, and under the compression within the duct, the velocity is further increased to 10 m/s. In Figure 4b, it can be observed that the air flowing out of the duct becomes turbulent, forming vortices at the upper part of the battery cabinet.

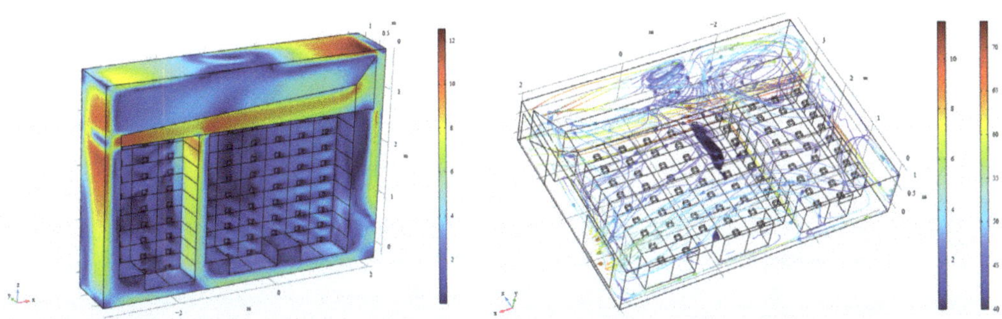

(a) (b)

Figure 4. Flow field simulation results before optimization. (**a**) Flow field contour plot; (**b**) Flow field streamline results.

As shown in Figure 5a, it can be observed that the airflow exiting the duct entrance is significantly slowed down due to the vortex effect in the middle section of the battery cabinet. As a result, the airflow velocity over the top of Battery Pack 2 is less than 2 m/s,

 https://doi.org/10.3390/sym18010107

leading to poor forced convection heat dissipation. By outputting the flow velocity and pressure along the cross-sectional line, it is found that the airflow from the duct outlet and the airflow escaping from the battery cabinet entrance converge, forming a vortex that causes a sudden drop in airflow velocity, with a velocity of only 1.5 m/s. Figure 5b also indicates that the lower passage of the battery cabinet is too long, causing the airflow to form steady laminar flow. This leads to the formation of a heat accumulation area at the very bottom of the contact surface between Battery Pack 1 and Battery Pack 2, where the temperature reaches 42.5 °C.

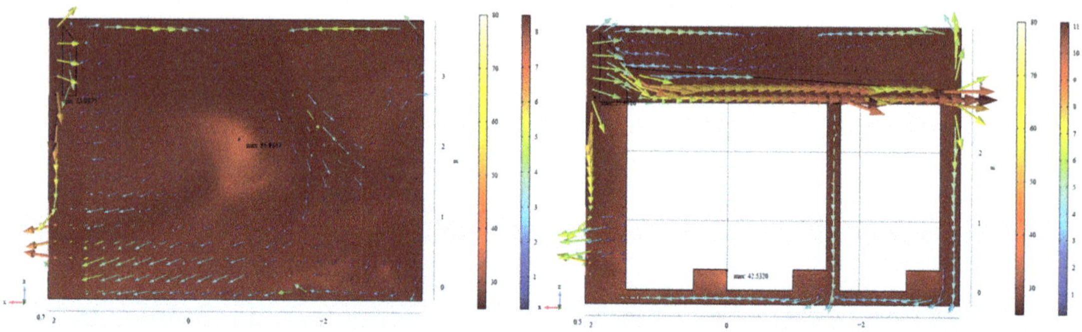

(a) (b)

Figure 5. Temperature contour plots of the battery cabinet cross-section. (**a**) Temperature distribution at the top of the battery pack; (**b**) Temperature distribution at the middle section of the battery pack.

The airflow exiting at the inlet and the airflow exiting at the duct outlet collide and compress at the top of Battery Pack 2. In addition to forming vortices, the two airflows flow toward Battery Packs 1 and 3, resulting in a reduced airflow and lower velocity at the top of Battery Pack 2, thereby leading to poor convective heat transfer and the formation of heat accumulation areas. The temperature distribution contour plots for the top and middle sections of the battery pack along the X-Z plane are shown in Figure 6a,b, with arrows indicating the direction and magnitude of the airflow velocity. In Figure 6a, it can be observed that the air velocity exiting the duct outlet is relatively high, and under the influence of the airflow exiting the duct entrance, it deflects towards the top of Battery Pack 3. Subsequently, under the pressure differential, it flows through the lower region of the battery cabinet.

From Figure 6, it can be seen that the airflow exiting the duct has both wind speed and pressure that are twice those of the airflow escaping from the battery cabinet entrance. The two airflows collide in the middle section, forming a vortex, which significantly reduces the airflow velocity.

Based on the above results, it can be concluded that under the current design, the heat accumulation areas in the battery cabinet are primarily located at the top of Battery Pack 2 and at the bottom contact surface between Battery Packs 1 and 2. Therefore, it is necessary to optimize the geometric design parameters of the airflow duct to improve the airflow velocity and direction in these two areas. Additionally, the airflow exiting the duct forms a strong recirculation vortex under the influence of the battery cabinet's outer shell, which causes a reduction in airflow velocity at the top of the battery packs, resulting in poor heat dissipation and the formation of heat accumulation areas. Therefore, optimizing the geometric parameters of the airflow duct to reduce the airflow volume at the duct outlet and adjust the impact area and strength of the recirculation vortex is one of the effective optimization directions.

https://doi.org/10.3390/sym18010107

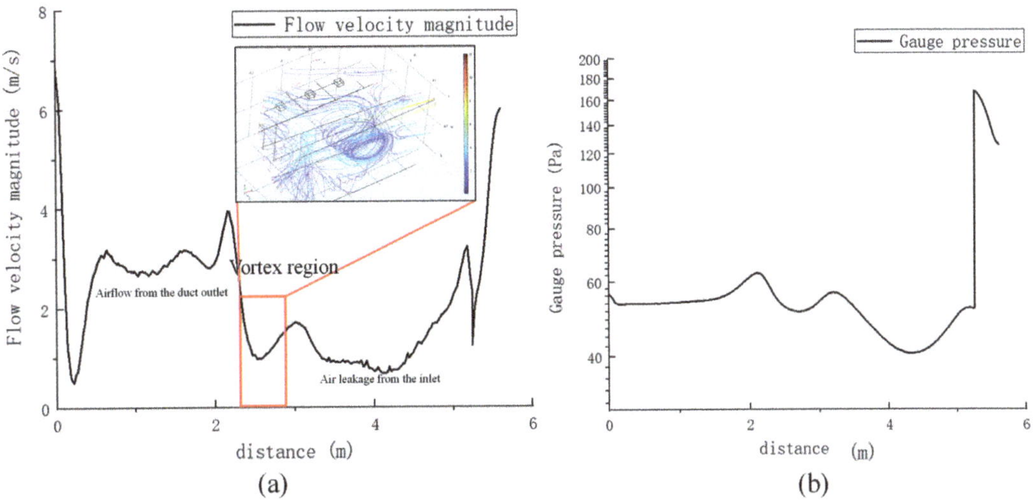

Figure 6. Flow field results along the cross-sectional line. (**a**) Flow velocity; (**b**) Relative pressure.

3.2. Enhanced Heat Dissipation Performance of the Cabinet After Optimization

Through the BOBYQA optimization method, four modifications were made to the geometry of the battery energy storage cabinet: (1) The slope at the airflow duct inlet was increased; (2) The slope of the main duct was increased, and the outlet area was reduced; (3) An additional airflow duct was added between Battery Pack 1 and Battery Pack 2; (4) The outlet height of the battery energy storage cabinet was increased. The detailed comparison between the geometric parameters of the battery energy storage cabinet before and after optimization is shown in Table 1.

Table 1. Comparison of the geometric structure before and after optimization.

Geometric Parameters	Td_1 (m)	Td_2 (m)	Ydh (m)	Jdh (m)	Fdh (m)	Ckh (m)
before optimization	0	0.2	0.2	0.5	1	1
after optimization	0.1	0.2	0.119	0.5	1	1

As shown in Figure 7, the temperature contour plot of the optimized battery energy storage cabinet structure reveals that the maximum temperature in the hotspot area decreases from 72.9 °C to 57.6 °C, and the temperature distribution at the top of the battery packs becomes more uniform, reducing the likelihood of failures caused by high temperature gradients. In addition, the overall average temperature of the battery energy storage cabinet is 29.3 °C. After optimization, the maximum temperature decreases by 20.99%, while the average temperature decreases by only 1.35%. In addition, Figure 7b shows that the middle section of Battery Pack 2 experiences a temperature drop of approximately 40 °C after optimization, indicating a substantial improvement in the hotspot area.

The cycle life model is used to evaluate the lifespan of the battery packs in the battery energy storage cabinet before and after optimization, with the calculation formula given as [31]:

$$N(T) = N_{ref} \cdot e^{\left[\frac{E_a}{R}\left(\frac{1}{T} - \frac{1}{T_{ref}}\right)\right]} \tag{13}$$

where $N(T)$ is the cycle life (number of cycles) at temperature T, N_{ref} is the cycle life at the reference temperature T_{ref}, E_a is the activation energy for cycle aging, and R is the

 https://doi.org/10.3390/sym18010107

gas constant with a value of 8.314 J/(mol·K). Before optimization, the maximum surface temperature of the battery pack was 72.9 °C, and the average surface temperature was 29.644 °C; after optimization, the maximum surface temperature decreased to 57.6 °C, and the average surface temperature was 29.278 °C. Based on the above formula, the cycle life of the battery pack in the hotspot area increased by approximately 106.17%, while the average cycle life of the battery energy storage cabinet improved by 2.19%. A comparison of the battery life before and after optimization is shown in the Figure 8.

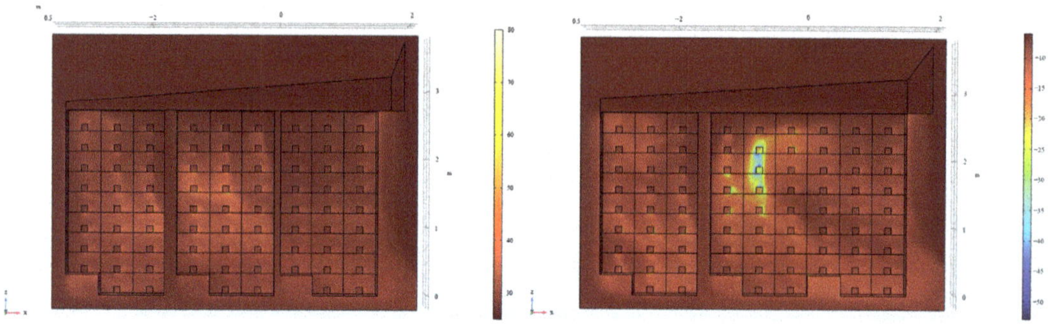

(a) (b)

Figure 7. Temperature distribution of the battery cabinet after optimization. (**a**) Temperature contour plot; (**b**) Distribution of the decrease in temperature after optimization.

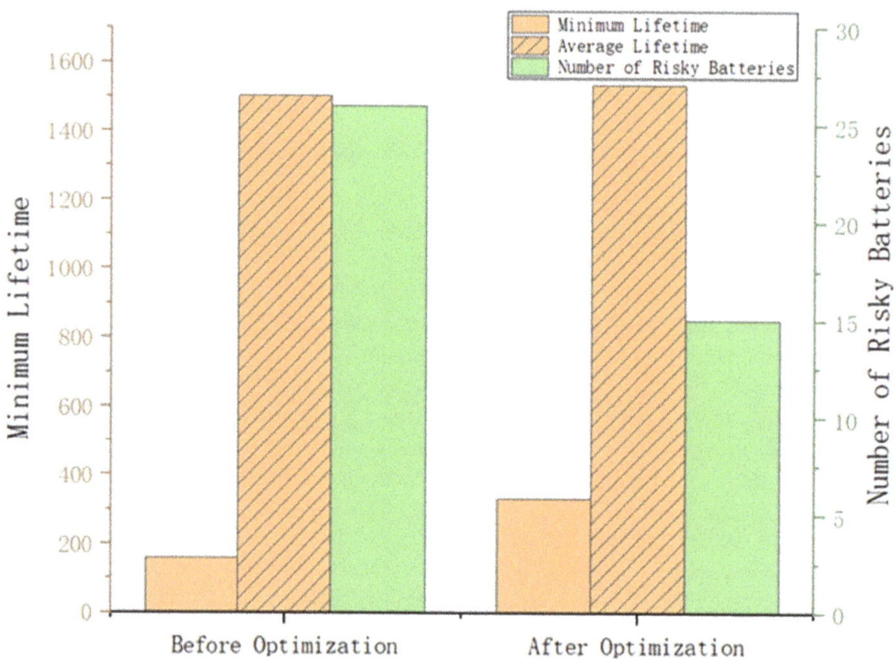

Figure 8. Minimum cycle life, average cycle life, and number of high-risk battery packs before and after optimization.

https://doi.org/10.3390/sym18010107

From Figure 8, it can be seen that after optimization, the average cycle life of the battery energy storage cabinet changes only slightly, while the minimum cycle life and the number of high-risk battery packs improve significantly.

Comparing the velocity contour plot of the battery energy storage cabinet structure before optimization, in Figure 9a, it can be seen that the geometric changes at the duct inlet lead to an increased airflow escaping from the battery cabinet entrance, with the airflow velocity reaching up to 14 m/s. Due to the addition of an airflow duct between Battery Pack 1 and Battery Pack 2, the velocity at the duct outlet decreases. Furthermore, in Figure 9b, it can be observed that the increased airflow escaping from the battery cabinet entrance weakens the airflow exiting the duct outlet. As a result, the airflow escaping from the battery cabinet entrance becomes dominant at the gathering point. Due to the collision and compression of the two opposing airflows, the airflow velocity drops to 5.4 m/s and flows towards the tops of Battery Packs 2 and 3.

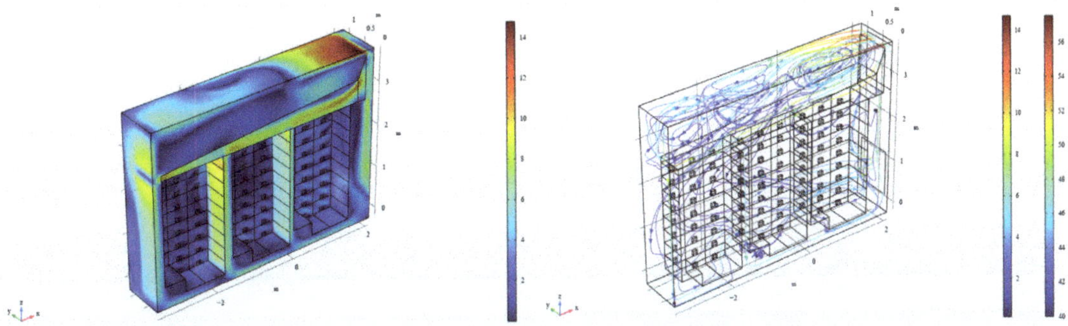

(a) (b)

Figure 9. Flow field simulation results after optimization. (**a**) Flow field contour plot; (**b**) Flow field streamline results.

Unlike the airflow before optimization, the vortex at the top of the battery packs in the optimized battery energy storage cabinet has changed from a strong single vortex into two weaker vortices, with the width of the vortices increasing. By combining Figures 9a and 10a, it can be observed that after geometric optimization, the intensity of the airflow from the duct outlet has significantly decreased. Additionally, part of the airflow forms a vortex rotating around the y-axis, while the remaining airflow compresses the y-directional width of the original vortex and converges with the airflow escaping from the battery storage cabinet entrance, flowing towards the top of the battery packs. Figure 10b shows the contour surface distribution of relative pressure in the region, where it can be seen that the newly formed vortices supplement the original vortices, increasing their flow velocity.

The temperature distribution and flow direction within the section of the battery energy storage cabinet are shown in Figure 11. Compared to the geometric structure before optimization (Figure 5), the optimized geometric structure exhibits a more uniform temperature distribution. Under the influence of the vortex at the top of Battery Pack 2, the airflow direction is deflected, flowing over the top of Battery Pack 2 and eventually towards Battery Pack 3. Finally, constrained by the geometry of the battery cabinet, the airflow flows through the lower sections of Battery Packs 3, 2, and 1, towards the cabinet outlet.

 https://doi.org/10.3390/sym18010107

After geometric optimization, the main cooling airflow now covers the primary areas of all the battery packs, with the airflow direction roughly forming a "U" shape, avoiding the velocity dissipation caused by abrupt directional changes.

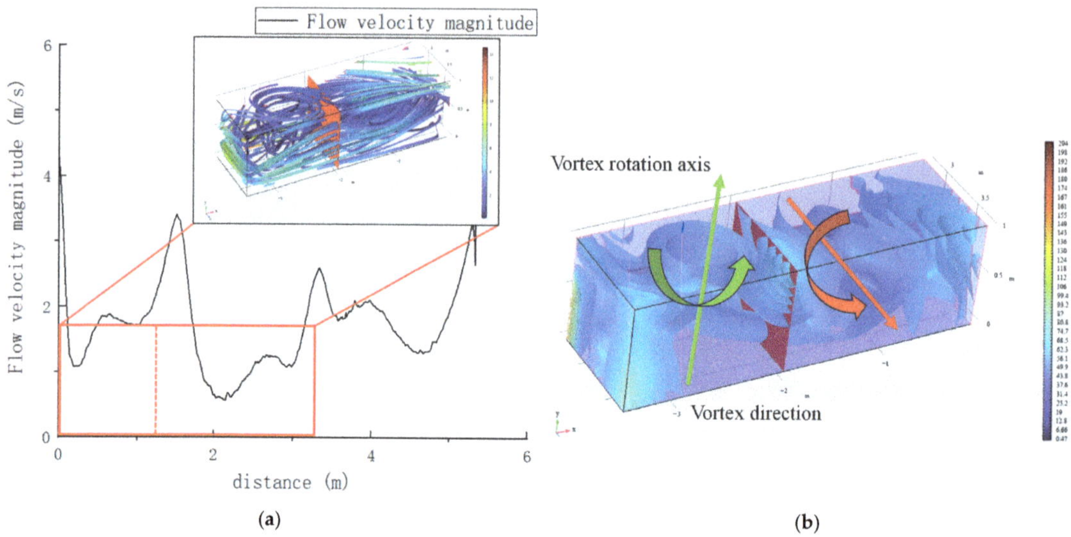

Figure 10. Flow field results along the cross-sectional line. (**a**) Flow velocity; (**b**) Contour surface of relative pressure.

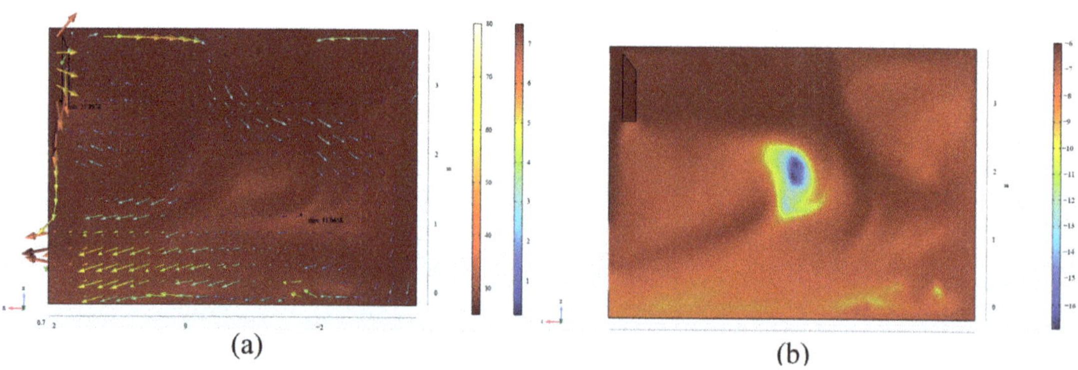

Figure 11. Temperature distribution at the top section of the battery pack. (**a**) Temperature contour plot; (**b**) Distribution of the decrease in temperature after optimization.

To analyze the differences in the cooling airflow formed within the plane before and after optimization, the gas streamlines within the plane were extracted, and the velocity magnitude was represented using color values along with the deformation in the *y*-axis direction. Additionally, the heat flux magnitude was plotted within the plane. By comparing Figure 12a with Figure 12c, it can be observed that the change in the geometric structure led to a modification in the airflow characteristics within the plane.

Before optimization, the two opposing airflows from the duct inlet and duct outlet collided at the top of Battery Pack 2. The airflow from the duct inlet significantly decreased in velocity, and the direction of the airflow from the duct outlet changed, flowing towards Battery Pack 3. This resulted in poor heat dissipation in the main section of Battery Pack 2,

 https://doi.org/10.3390/sym18010107

forming a heat accumulation area (Figure 12b). Furthermore, as the airflow passed near the battery cabinet inlet and outlet, the airflow velocity increased when passing over Battery Pack 1. As a result, the average temperature of Battery Pack 1 was significantly lower than that of the other battery packs. In the heat flux contour plot, distinct isolated high-heat spots (T > 40 °C) are observed, with arrows showing disordered directions and large differences in length, indicating that the heat dissipation path is locally "blocked" and the recirculation zone becomes a potential hotspot for thermal runaway risks.

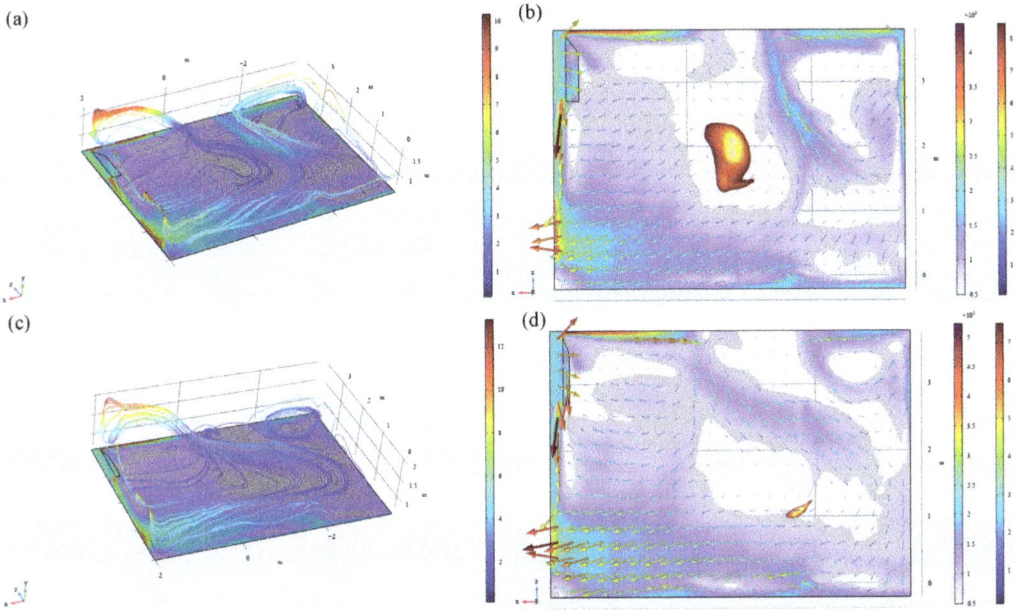

Figure 12. Temperature contour plots of the top section of the battery pack before and after optimization. (**a**) Flow field streamlines before optimization; (**b**) Heat flux contour plot before optimization (arrows represent flow direction and magnitude); (**c**) Flow field streamlines after optimization; (**d**) Heat flux contour plot after optimization. Arrows represent flow direction and magnitude.

After geometric optimization, the airflow volume from the duct inlet increased, and the airflow volume from the duct outlet decreased. This resulted in the inlet airflow maintaining continuous wall-attached laminar flow, and the original recirculation vortex was significantly weakened. The bottom supplemental airflow, guided by the flow channel, was lifted and merged with the main flow, forming a stable cooling channel that spans across the left and right sides. Figure 12d shows that both the area and peak of the high heat flux region have decreased, with the hot spots sinking and dissipating. The direction of the vector arrows is more consistent, and the lengths are more balanced, indicating that the heat at the top is quickly carried away by the main airflow.

Based on the simulation results, it can be observed that heat accumulation areas in the battery energy storage cabinet predominantly appear in the top regions of the battery packs. Therefore, the flow field results of this cross-section were extracted and the flow field characteristics were analyzed, as shown in Figure 13. The airflow within the section is classified into five distinct flow categories (Airflow I, Airflow II, Airflow III, Airflow IV, and Airflow V). Airflow I represents the airflow escaping from the battery storage cabinet entrance, Airflow II represents the airflow from the duct outlet of the battery storage cabinet,

 https://doi.org/10.3390/sym18010107

Airflow III is the combined airflow formed by the convergence of Airflow I and Airflow II, Airflow IV is a portion of the airflow escaping from the battery storage cabinet entrance, and Airflow V is the airflow formed by the pressure differential at the outlet of the cabinet and the geometric effects of the cabinet.

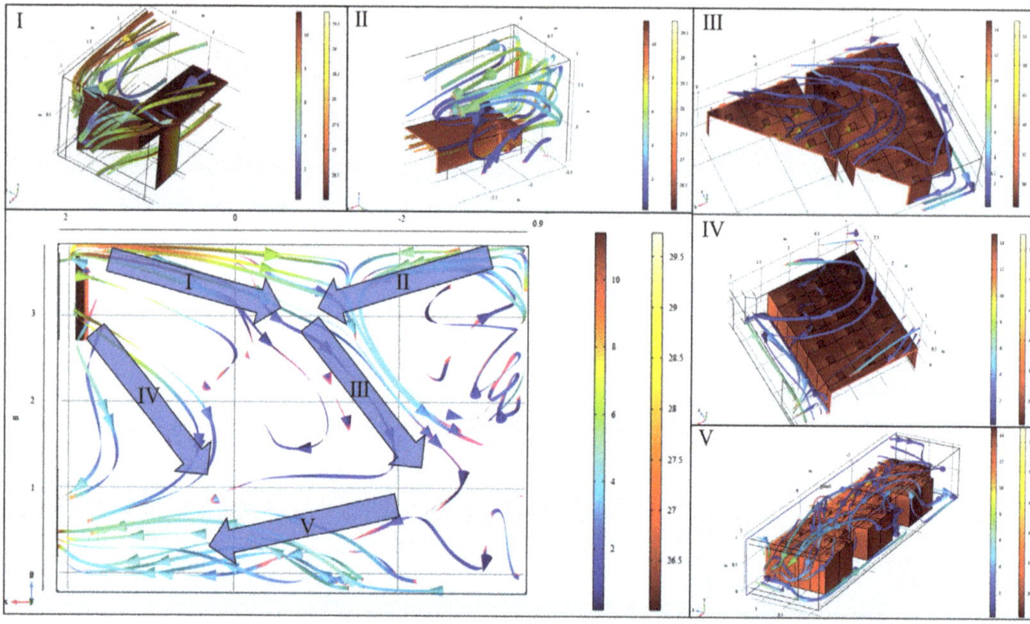

Figure 13. Flow field direction characteristics analysis at the top of the battery pack. (I) the airflow escaping from the battery storage cabinet entrance; (II) the airflow from the duct outlet of the battery storage cabinet; (III) the combined airflow formed by the convergence of Airflow I and Airflow II; (IV) a portion of the airflow escaping from the battery storage cabinet entrance; (V) the airflow formed by the pressure differential at the outlet of the cabinet and the geometric effects of the cabinet.

3.3. Evaluation of Geometric Effects on Cooling Performance

It can be observed that Airflow I exhibits the highest velocity near the entrance of the battery energy storage cabinet and the duct inlet because of flow contraction; however, its volume flow rate is relatively small, and the velocity decays rapidly downstream. The airflow entering the duct is subsequently diverted by the channels between the battery packs and exits through the duct outlet, forming Airflow II. In general, the velocity of Airflow II is lower than that of Airflow I. Nevertheless, if the volume flow rate of Airflow II becomes too large, it impinges on Airflow I, generating a recirculation vortex and causing substantial dissipation of kinetic energy, which is detrimental to heat removal. Therefore, enlarging the duct entrance to increase the volume flow rate of Airflow I and widening the channels between the battery packs to reduce the velocity of Airflow II can enhance the heat dissipation performance of the battery energy storage cabinet. The formation of Airflow IV and Airflow V is primarily driven by the pressure difference at the cabinet outlet, and geometric modifications of the duct generally have no significant influence on their direction or velocity.

According to the above analyses, the thermal performance of this class of battery energy-storage cabinets is governed primarily by the strength of the skimming flow over

 https://doi.org/10.3390/sym18010107

the top of the battery pack. We therefore propose the following quadratic regression to rapidly evaluate design quality using a few key geometric descriptors:

$$\Phi_{top} = c_0 + c_1\sigma_{in} + c_2\sigma_m - d_1\sigma_{in}^2 - d_2\sigma_m^2 - e\chi_r^2 \quad (14)$$

Here Φ_{top} denotes the skimming-flow intensity at the cell tops. The terms are interpreted as follows: c_0—baseline term that aggregates geometry–independent contributions such as total inlet flow rate, cabinet pressure drop, and local wall roughness. It sets the reference skimming level "near zero variations" of the design variables; $c_1\sigma_{in}$—linear inlet-slope term. A larger (but not excessive) inlet slope σ_{in} improves guidance, concentrates the momentum of Airflow I, and increases the tangential velocity along the top surface; $-d_1\sigma_{in}^2$—quadratic penalty for inlet over-slope. An overly steep inlet promotes separation and recirculation; the quadratic term attenuates the net gain; $c_2\sigma_m$—linear main-duct slope term. A slight upward/diffuser-type inclination of the main duct helps deflect the core stream toward the top region; $-d_2\sigma_m^2$—quadratic penalty for excessive main-duct slope, which likewise triggers separation and is suppressed by the second-order term; $-e\chi_r^2$—maldistribution penalty. As the channel maldistribution parameter χ_r deviates further from balance, the confluence of Airflows I and II becomes less uniform, promoting mutual impingement and recirculation cores that weaken the top skimming flow.

To enhance numerical stability, all predictors except the intercept are column-standardized; ordinary least squares (OLS) is then fitted in the standardized space with covariance estimated from the residuals. The regression coefficients and their uncertainties are rigorously back-transformed to the physical space, yielding (c_0, c_1, c_2, d_1, d_2, e) together with standard errors and 95% confidence intervals that can be used directly in the governing formula.

By extracting eighty sets of design-parameter combinations and the corresponding top skim-flow measurements of the battery pack from the iterations of the BOBYQA optimization, we fitted the parameters of the above equation as listed in Table 2.

Table 2. Values of design parameters.

Coefficient	c_0	c_1	c_2	d_1	d_2	e
Value	−0.2558	1.0136	−6.8843	0.1225	95.3327	0.4981

The fitted model provides moderate-to-good explanatory power for the observations ($R^2 \approx 0.66$, $RMSE \approx 0.28$, $MAE \approx 0.22$).

The fitted scatter plot and the residual distribution of the model are shown in Figure 14. In the scatter plot, the horizontal axis denotes the observed values and the vertical axis denotes the predicted values, with the red dashed line representing the ideal case of $y = \hat{y}$. The data points are generally concentrated around the ideal line, indicating that the model captures the overall trend of the data; however, greater dispersion is observed at higher observed values, suggesting the presence of heteroscedasticity. The performance metrics report an R^2 of 0.6564, implying that the model explains approximately 65.6% of the variance in the target variable and thus achieves a moderate level of fit. The root mean square error (*RMSE*) is 0.2766 and the mean absolute error (*MAE*) is 0.2236, with *RMSE* exceeding *MAE*, which suggests the existence of a small number of relatively large prediction errors. The residual histogram exhibits an approximately bell-shaped distribution centered near zero but slightly left-skewed, indicating a general tendency toward overestimation. Moreover, the increased dispersion of residuals at higher value points points to potential systematic bias. Overall, the model adequately reflects the relationship between the dependent and

https://doi.org/10.3390/sym18010107

independent variables, yet issues such as bias and heteroscedasticity remain, highlighting the need for further refinement through feature optimization, model enhancement, or the adoption of robust estimation methods.

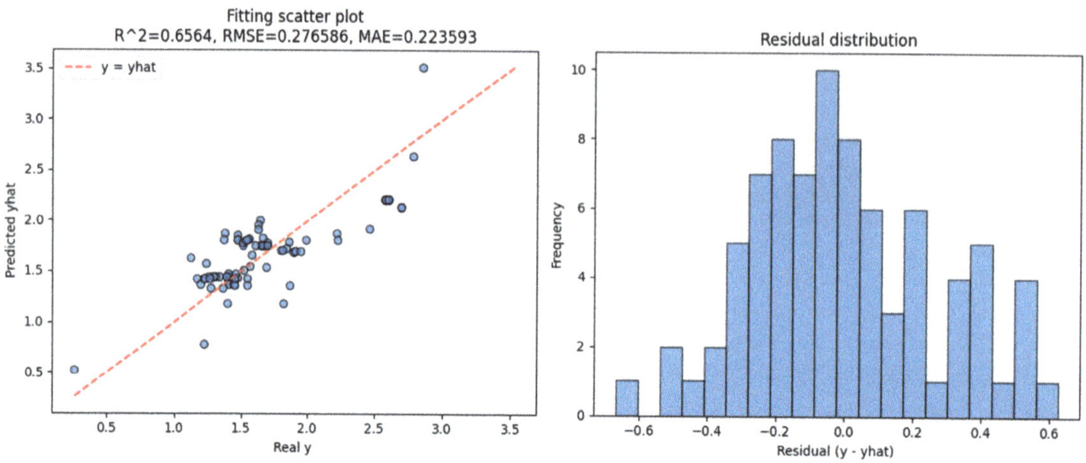

Figure 14. Fitting Scatter Plot and Residual Distribution.

From a physical perspective, the inlet linear effect is positive and statistically significant ($c_1 > 0$), indicating that a moderate increase in inlet slope strengthens the top skim flow. Both quadratic penalty coefficients are positive and significant (d_1, $d_2 > 0$), confirming the expected diminishing-returns behavior at larger angles. By contrast, the main-channel linear effect is negative and significant ($c_2 < 0$), implying that, under the current sampling distribution, main-channel lift/expansion is overall unfavorable. The penalty associated with channel non-uniformity is positive but statistically insignificant ($e > 0$), suggesting insufficient identifiability of r. A simple magnitude check shows that the linear gains and quadratic penalties are of comparable size, reproducing the observed range of Φ in a physically consistent manner. A more detailed explanation is provided in Section S3 of the Supplementary Materials.

According to the above formula, the results before and after optimization are 0.2413 and 0.3535, respectively, corresponding to an improvement of 46.56%. The optimized scheme is evidently superior, with the primary benefit arising from the significant reduction in χ^2. This is consistent with the thermal flow field results and the analysis of the optimization rationale presented earlier. The fitted formula demonstrates high reliability and provides valuable guidance for the design of this type of battery energy storage cabinet.

4. Conclusions and Outlook

4.1. Conclusions

This study optimized the airflow duct geometry of a battery energy storage cabinet using the BOBYQA optimization algorithm and analyzed its flow and thermal field characteristics. The conclusions are as follows:

(1) The airflow resulting from the vortex creates low local airflow and velocity at the top of the battery pack, leading to the formation of local heat accumulation area.

(2) The excessive airflow and velocity in the duct create a strong recirculation vortex at the duct outlet, causing the cooling airflow at the top of the battery pack to be-

https://doi.org/10.3390/sym18010107

come turbulent. This leads to the formation of local heat accumulation areas and significantly increase the risk of thermal failure.

(3) By adjusting the geometric features of the structure, such as the duct inlet and main duct, through the BOBYQA optimization algorithm, the heat dissipation airflow was optimized. The maximum temperature in the hotspot area was reduced from 72.9 °C to 57.6 °C; the cycle life of the battery pack in the hotspot area increased by approximately 106.17%; the average cycle life of the battery energy storage cabinet improved by 2.19%.

(4) Adjusting the geometry of the airflow duct to change the airflow volume and velocity improved the flow disruption caused by the recirculation vortex at the duct outlet. The intensity of the vortex was weakened, allowing it to supplement the airflow entering the duct and form a stable cooling channel.

(5) The main heat dissipation airflow is formed by the convergence of the airflows from the battery storage cabinet entrance and the duct outlet. Increasing the airflow of Airflow I and reducing the velocity of Airflow II helps improve the heat dissipation of the battery energy storage cabinet. Excessive airflow in the duct is not conducive to forming stable cooling airflow, while appropriately increasing the airflow from the cabinet entrance and weakening the airflow at the duct outlet benefits the heat dissipation at the top of the battery packs.

4.2. Limitations and Future Work

The proposed design modifications demonstrate significant thermal benefits in the numerical study and have been validated through their commercial application in air-cooled lithium-ion battery energy storage containers manufactured by SDEE Contemporary Energy Technology Co., Ltd., located in Jinan, China. However, there are several practical challenges and limitations that must be addressed before these designs can be fully implemented in commercial battery energy storage cabinets.

One key challenge lies in the manufacturability, assembly clearance, and maintenance-access requirements of the optimized flow-path features. While these modifications improve thermal performance, their practical implementation in a real-world setting requires careful consideration of how easily they can be manufactured, assembled, and maintained. The performance of these features must also be evaluated in conjunction with the actual fan curve to ensure an efficient operating point, accounting for trade-offs between enhanced cooling and potential increases in pressure drop, fan power consumption, and noise levels.

Furthermore, ensuring the long-term reliability of these designs is critical. The system must be robust against degradation mechanisms such as dust accumulation, filter aging, and partial blockage, all of which can alter flow distribution and affect thermal uniformity over time. These factors are often overlooked in idealized models but are crucial for the performance of battery energy storage systems in real-world operating conditions.

From a modeling perspective, this work has several limitations that need to be acknowledged. The steady-state assumption used in the simulations does not account for transient behaviors, which can be important in real-world applications where temperatures fluctuate over time. Additionally, the external thermal boundary conditions used in the model involve some uncertainty, and the simplified heat-generation representation may not fully capture the complexity of actual battery systems. Another limitation is the use of BOBYQA, a derivative-free local optimizer. While this method is effective for certain problems, the final design can still depend on the chosen parameterization, bounds, and initial point, which means it may not guarantee global optimality.

 https://doi.org/10.3390/sym18010107

Although numerical verification, such as mesh sensitivity and convergence monitoring, supports the credibility of the CFD results, further validation through experimental or field studies under various operating conditions is essential. Future work should, therefore, focus on multi-condition robust optimization, explicitly coupling the fan and system performance with energy and noise objectives, and incorporating uncertainty quantification into the design process. Additionally, experimental calibration and validation are crucial steps for ensuring the applicability of the proposed designs in real-world scenarios. Surrogate modeling could also be explored as a way to accelerate the optimization process by capturing nonlinear interactions among geometric parameters more effectively.

In conclusion, while the proposed design modifications show clear potential for improving the thermal performance of battery energy storage systems, overcoming the practical challenges outlined above will be crucial for their successful implementation in commercial applications. Future research should address these challenges to ensure the robustness and efficiency of the designs across a range of real-world conditions.

Supplementary Materials: The following supporting information can be downloaded at: https://www.mdpi.com/article/10.3390/sym18010107/s1, Figure S1. Temperature-Dependent Properties of Air (a) Dynamic Viscosity (b) Specific Heat Capacity (c) Thermal Conductivity (d) Density of the air, Figure S2. Meshes with different average element sizes: (a) 0.134 m, (b) 0.106 m, (c) 0.074 m, and (d) 0.046 m, Figure S3. Velocity fields obtained with different average mesh sizes: (a) 0.134 m, (b) 0.106 m, (c) 0.074 m, and (d) 0.046 m, Figure S4. (a) Sampling line; (b) velocity magnitude along the sampling line; Table S1. Turbulence Model Parameters; S1. Details of the Steady-State Conjugate Heat Transfer Finite Element Model Setup, S2. Mesh Independence Verification, S3. Model Regression Explanation.

Author Contributions: Conceptualization, L.W. and Y.T.; methodology, Y.T.; software, D.L.; validation, J.Y., W.Q. and Y.Z.; formal analysis, L.W.; investigation, J.Y.; resources, Y.C., G.W. and Z.L.; data curation, L.W. and G.W.; writing original draft preparation, L.W.; writing review and editing, Y.T.; visualization, L.W.; supervision, Z.L.; project administration, Y.T. and Z.L.; funding acquisition, Y.T. All authors have read and agreed to the published version of the manuscript.

Funding: This work was supported by National Natural Science Foundation of China (No. 52305203), Guangxi Key Laboratory of Manufacturing System & Advanced Manufacturing Technology (No. GUIKE AD25069080, No. 24354S012), Shandong Excellent Young Scientists Fund Program (Overseas) (No. 2024HWYQ-003), Young Taishan Scholars Program of Shandong Province (No. tsqn202306087), "Qilu Young Scholar" program of Shandong University (No. 31360082363167), Central Guidance on Local Science and Technology Development Fund of Shandong Province (YDZX2024084), during the course of this work.

Data Availability Statement: The original contributions presented in this study are included in the article. Further inquiries can be directed to the corresponding authors.

Conflicts of Interest: Authors Jianbin Yu, Wei Qin, Yangyang Zhang, and Guoyan Wang were employed by the company SDEE Contemporary Energy Technology Co., Ltd. The remaining authors declare that the research was conducted in the absence of any commercial or financial relationships that could be construed as a potential conflict of interest.

References

1. Ngoy, K.R.; Lukong, V.T.; Yoro, K.O.; Makambo, J.B.; Chukwuati, N.C.; Ibegbulam, C.; Eterigho-Ikelegbe, O.; Ukoba, K.; Jen, T.-C. Lithium-ion batteries and the future of sustainable energy: A comprehensive review. *Renew. Sustain. Energy Rev.* **2025**, *223*, 115971. [CrossRef]
2. Chen, M. Thermal Safety of Lithium-Ion Batteries: Current Status and Future Trends. *Batteries* **2025**, *11*, 112. [CrossRef]

https://doi.org/10.3390/sym18010107

3. Bandhauer, T.M.; Garimella, S.; Fuller, T.F. A critical review of thermal issues in lithium-ion batteries. *J. Electrochem. Soc.* **2011**, *158*, R1. [CrossRef]
4. Troxler, Y.; Wu, B.; Marinescu, M.; Yufit, V.; Patel, Y.; Marquis, A.J.; Brandon, N.P.; Offer, G.J. The effect of thermal gradients on the performance of lithium-ion batteries. *J. Power Sources* **2014**, *247*, 1018–1025. [CrossRef]
5. Ma, S.; Jiang, M.; Tao, P.; Song, C.; Wu, J.; Wang, J.; Deng, T.; Shang, W. Temperature effect and thermal impact in lithium-ion batteries: A review. *Prog. Nat. Sci. Mater. Int.* **2018**, *28*, 653–666. [CrossRef]
6. Wang, Y.; Zhang, X.; Chen, Z. Low temperature preheating techniques for Lithium-ion batteries: Recent advances and future challenges. *Appl. Energy* **2022**, *313*, 118832. [CrossRef]
7. Na, X.; Kang, H.; Wang, T.; Wang, Y. Reverse layered air flow for Li-ion battery thermal management. *Appl. Therm. Eng.* **2018**, *143*, 257–262. [CrossRef]
8. Motloch, C.G.; Christophersen, J.P.; Belt, J.R.; Wright, R.B.; Hunt, G.L.; Sutula, R.A.; Duong, T.; Tartamella, T.J.; Haskins, H.J.; Miller, T.J. High-power battery testing procedures and analytical methodologies for HEV's. *SAE Trans.* **2002**, 797–802.
9. Zhang, X.; Li, Z.; Luo, L.; Fan, Y.; Du, Z. A review on thermal management of lithium-ion batteries for electric vehicles. *Energy* **2022**, *238*, 121652. [CrossRef]
10. Kim, J.; Oh, J.; Lee, H. Review on battery thermal management system for electric vehicles. *Appl. Therm. Eng.* **2019**, *149*, 192–212. [CrossRef]
11. Liu, H.; Wei, Z.; He, W.; Zhao, J. Thermal issues about Li-ion batteries and recent progress in battery thermal management systems: A review. *Energy Convers. Manag.* **2017**, *150*, 304–330. [CrossRef]
12. Wang, C.; Xi, H.; Wang, M. Investigation on forced air-cooling strategy of battery thermal management system considering the inconsistency of battery cells. *Appl. Therm. Eng.* **2022**, *214*, 118841. [CrossRef]
13. Zhang, F.; Wang, P.; Yi, M. Design optimization of forced air-cooled lithium-ion battery module based on multi-vents. *J. Energy Storage* **2021**, *40*, 102781. [CrossRef]
14. Xin, S.; Wang, C.; Xi, H. Thermal management scheme and optimization of cylindrical lithium-ion battery pack based on air cooling and liquid cooling. *Appl. Therm. Eng.* **2023**, *224*, 120100. [CrossRef]
15. Xie, J.; Ge, Z.; Zang, M.; Wang, S. Structural optimization of lithium-ion battery pack with forced air cooling system. *Appl. Therm. Eng.* **2017**, *126*, 583–593. [CrossRef]
16. Xu, X.M.; He, R. Research on the heat dissipation performance of battery pack based on forced air cooling. *J. Power Sources* **2013**, *240*, 33–41. [CrossRef]
17. Du, J.; Tao, H.; Chen, Y.; Yuan, X.; Lian, C.; Liu, H. Thermal management of air-cooling lithium-ion battery pack. *Chin. Phys. Lett.* **2021**, *38*, 118201. [CrossRef]
18. Sahin, R.C.; Gocmen, S.; Cetkin, E. Thermal management system for air-cooled battery packs with flow-disturbing structures. *J. Power Sources* **2022**, *551*, 232214. [CrossRef]
19. Li, W.; Wang, X.; Cen, P.Y.; Chen, Q.; Cordeiro, I.M.D.C.; Kong, L.; Lin, P.; Li, A. A Comparative Numerical Study of Lithium-Ion Batteries with Air-Cooling Systems towards Thermal Safety. *Fire* **2024**, *7*, 29. [CrossRef]
20. Lin, X.; Shao, K.; Wang, C. Optimization and numerical simulation of novel air-cooling system for the thermal management of lithium-ion battery pack. *Int. J. Electrochem. Sci.* **2022**, *17*, 220141. [CrossRef]
21. Feng, X.H.; Li, Z.Z.; Gu, F.S.; Zhang, M.-L. Structural design and optimization of air-cooled thermal management system for lithium-ion batteries based on discrete and continuous variables. *J. Energy Storage* **2024**, *86*, 111202. [CrossRef]
22. Argade, S.; De, A. Optimization study of a Z-type airflow cooling system of a lithium-ion battery pack. *Phys. Fluids* **2024**, *36*, 067119. [CrossRef]
23. Dai, H.; Wang, Y. Study on the Influence of Air Inlet and Outlet on the Heat Dissipation Performance of Lithium Battery. *World Electr. Veh. J.* **2023**, *14*, 113. [CrossRef]
24. Gogoi, B.; Deka, H.; Bora, B.J.; Paramasivam, P.; Kamangar, S.; Arabi, A.I.A.; Razak, A.; Wodajo, A.W. Optimizing thermal performance in air-cooled Li-ión battery packs with vortex generators for cleaner energy storage. *Sci. Rep.* **2025**, *15*, 25477. [CrossRef]
25. Yang, C.; Xi, H.; Wang, M. Structure optimization of air cooling battery thermal management system based on lithium-ion battery. *J. Energy Storage* **2023**, *59*, 106538. [CrossRef]
26. Dong, Y.; Ding, Y.; Elfeky, K.E.; Qi, Y.; Chu, W.; Wang, Q. Thermal Management for a Stadium Power Supply Container Using a Rack-Level Air Cooling Strategy. *Energies* **2024**, *17*, 1654. [CrossRef]
27. Lv, Z.; Sun, Z.; Wang, L.; Liu, Q.; Zhang, J. Multi-Level Thermal Modeling and Management of Battery Energy Storage Systems. *Batteries* **2025**, *11*, 219. [CrossRef]
28. Permana, I.; Agharid, A.P.; Wang, F.; Lin, S.H. Performance investigation of thermal management system on battery energy storage cabinet. *Therm. Sci.* **2023**, *27*, 4389–4400. [CrossRef]

29. Galdi, G. *An Introduction to the Mathematical Theory of the Navier-Stokes Equations: Steady-State Problems*; Springer Science & Business Media: New York, NY, USA, 2011.
30. Powell, M.J.D. *The BOBYQA Algorithm for Bound Constrained Optimization Without Derivatives*; Cambridge NA Report NA2009/06; University of Cambridge: Cambridge, UK, 2009; Volume 26, pp. 26–46.
31. Wang, Q.; Ping, P.; Zhao, X.; Chu, G.; Sun, J.; Chen, C. Thermal runaway caused fire and explosion of lithium ion battery. *J. Power Sources* **2012**, *208*, 210–224. [CrossRef]

https://doi.org/10.3390/sym18010107

MDPI AG
Grosspeteranlage 5
4052 Basel
Switzerland
Tel.: +41 61 683 77 34

Symmetry Editorial Office
E-mail: symmetry@mdpi.com
www.mdpi.com/journal/symmetry

www.ingramcontent.com/pod-product-compliance
Lightning Source LLC
Chambersburg PA
CBHW061840210926
51850CB00006B/933